NAVIGATING SOCIAL–ECOLOGICAL SYSTEMS

In the effort towards sustainability, it has become increasingly important to develop new conceptual frames to understand the dynamics of social and ecological systems. Drawing on complex systems theory, this book investigates how human societies deal with change in linked social–ecological systems, and build capacity to adapt to change. The concept of resilience is central in this context. Resilient social–ecological systems have the potential to sustain development by responding to and shaping change in a manner that does not lead to loss of future options. Resilient systems also provide capacity for renewal and innovation in the face of rapid transformation and crisis. The term navigating in the title is meant to capture this dynamic process.

Navigating Social–Ecological Systems deliberately transcends academic disciplines, because the issues in focus require collaboration over the boundaries of the natural sciences, social sciences, and the humanities. Case studies and examples from several geographic areas, cultures, and resource types are included, merging forefront research from different disciplines into a common framework for new insights into sustainability.

FIKRET BERKES is Professor and former Director, Natural Resources Institute, University of Manitoba, Canada. He holds the Canada Research Chair in Community-Based Resource Management.

JOHAN COLDING is at the Centre for Research on Natural Resources and the Environment, Stockholm University, and the Beijer International Institute of Ecological Economics at the Royal Swedish Academy of Sciences, Stockholm, Sweden.

CARL FOLKE is Director of the Centre for Research on Natural Resources and the Environment (CNM), and a Professor in the Department of Systems Ecology, at Stockholm University, Sweden. He is also a Professor at the Beijer International Institute of Ecological Economics at the Royal Swedish Academy of Sciences, Stockholm, Sweden.

NAVIGATING SOCIAL–ECOLOGICAL SYSTEMS

Building Resilience for Complexity and Change

Edited by

FIKRET BERKES
*The University of Manitoba,
Winnipeg, Canada*

JOHAN COLDING
*The Royal Swedish Academy of Sciences, Stockholm, Sweden, and
Stockholm University, Sweden*

and

CARL FOLKE
*Stockholm University, Sweden, and
The Royal Swedish Academy of Sciences, Stockholm, Sweden*

CAMBRIDGE UNIVERSITY PRESS
Cambridge, New York, Melbourne, Madrid, Cape Town, Singapore, São Paulo

Cambridge University Press
The Edinburgh Building, Cambridge CB2 8RU, UK

Published in the United States of America by Cambridge University Press, New York

www.cambridge.org
Information on this title: www.cambridge.org/9780521815925

First published 2003
Reprinted 2006
This digitally printed version 2008

A catalogue record for this publication is available from the British Library

Library of Congress Cataloguing in Publication data

Navigating social–ecological systems : building resilience for complexity and change /
edited by Fikret Berkes, Johan Colding, and Carl Folke.
p. cm.
Includes bibliographical references and index.
ISBN 0 521 81592 4
1. Social ecology – Congresses. I. Berkes, Fikret. II. Colding, Johan. III. Folke, Carl.
HM861 N38 2002
304.1–dc21 2002022268

ISBN 978-0-521-81592-5 hardback
ISBN 978-0-521-06184-1 paperback

Contents

Contributors

Janis B. Alcorn
World Resources Institute, Institutions and Governance Program, c/o 3508 Woodbine Street, Chevy Chase, MD 20815, USA

John Bamba
Institut Dayakologi, Jalan Budi Utomo, Block A3 No. 4, Pontianak 78241, West Kalimantan, Indonesia

Fikret Berkes
Natural Resources Institute, The University of Manitoba, Winnipeg, Manitoba, Canada R3T 2N2

Kristen Blann
Department of Fisheries, Wildlife, and Conservation Biology, University of Minnesota, 200 Hodson Hall, 1980 Folwell Avenue, St Paul, Minnesota 55108, USA

Lars Carlsson
Department of Political Science, Luleå University of Technology, 97187 Luleå, Sweden

Johan Colding
The Beijer International Institute of Ecological Economics, The Royal Swedish Academy of Sciences, Box 50005, S-104 05 Stockholm, Sweden, and Centre for Research on Natural Resources and the Environment, Stockholm University, S-106 91 Stockholm, Sweden

Iain J. Davidson-Hunt
Natural Resources Institute, The University of Manitoba, Winnipeg, Manitoba, Canada R3T 2N2

Thomas Elmqvist
Department of Systems Ecology, Stockholm University, S-106 91 Stockholm, Sweden

Carl Folke
Centre for Research on Natural Resources and the Environment and Department of Systems Ecology, Stockholm University, S-106 91 Stockholm, Sweden, and The Beijer International Institute of Ecological Economics, The Royal Swedish Academy of Sciences, Box 50005, S-104 05 Stockholm, Sweden.

Madhav Gadgil
Centre for Ecological Sciences, Indian Institute of Science, Bangalore 560012, India

Lance H. Gunderson
Department of Environmental Studies, Emory University, Atlanta, Georgia 30322, USA

Monica Hammer
Department of Systems Ecology, Stockholm University, S-106 91 Stockholm, Sweden

C. S. (Buzz) Holling
16871 Sturgis Circle, Cedar Key, Florida 32625, USA

Anne Kendrick
Natural Resources Institute, The University of Manitoba, Winnipeg, Manitoba, Canada R3T 2N2

Steve Light
Institute for Agriculture and Trade Policy, 2105 1st Ave S., Minneapolis, Minnesota 55404–2505, USA

Bobbi Low
School of Natural Resources and Environment, University of Michigan, Ann Arbor, Michigan 48109, USA

Stefanus Masiun
Jl. Boedi Utomo, Blk A-III/5, Komp. Bumi Indah Khatulistiwa, Pontianak, 78241, Kalimantan Barat, Indonesia

Jo Ann Musumeci
Three Oaks Research, 2606 Pleasant Ave. South, Minneapolis, Minnesota 55408–1441, USA

Ita Natalia
KpSHK (Consortium for Supporting Community Based Forest System Management) – NTFP Focal Point. Jl. Citarum Blok B X/23, Bogor Baru, Bogor 16152, Indonesia

Per Olsson
Department of Systems Ecology, Stockholm University, S-106 91 Stockholm, Sweden

Elinor Ostrom
Center for the Study of Institutions, Population, and Environmental Change

(CIPEC), Indiana University, 513 North Park Street, Bloomington, Indiana 47408, USA

Antoinette G. Royo
The BSP-KEMALA, Ratu Plaza–17th Floor, Jalan Sudirman 9, Jakarta 10270, Indonesia

Cristiana S. Seixas
Natural Resources Institute, The University of Manitoba, Winnipeg, Manitoba, Canada R3T 2N2

Carl Simon
Center for Study of Complex Systems, University of Michigan, Ann Arbor, Michigan 48109, USA

Maria Tengö
Department of Systems Ecology, Stockholm University, S-106 91 Stockholm, Sweden

Ronald L. Trosper
College of Ecosystem Science and Management, Northern Arizona University, Flagstaff, Arizona, PO Box 15018, 86011–5018, USA

James Wilson
School of Marine Sciences, University of Maine, Orono, Maine 04469–5782, USA

Preface

It is evident that the dominant worldview in resource and environmental management of 'systems in equilibrium' is incompatible with observations of the complex dynamics of social and ecological systems. In the effort towards sustainability, it has become increasingly important to develop new conceptual frames to understand these dynamics. The framework underlying the book is complex systems theory, with the explicit objective of examining ways of building social–ecological resilience to enhance the capacity to deal with complexity and change. In particular, we look for effective ways of analyzing the phenomenon of change and how to respond to change in a manner that does not lead to loss of future options. The 14 chapters of the volume investigate how human societies deal with change in coupled social–ecological systems and build capacity to adapt to change. The term navigating in the title of the book is meant to capture this dynamic process.

It is an edited volume, but it is different from most edited volumes. We have used a common framework for the syntheses and the case-study analyses of a diversity of resource management systems. The chapters, written by scholars from several disciplines, have been developed on the basis of the common framework. The Introduction presents the framework and direction of the volume followed by four major sections: perspectives on resilience; building resilience in local management systems; social–ecological learning and adaptation; and cross-scale institutional response to change. In the final chapter we synthesize the lessons of the volume, emphasizing the need to learn to live with change and uncertainty; to nurture diversity for resilience; to combine different types of knowledge for learning about complex systems; and to create opportunity for self-organization towards social–ecological sustainability. The volume deliberatively transcends disciplinary boundaries, because the issues in focus require collaboration over the boundaries of the natural sciences, social sciences, and the humanities.

The work with the volume was initiated as a project of the Resilience Network, a research program of the Beijer International Institute of Ecological Economics in Stockholm, Sweden, and University of Florida, USA. We are forever indebted to C.S. (Buzz) Holling, the founder of the Resilience Network, to Karl-Göran Mäler, the Director of the Beijer Institute, and to the program director of the network, Lance Gunderson, who is also a chapter author of this volume, for providing support to our work on understanding the dynamics of social–ecological systems in the context of this 3-year research program.

The project, *Dynamics of Ecosystem–Institution Linkages for Building Resilience,* started in early 1998. Project members and potential chapter authors were sent an invitation, along with a Beijer Discussion Paper providing a tentative common framework for the project. Two workshops were held, the first in the fall of 1998 at the Beijer Institute in Stockholm, in which the framework and possible contributions were discussed and improved and chapter outlines constructed. Draft chapters were presented and discussed at a second workshop held in the fall of 1999 at the University of Manitoba, Winnipeg. In June 2000, many of the papers were presented at Indiana University at a conference of the International Association for the Study of Common Property (IASCP). The editors examined the second drafts, they were revised and sent for peer review in August/September 2000. Three or four scientific experts, two of them external to the project group, have reviewed each chapter. The work of the project and the production of the book have been a joint effort, and consequently the editorial author order is alphabetic.

Fikret Berkes, Johan Colding, and Carl Folke
Stockholm and Winnipeg
October 2001

Acknowledgements

There are many that have supported us during the work with this book. Special thanks go to the referees who reviewed the chapters. In addition to those who were also chapter authors, we are most grateful for the constructive comments by Alpina Begossi, Christo Fabricius, Clark Gibson, C.S. Holling, Gary Kofinas, Simon Levin, Robin Mearns, Nick Menzies, Richard Norgaard, Uygar Ozesmi, Kathryn Papp, P.S. Ramakrishnan, Steve Sanderson, John Sinclair, Nancy Turner, Paulo Vieira, Brian Walker, and Oran Young.

The content of the volume was strongly inspired by the work of the Resilience Alliance. Buzz Holling was instrumental in creating the alliance, now under the enthusiastic leadership of Brian Walker. Chapters of the book were discussed in meetings of the alliance, and we gratefully acknowledge the support of all of our exceptional colleagues at the Resilience Alliance for their fundamental role in providing inspiration and improving the content of the volume. We are also indebted to the valuable discussions and comments of our colleagues, Jan Bengtsson, Kristen Bingeman, Line Gordon, Don Ludwig, Phil Lyver, Fredrik Moberg, Magnus Nyström, Lowell (Rusty) Pritchard, and Max Troell.

We express our genuine thanks to Christina Leijonhufvud and Astrid Auraldsson of the Beijer Institute for their assistance with the Stockholm workshop, and to Anne Kendrick and Viviane Weitzner of the Natural Resources Institute, University of Manitoba, for the Winnipeg workshop. Thanks are due to Peter Vandeburg and Magnus Anderson for producing the illustrations, and to Amy Rader and Miriam Huitric for improving readability. The chapter authors deserve special thanks for their patience with several rounds of revising and refining manuscripts and handling numerous inquiries and comments from the editors.

The Dynamics of Ecosystem–Institution Linkages for Building Resilience project was supported by a grant from the John D. and Catherine T. MacArthur Foundation, which we gratefully acknowledge. The work of Johan Colding and

Carl Folke reported in this book was also supported by the Swedish Council for Planning and Coordination of Research (FRN), the Rockefeller Foundation, and the Swedish Man and the Biosphere Program with funds from the Swedish Research Council (Vetenskapsrådet). Fikret Berkes' work was supported by the Social Sciences and Humanities Research Council of Canada (SSHRC), and the project participation of the University of Manitoba researchers was supported by a grant from the Research Development Initiatives (RDI) program of the SSHRC.

Foreword: The backloop to sustainability

C.S. HOLLING

Introduction

I hazard a guess that people know enough about growth to know how to nurture it – mostly. But when growth stops or collapses, they do not know enough about protection or about novelty to know how to renew confidently for the next phase of growth. And they do not know how the two – growth and novelty – interact. As one consequence, economic forecasters, for example, do well in predicting rates of growth while on a growth path. They do a poor job at times of recession, or even worse at times of looming depression.

That is why I said 'mostly'. Growth of a cell or a society occurs gradually. It builds potential that accumulates slowly and it creates two conflicting attributes – increasing potential but also increasing vulnerability. Increase in potential roughly represents an increase in wealth represented in those structures that acquire, store, maintain and use potential. Increase in wealth gives potential for alternative futures. The increase in vulnerability comes from increase in structure that adds complexity but also vulnerability. As a consequence, eventually cells can die and societies can revolt. Growth then stops or reverses.

But cells and societies also reproduce and reinvent in the process of cyclic transformations. That is when evolution and deep changes are created. The bewildering, entrancing, unpredictable nature of nature and people, the richness, diversity and changeability of life come from that evolutionary dance generated by cycles of growth, collapse, reorganization, renewal and re-establishment.

We call that the adaptive cycle, as noted in Figure 1.2 in the introductory chapter, where its essential features are described. The 'front-loop' of that cycle is the loop of growth. The 'back-loop' is the loop of reorganization.

The editors of this book, Carl Folke, Fikret Berkes and Johan Colding, are interested in sustainable systems. Those are systems that persist, but also that evolve and change. Growth is important, but even more so are the forces in a healthy system that dominate during episodes when growth is halted or reversed, when deep uncertainty explodes, when several alternative futures become suddenly perceived and unpredictability explodes. It is a time of crisis, but also of opportunity. Unexpected interactions can occur among previously separate properties that can nucleate an inherently novel and unexpected focus for future good or ill.

At such times, the future can also be suddenly shaped by externally triggered events such as those from slowly changing climate, from entrants of invasive species, from human immigrants driven by geopolitical changes or from unexpected terrorist events. Such apparently external events can launch future development along an unpredictable path.

During such times, uncertainty is high, control is weak and confused, and unpredictability is high. But space is also created for reorganization and innovation. It is therefore also a time when individual cells, individual organisms or individual people have the greatest chance of influencing events. There is opportunity with low costs of failure possible. The future can be mapped by experiments rather than by long-term plans. It is the time when a Gandhi or a Hitler can use events of the past to transform the future for great good or great ill. In a biological evolutionary setting, it is a time when mammals can replace dinosaurs as the dominant life form. It is the time of the 'Long Now' (Brand, 1999).

The editors therefore break a tradition in this book. They see that the essence of sustainability cannot be defined from metaphors of growth, equilibrium and stability. Rather it is defined from metaphors of novelty, memory and instability. They reverse existing traditions of exploration and analysis by focusing on the back-loop of collapse and reorganization, rather than on the front-loop of growth and predictability. They therefore focus on foundations for change. They focus on forces of evolution from biology, ecology, society and culture.

Their approach is integrative, merging the natural and social sciences. And they do that by choosing largely an analysis of existing contemporary and traditional societies around the world and by exploring the responses of such systems to crises and change. That focus emerged as one of the four central themes of the Resilience Project, a 5-year international project to develop integrative theory for sustainable systems and to propose integrative practice that can be tested within developed and developing regions.

The overall theories of the Resilience Project emerged from selecting, expanding and integrating existing theories in economics, ecosystem science, institutional research and adaptive complex system theory. The practice emerged from experience in regions where there is significant multiple use of renewable resources: agriculture, forestry, fisheries, rangeland grazing, wildlife and eco-tourism. Specifically, the regions included semi-arid grasslands and savannas in Africa, Australia and North America, coral reefs in tropical regions, boreal forested regions in Canada, the USA and Europe, enclosed seas of the Baltic region of Europe and in south Florida, and wetlands of Wisconsin, Minnesota, Florida and Europe. The research has been an effort of synthesis through cooperation among a wonderful international group of scientists, scholars and practitioners, together with their students and collaborators.

This book is one of four that the project has created. One concerns nonlinear economics (Mäler and Starrett, in process) and the breakdown of traditional linear economics under certain conditions when resources are exploited in ecosystems. Another explores different large-scale ecosystems and identifies their structure and function (Gunderson and Pritchard, 2002), particularly the causes of multi-stable states and the surprises that result. Still another is the central integrative volume (Gunderson and Holling, 2002) that presents the integrative theory called panarchy. It is that theory that this book chooses as its base in its delightful examination of the structures formed by people and nature, particularly at times of fundamental crisis or transformation. Its message is therefore of deep significance at these times of national and international transformations in economics, society and security.

The terrorist attacks on September 11, 2001 in New York and Washington are the events that make this period one recognized as a time of crisis and transformation by the peoples of the world. But those events emerge from slower processes that have paced changes in development, politics and our natural endowments since the Second World War – locally and globally. This book and indeed all four of the books provide a foundation to develop and evaluate responses of nations and people to such profound changes. We do not do that here, because the terrorist events are so recent. But the shape of the influence of these works is becoming clearer, and will be the foundation for the next immediate target of thought and action. We encourage readers to do the same, enriching the effort with their own experiments in enquiry and invention.

The pathology of regional development

Our resilience work focused particularly on regions where local history and status interact with global and international processes. It was launched by the following pattern that was observed in several dozen examples of development and resource management policies initiated in both developed and developing nations (Gunderson, Holling and Light, 1995; Holling and Meffe, 1996). That pattern exposed an intriguing paradox in regions dominated by the 'modern' context of the developed nations since the Second World War. It consistently emerged as those regional systems experienced a crisis or policy change.

The Regional Resource and Development Pathology has the following features:

1. The new policies and development initially succeed in reversing the crisis or in enhancing growth.
2. Implementing agencies initially are responsive to the ecological, economic and social forces, but evolve to become narrow, rigid and myopic. They become captured by economic dependents and the perceived needs for their own survival.
3. Economic sectors affected by the resources grow and become increasingly dependent on perverse subsidies.
4. The relevant ecosystems gradually lose resilience to become fragile and vulnerable and more homogeneous as diversity and spatial variability are reduced.
5. Crises and vulnerabilities begin to become more likely and evident and the public begin to loose trust in governance.

In rich regions the result is spasmodic lurches of learning with expensive actions directed to reverse the worst of the consequences of past mistakes. An example is the present effort to restore the Everglades ecosystem in south Florida – the largest effort of restoration that has ever been attempted (Gunderson, 1999).

In poor regions the result is dislocation of people, increasing uncertainty, impoverishment and a poverty trap. Rarely, a radical new approach to development is invented that depends more on people's inventiveness and the transformation of strategic goals than on money. An example is the invention of community and economic utilization of biodiversity in Zimbabwe after the catastrophic droughts of the 1980s exposed the unsustainability of past development (Lynam, 1999). But that transformation is now being destroyed as its vulnerability to national political corruption is exposed.

Diagnosis of the pathology

Sustainable development and management of global and regional resources are not an ecological problem, nor an economic one, nor a social one. They are a combination of all three. And yet actions to integrate all three in the developed nations have short-changed one or more. Sustainable designs driven by conservation interests ignore the needs for an adaptive form of economic development that emphasizes individual enterprise and flexibility. Those driven by industrial interests act as if the uncertainty of nature can be replaced with human engineering and management controls, or ignored altogether. Those driven by social interests act as if community development and empowerment alone can surmount any constraints of nature or of external forces. As investments fail, the policies of government, private foundations, international agencies and non-governmental organizations (NGOs) flop from emphasizing one kind of partial solution to another. Over the last three decades, such policies have flopped from large investment schemes, to narrow conservation ones, to equally narrow community development ones, to libertarian market solutions.

There has been lots of despair over failures but little benefit from the learning that has occurred. And little sharing of learning across regions.

Each spasm of policy change builds on theory, though many would deny anything but the most pragmatic and non-theoretical foundations to their proposed actions. The conservationists depend on theories of ecology and evolution, the developers on variants of free market models, the community activists on theories of community and social organization. All these theories are correct. Correct in the sense of being partially tested and credible representations of one, but only one, part of reality. The problem is that they are partial. Each misses a critical dimension. Economic theory deals poorly with slow variables that form cultural and ecological foundations for sustainability. Ecological theory ignores the richness of people's needs and inventiveness. Social theory is fragmented and static.

But our integrated theory has now been developed by a leading group of ecologists, economists and social scientists drawing upon extensive regional experience. It is a theory that recognizes the synergies and constraints among nature, economic activities, and people – a theory that informs and emerges from empirical practice.

Even the most ruthlessly pragmatic goals for developing policies and investments for sustainability need such a theoretical foundation that integrates ecological with economic with institutional with evolutionary theory – that overcomes the disconnect rooted in current theoretical limitations within each field. It is that integrative theoretical foundation and the practical consequences of it that have been the focus of the Resilience Project supported by The MacArthur Foundation. It is that integrative theory that was expanded by the discoveries in this book.

A prescription

The failures of the past have not been complete: there have been partial successes. This mixed picture comes because theories, trials and projects were not wrong, just too partial. The recent fad for community-based development alone is another such correct, partial solution that will fail. The gales of change internationally (international financial contagion, migration, the emergence of the Internet), globally (climate change, ozone depletion, novel diseases) and regionally (conflicts and politics of sustainability, terrorism, biodiversity and resilience loss) create opportunity and a potential for constructive change. Now is the time to protect and integrate the good experience, ignore the bad and launch and communicate safe-fail experiments.

Oddly, the present recognition of global crises makes this the time to share the fruits of innovative development widely between North and South as it emerges, not just among those of the North or those of the South after it has in part failed. The Internet provides an arena to invent and communicate ways of learning and doing that are discovered in local regions around the world.

These gales of change suggest that the window for constructive change has opened at several scales. It is a time when conditions of the back-loop of the adaptive cycle dominate. Under those conditions, the elements of a prescription for facilitating constructive change are:

- Identify and reduce destructive constraints and inhibitions on change, such as perverse subsidies.
- Protect and preserve the accumulated experience on which change will be based.
- Stimulate innovation and communicate the results in a variety of safe-fail experiments that probe possible directions, in a way that are low in costs for people's careers and organizations' budgets.
- Encourage new foundations for renewal that build *and sustain* the capacity of people, economies and nature for dealing with change.
- Encourage new foundations to expand and communicate understanding of change.

Lessons that derive from exploration of these backloop studies include the expectation that dynamics of social–ecological systems will have multiple domains of attraction and that the system can flip from one to another, with large consequences for people (Berkes and Folke, 1998). The delightful simplified models of Carpenter and Brock show the consequences are real when integration is at the heart of the models (Carpenter, Brock, and Hanson, 1999). Resilience, multi-stable states, and learning about slow and spatially remote variables are revealed as a key to sustainability (Holling 2001). That is where the social–ecological memory plays a central role, as shown in several of the studies of this book. For some time prior to a domain flip, the impending collapse can be evident to some participants in the system and the system itself becomes an accident waiting to happen. Breakdown is inevitable. After collapse, innovation and experimentation can be favoured. Participants find themselves asking how learning can be stimulated in ways that enhance sustainability. In the end, we find that we need to create excitement, identify options in the form of alternative visions of the future, and build hope.

References

Berkes, F. and Folke, C., eds. 1998. *Linking Social and Ecological Systems. Management Practices and Social Mechanisms for Building Resilience.* Cambridge: Cambridge University Press.

Brand, S. 1999. *The Clock of the Long Now.* New York: Basic Books.

Carpenter, S., Brock, W., and Hanson, P. 1999. Ecological and social dynamics in simple models of ecosystem management. *Conservation Ecology* 3(2): 4. www.consecol.org/vol3/iss2/art4

Gunderson, L.H. 1999. Resilience, flexibility and adaptive management – antidotes for spurious certitude? *Conservation Ecology* 3(1): 7 www.consecol.org/vol3/iss1/art7

Gunderson, L.H. and Holling, C.S., eds. 2002. *Panarchy: Understanding Transformations in Systems of Humans and Nature.* Washington DC: Island Press.

Gunderson, L.H., Holling, C.S., and Light, S.S., eds. 1995. *Barriers and Bridges to the Renewal of Ecosystems and Institutions.* New York: Columbia University Press.

Gunderson, L.H. and Pritchard, L., eds. 2002. *Resilience and the Behavior of Large-Scale Ecosystems.* Washington DC: Island Press.

Holling, C.S. 2001. Understanding the complexity of economic, ecological and social systems. *Ecosystems* 4: 390–405.

Holling, C.S. and Meffe, G.K. 1996. Command and control and the pathology of natural resource management. *Conservation Biology* 10: 328–37.

Lynam, T. 1999. Adaptive analysis of locally complex systems in a globally complex world. *Conservation Ecology* 3(2): 13. www.consecol.org/vol3/iss2/art13

Mäler, K.-G. and Starrett, D., eds. in process. *Ecological and Economic Modelling.*

1

Introduction

FIKRET BERKES, JOHAN COLDING, AND CARL FOLKE

1.1 Building capacity to adapt to change: the context

A common perspective until recently was that our problem-solving abilities have been improving over the years. In the area of resource and environmental management, for example, there was a great deal of faith in our growing scientific understanding of ecosystems, our bag of increasingly sophisticated tools and technologies, and the application of market mechanisms to problems such as air pollution control and fishery management through individually allocated quotas. However, the experience over the last few decades does not support such optimism (e.g., Clark and Munn, 1986; Ludwig, Hilborn, and Walters, 1993; Gunderson, Holling, and Light, 1995). Many of our resource and environmental problems are proving resistant to solutions. A gap has developed between environmental problems and our lagging ability to solve them. This is coming at a time when the Earth has become an increasingly human-dominated system. Many of the changes in the biosphere, including the modification of landscapes, loss of biodiversity and, according to some, climate change, are driven by human activities. Furthermore, changes are occurring at an increasingly faster rate than previously experienced in human history.

There is an emerging consensus regarding the need to look for broader approaches and solutions, not only with resource and environmental issues but along a wide front of societal problems. A survey of senior American Association for the Advancement of Science (AAAS) scientists revealed an intriguing insight. When asked about the most urgent challenges facing science and society, scientists identified many items, but a common thread was that each issue 'seemed to have radically outgrown its previously accepted conceptual framing' (Jasanoff *et al.*, 1997). For each of the issues identified, there were new theories and explanations appearing on the horizon, many calling for more creative forms of collaboration between scientists and society, involving a broader range of disciplines and skills needed for the process.

Broader public participation was also important. Scientific solutions were being undertaken with greater attention to their social context, and the interaction between science and society was increasingly seen as important (Jasanoff *et al.*, 1997). The kind of research that is needed may be 'created through processes of co-production in which scholars and stakeholders interact to define important questions, relevant evidence, and convincing forms of argument' (Kates *et al.*, 2001).

There is also an emerging consensus on the nature of the problem. Many of our resource and environmental problems are seen as *complex systems* problems (Levin, 1999a). Natural systems and social systems are complex systems in themselves; furthermore, many of our resource and environmental problems involve the additional complexity of interactions between natural and social systems (Norgaard, 1994; Berkes and Folke, 1998). Such complexity creates a huge challenge for disciplinary approaches. 'Phenomena whose causes are multiple, diverse and dispersed cannot be understood, let alone managed or controlled, through scientific activity organized on traditional disciplinary lines' (Jasanoff *et al.*, 1997). Complex systems thinking is therefore used to bridge social and biophysical sciences to understand, for example, climate, history and human action (McIntosh, Tainter, and McIntosh, 2000). It is at the basis of many of the new integrative approaches, such as sustainability science (Box 1.1) and ecological economics (Costanza *et al.*, 1993; Arrow *et al.*, 1995). It has led to the recognition that much of conventional thinking in resource and environmental management may be contributing to problems, rather than to solutions (Holling and Meffe, 1996).

In this volume, our ultimate objective is to contribute to efforts towards *sustainability*, that is, the use of environment and resources to meet the needs of the present without compromising the ability of future generations to meet their own needs. We consider sustainability as a process, rather than an end product, a dynamic process that requires adaptive capacity for societies to deal with change. Rather than assuming stability and explaining change, as often done, one needs to assume change and explain stability (van der Leeuw, 2000). For our purposes, sustainability implies maintaining the capacity of ecological systems to support social and economic systems. Sustaining this capacity requires analysis and understanding of feedbacks and, more generally, the dynamics of the interrelations between ecological systems and social systems.

Social systems that are of primary concern for this volume include those dealing with governance, as in property rights and access to resources. Also of key importance are different systems of knowledge pertinent to the dynamics

Box 1.1 Sustainability science

By structure, method, and content, sustainability science must differ fundamentally from most science as we know it. Familiar approaches to developing and testing hypotheses are inadequate because of nonlinearity, complexity, and long time lags between actions and consequences. Additional complications arise from the recognition that humans cannot stand outside the nature–society system. The common sequential analytical phases of scientific inquiry such as conceptualizing the problem, collecting data, developing theories, and applying the results will become parallel functions of social learning, which incorporate the elements of action, adaptive management, and policy as experiment. Sustainability science will therefore need to employ new methodologies that generate the semi-quantitative models of qualitative data, build upon lessons of case studies, and extract inverse approaches that work backwards from undesirable consequences to identify pathways that can avoid such outcomes. Scientists and practitioners will need to work together with the public at large to produce trustworthy knowledge and judgement that is scientifically sound and rooted in social understanding.

Source: http://sustsci.harvard.edu/keydocs/friibergh.htm

of environment and resource use, and world views and ethics concerning human–nature relationships. *Ecological systems* (ecosystems) refer to self-regulating communities of organisms interacting with one another and with their environment. When we wish to emphasize the integrated concept of humans-in-nature, we use the terms *social–ecological systems* and *social–ecological linkages*, consistent with our earlier work (Berkes and Folke, 1998). We hold the view that social and ecological systems are in fact linked, and that the delineation between social and natural systems is artificial and arbitrary. The specific objectives of the volume are to investigate:

- how human societies deal with change in social–ecological systems, and
- how capacity can be built to adapt to change and, in turn, to shape change for sustainability.

Figure 1.1 sketches the scope of the inquiry. We consider change and the impact of change as universal givens. The social–ecological system is impacted by change and deals with it as a function of its capacity to adapt to change and shape it. We look for effective ways of analyzing the phenomenon

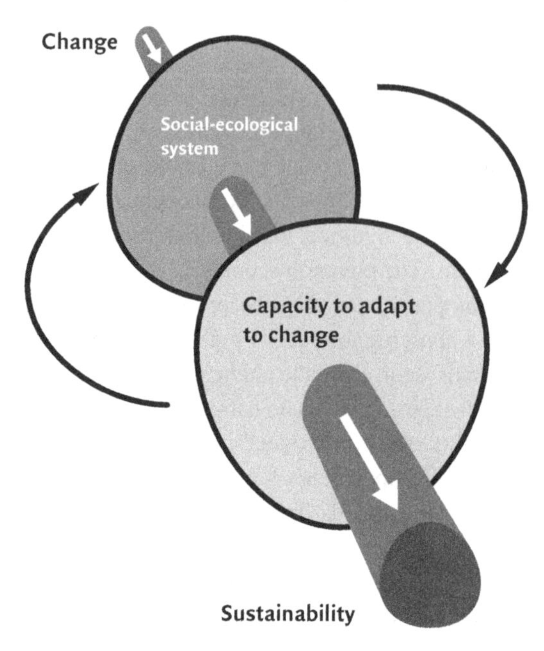

Figure 1.1 The focus on adaptive capacity for sustainability. Sustainability is viewed as a process, rather than an end-product, a dynamic process that requires adaptive capacity in resilient social–ecological systems to deal with change.

of change and how to respond to change in a manner that does not lead to loss of future options. We seek to analyze social–ecological system adaptability to meet novel challenges without compromising sustainability. The approach used in the volume is novel in that we are not focusing merely on environmental change or on social change but rather on social–ecological system change.

This chapter starts with the investigation of some of the implications of complexity in natural systems and in resource and environmental management systems. This is followed by a section that provides an overview of several integrative fields, such as common property and ecological economics that deal with integrated social–ecological systems and provide the starting point for many of the chapters in this volume. We then turn to explaining the rationale of the resilience approach. The systems we deal with are complex, but, as C.S. Holling points out, not *infinitely* complex. In seeking to integrate the two streams of thought, ecological system complexity and social system complexity, we use the idea of *resilience* as our organizing concept and scoping device. Thus, we deal with the issue of change and adaptation through the lens of resilience, which is the subject of the fourth section of this chapter.

1.2 Complex systems: ecology and resource management

A major change in the science of the last few decades has been the recognition that nature is seldom linear and predictable. Processes in ecology, economics and many other areas are dominated by nonlinear phenomena and an essential quality of uncertainty. These observations have led to the notion of *complexity*, developed through the work of many people and groups, notably the Santa Fe Institute (2002). Earlier challenges to the idea of linear causality and reductionistic science go back to general systems theory developed in the 1930s and 1940s (von Bertalanffy, 1968). General systems theory is concerned with the exploration of *wholes* and *wholeness*. It emphasizes connectedness, context and feedback, a key concept that refers to the result of any behavior that may reinforce (positive feedback) or modify (negative feedback) subsequent behavior. It argues that the understanding of the essential properties of the parts of a system comes from an understanding of not only these components but of their interrelations as well. Understanding comes from the examination of how the parts operate together, and not from the examination of the parts themselves in isolation.

With the science of complexity (Costanza *et al.*, 1993; Kauffman, 1993; Holland, 1995; Levin, 1999a), a new understanding of systems is emerging to augment general systems theory. A complex system can be distinguished from one that is simple – one that can be adequately captured using a single perspective and a standard analytical model, as in Newtonian mechanics and gas laws. By contrast, a complex system often has a number of attributes not observed in simple systems, including nonlinearity, uncertainty, emergence, scale, and self-organization.

Nonlinearity is related to inherent uncertainty. Mathematical solutions to nonlinear equations do not give simple numerical answers but instead produce a large collection of values for the variables that satisfy an equation. The solutions produce not one simple equilibrium but many equilibria, sometimes referred to as stable states or stability domains, each of which may have their own threshold effects (Scheffer *et al.*, 2001). Complex systems organize around one of several possible equilibrium states or attractors. When conditions change, the system's feedback loops tend to maintain its current state – up to a point. At a certain level of change in conditions (threshold), the system can change very rapidly and even catastrophically (called a flip). Just when such a flip may occur, and the state into which the system will change, are rarely predictable. If so, Holling (1986) pointed out, phenomena such as climate change would hardly be expected to proceed smoothly and predictably, and he drew attention to a system's resilience as a critical factor in environmental management. Resilience may be considered an emergent property of a system, one that cannot be predicted or understood simply by examining the system's parts. Resilience absorbs change and provides the capacity to adapt to change, as defined later and as illustrated in several chapters of this volume.

Scale is important in dealing with complex systems. A complex system is one in which many subsystems can be discerned. Many complex systems are hierarchic – each subsystem is nested in a larger subsystem, and so on (Allen and Starr, 1982). For example, a small watershed may be considered an ecosystem, but it is part of a larger watershed that can also be considered an ecosystem and a larger one that encompasses all the smaller watersheds. Similarly, institutions may be considered hierarchically, as a nested set of systems from the local level, through regional and national, to the international. Phenomena at each level of the scale tend to have their own emergent properties, and different levels may be coupled through feedback relationships (Gunderson and Holling, 2002). Therefore, complex systems should be analyzed or managed simultaneously at different scales. Consider, for example, biodiversity conservation. Problems and solutions of conservation at the genetic level are considerably different from those at the species level or the landscape level. Different groups of conservationists focus on different levels; they may use different research approaches and may recommend different policies. Biodiversity can be considered at different levels of the scale. However, because there are strong feedbacks among the genetic, species, and landscape levels, there is coupling between different levels, and the system should be analyzed simultaneously across scale.

Self-organization is one of the defining properties of complex systems. The basic idea is that open systems will reorganize at critical points of instability. Holling's adaptive renewal cycle, discussed later in the section on resilience, is an illustration of reorganization that takes place within cycles of growth and renewal (Gunderson and Holling, 2002). The self-organization principle, operationalized through feedback mechanisms, applies to many biological systems, social systems and even to mixtures of simple chemicals. High-speed computers and nonlinear mathematical techniques help simulate self-organization by yielding complex results and yet strangely ordered effects. For example, for many complex systems such as genes, Kauffman (1993) argues that spontaneous self-organization is not random but tends to converge towards a relatively small number of patterns or attractors. At each point at which new organization emerges, the system may branch off into one of a number of possible states. The direction of self-organization will depend on such things as the system's history; it is path dependent and difficult to predict.

These characteristics of complex systems have a number of rather fundamental implications for resource and environmental management. In this chapter we deal with three of them: (1) the essential inadequacy of models and perspectives based on linear thinking; (2) the recognition of the significance of qualitative analysis as a complement to quantitative approaches; and (3) the importance of using a multiplicity of perspectives in the analysis and management of complex systems.

The inadequacy of conventional resource management models and output objectives, such as the maximum sustainable yield (MSY) in fisheries, has been discussed for some time. For example, Larkin (1977) pointed out in a seminal paper that MSY assumes away such complexity as food-web relations in trying to predict single species yields. These models often do not work. However, the issue is more than the ecological shortcomings of a few management tools such as MSY. There is a more fundamental problem. The conventional wisdom in much of twentieth-century ecology is based on the idea of single equilibria. Although most ecologists no longer hold the popular idea of a 'balance of nature,' many of them consider population phenomena in the framework of equilibria and consider population numbers, and ecosystem behavior in general, to be predictable, at least in theory. To be sure, very few ecologists would consider predictive models in ecology as easy to achieve. But there is a fundamental difference between the view that quantitative prediction is *difficult* and data intensive ('we need more research') and the view that nature is *not* equilibrium centered and *inherently* unpredictable. For much of ecology and resource management science, complexity is a subversive idea that challenges the basis of population and yield models.

Recognizing the importance of qualitative analysis is one consequence of the recognition of complex system phenomena for natural resource management (Box 1.1). By qualitative analysis we mean the understanding of the system's behavior to help guide management directions. Qualitative analysis follows from the nature of nonlinearity. Because there are many possible mathematical solutions to a nonlinear model and no one 'correct' numerical answer, simple quantitative output solutions are not very helpful (Capra, 1996). This does not imply that quantitative analysis is not useful. Rather, it means that there is an appropriate role for both quantitative and qualitative analyses, which often complement each other.

Some of this qualitative management thinking has been put to work. Managers may specify objectives in the form of management directions and the understanding of key processes for sustainability. For example, Lugo (1995) pointed out that trying to quantify supposedly sustainable levels of yield in tropical forests rarely leads to ecosystem sustainability. If the objective is conservation, a strategy of focusing on resilience, through an understanding of regeneration cycles and ecological *processes* such as plant succession, may be the key to tropical forest sustainability.

In the area of fisheries, some managers are beginning to experiment with the use of reference directions (e.g., increasing the number of sexually mature year-classes in the population or reducing the proportion of immature individuals in the catch) instead of the conventional target reference points (e.g., a catch of 1000 tons of a particular species). Note that using reference directions, rather than targets, still requires quantitative data, but the choice of the management direction itself is a qualitative decision. This alternative approach shifts the focus of management action from the exacting and difficult question 'where do we want to be?' to the simpler and more manageable 'how do we move from here towards the desired direction?' (Berkes *et al.*, 2001: 131).

The need to use a multiplicity of perspectives follows from complex systems thinking. Because of a multiplicity of scales, there is no one 'correct' and all-encompassing perspective on a system. One can choose to study a particular level of biodiversity conservation; but the perspective from that particular level will be different from the perspective from another. In complex systems, time flows in one direction, i.e., time's arrow is not reversible. Especially with social systems, it is difficult or impossible to understand a system without considering its history, as well as its social and political contexts. For example, each large-scale management system (e.g., Gunderson *et al.*, 1995) or each local-level common property system (e.g., Ostrom, 1990) will have its unique history and context. A complex social–ecological system cannot be captured using a single perspective. It can be best understood by the use of a multiplicity of perspectives.

These considerations provide an insight into the reasons that conventional scientific and technological approaches to resource and ecosystem management are not working well, and in some cases making problems worse. In part, this failure is related to the focus on wrong kinds of sustainability and on narrow types of scientific practice (Holling, Berkes, and Folke, 1998). In part, it is related to the ideology of a strongly positivist resource management science, with its emphasis on centralized institutions and command-and-control resource management. Such management is based on a thinking of linear models and mechanistic views of nature. It aims to reduce natural variation in an effort to make an ecosystem more productive, predictable, economically efficient, and controllable. But the reduction of the range of natural variation is the very process that may lead to a loss of resilience in a system, leaving it more susceptible to resource and environmental crises (Holling and Meffe, 1996).

Taken together, these implications of complex systems thinking suggest the need for a new kind of resource and environmental management science that takes a critical view of the notions of control and prediction. Holling (1986) called it the 'science of surprise.' An appropriate metaphor may be the message on the sign that appears on some remote logging roads on Vancouver Island in Canada: 'Be prepared for the unexpected.'

The lesson from complex systems thinking is that management processes can be improved by making them adaptable and flexible, able to deal with uncertainty and surprise, and by building capacity to adapt to change. Holling (1978) recognized early on that complex adaptive systems required adaptive management. *Adaptive management* emphasizes learning-by-doing, and takes the view that resource management polices can be treated as 'experiments' from which managers can learn (Walters, 1986; Gunderson, 1999). Organizations and institutions can 'learn' as individuals do, and hence adaptive management is based on social and institutional learning. Adaptive management differs from the conventional practice of resource management by emphasizing the importance of *feedbacks* from the environment in shaping policy, followed by further systematic experimentation to shape subsequent policy, and so on. Thus, the process is iterative, based on feedback learning. It is co-evolutionary, involving two-way feedback between management policy and the state of the resource (Norgaard, 1994), and leading to self-organization through mutual feedback and entrainment (Colding and Folke, 1997).

1.3 Integrative approaches to social–ecological systems: an overview

Many of the principles of complex systems apply to both natural systems and social systems. Some of these principles or ideas, for example the importance of context and history in understanding a system, probably make more intuitive sense to social scientists than to natural scientists. Our effort in this volume is to seek principles and ideas which make sense to both natural scientists and social scientists and which can be mobilized towards our objective of examining how human societies deal with change in social–ecological systems, and how they can build capacity to adapt to change.

Until recent decades, the point of contact between social sciences and natural sciences was very limited in dealing with social–ecological systems. Just as mainstream ecology had tried to exclude humans from the study of ecology, many social science disciplines had ignored environment altogether and limited their scope to humans. The unity of biosphere and humanity had been sacrificed to a dichotomy of nature and culture. There were exceptions, of course, and some scholars were working to bridge the nature–culture divide (e.g., Bateson, 1979); we deal with some of them in Chapter 3. But, by and large, models of human societies in many social science disciplines did not include the natural environment. This changed in the 1970s and the 1980s with the rise of several subfields allied with the social sciences but explicitly including the environment in the framing of the issues. Six of these integrative areas are directly relevant to the perspectives of this volume: environmental ethics, political ecology, environmental history, ecological economics, common property, and traditional ecological knowledge. We describe each briefly here because many of the chapters in this volume borrow from the approaches and terminology of these fields.

Environmental ethics arose from the need to develop a philosophy of relations between humans and their environment, because conventional ethics only applied to relations among people. A number of schools of environmental ethics have emerged, including the ecosophy of Naess (1989). Particularly relevant to this volume, a discussion has developed on the subject of worldviews, pointing out that there is a wide diversity of spiritual and ethical traditions in the world that helps offer alternatives to the current views of the place of humans in the ecosystem (Callicott, 1994). Culturally different attitudes towards the environment have implications for the management of the environment, even though there is no clear correspondence between ethical traditions and their actual performance (Berkes, 2001). Some of the literature on environmental ethics emphasized belief systems (religion in the broad sense) as encoding wise environmental management. For example, Anderson (1996: 166) argued that 'all traditional societies that have succeeded in managing resources well, over time, have done it in part through religious or ritual representation of resource management.'

Political ecology grew out of the field of political economy, but it is different from political economy that tends to reduce everything to social constructions, disregarding ecological relations. 'Political ecology expands ecological concerns to respond to the inclusion of cultural and political activity within an analysis of ecosystems that are significantly but not always entirely socially constructed' (Greenberg and Park, 1994). The analysis of political ecology often starts by focusing on political–economic divisions among the actors. These may be divisions between local and international interests, between North and South; they may involve power relations based on differences of class, ethnicity, and gender (Blaikie and Jeanrenaud, 1996). The political ecology perspective compels the analyst to consider that there exist different actors who define knowledge, ecological relations, and resources in different ways and at different geographic scales. Actors will bring different cultural perspectives and experience, and may use different definitions in pursuit of their own political agendas (Blaikie, 1985; Blaikie and Jeanrenaud, 1996). With its explicit attention to the multiplicity of perspectives and to scale issues, political ecology fits well with systems thinking.

The rich accumulation of material documenting relationships between societies and their environment (Turner *et al.*, 1990) has given rise to a discipline identified as environmental history (Worster, 1988) or historical ecology (Balee, 1998). Investigating the root causes of environmental problems, environmental historians discussed, among others things, how ecological relations became more destructive as they became more distant, especially after the great transformation following the Industrial Revolution (Worster, 1988). They not only interpreted ancient landscapes but also analyzed the *dynamics* of these landscapes, making ecological sense of resource use practices, and their change that *resulted* in these landscapes. For example, Cronon (1983) studied the colonization of New England states, and found that the early European–Indian relationship could be characterized in terms of two competing economies. The Indian economy treated the environment as a portfolio of resources and services that supported livelihoods, whereas that of the colonists turned the environment into commodities, sequentially depleting one resource after another. Similarly, the push for valuable timber production under colonialism in India resulted in the commodification of resources serving diverse livelihood needs, and the depletion of certain species (Gadgil and Guha, 1992).

Ecological economics examines the link between ecology and economics. Taking issue with conventional economics that often downplays the role of the environment, and conventional ecology that ignores humans, ecological economics tries to bridge the two disciplines to promote an integrated view of economics within the ecosystem (Costanza, 1991). Among the defining characteristics of ecological economics are: the view of the economic system as a subset of the ecological system; a primary interest in natural capital; a greater concern with a wider range of values; and longer time horizons than those normally considered by economists. Ecological economics has helped reconceptualize systems problems such as conservation by shifting attention from the elements of the system to the structures and processes that perpetuate that system (Costanza, Norton, and Haskell, 1992). For example, biodiversity can be seen as providing ecosystem insurance, and redundancy as a mechanism to provide adaptive capacity in an ecosystem characterized by hierarchical organization, scale effects, and multiple equilibria (Barbier, Burgess, and Folke, 1994; Perrings *et al.*, 1995).

Ecological economics makes a distinction between *human-made capital,* generated through economic activity through human ingenuity and technological change, and *natural capital*, consisting of non-renewable resources extracted from ecosystems, renewable resources produced by the processes and functions of ecosystems, and ecological services sustained by the workings of ecosystems (Jansson *et al.*, 1994). To these, a third kind of capital may be added: *cultural capital* refers to the factors that provide human societies with the means and adaptations to deal with the natural environment and to actively modify it. Ecological knowledge and institutions, important for the arguments in this volume, are considered to be a part of this cultural capital (Berkes and Folke, 1994).

The field of common property examines the linkages between resource management and social organization, analyzing how institutions and property-rights systems deal with the dilemma of the 'tragedy of the commons' (McCay and Acheson, 1987; Berkes, 1989; Bromley, 1992; Ostrom *et al.*, 1999). The emphasis is on *institutions*, defined as 'humanly devised constraints that structure human interaction ... made up of formal constraints (rules, laws, constitutions), informal constraints (norms of behavior, conventions and self-imposed codes of conduct), and their enforcement characteristics' (North, 1994). Institutions are the set of rules actually used or the working rules or rules-in-use (Ostrom, 1992). However, they are also socially constructed, with normative and cognitive dimensions (Jentoft, McCay, and Wilson, 1998), particularly relevant to this volume in dealing with the nature and legitimacy of different kinds of knowledge.

Institutions of key importance are those that deal with property rights and common-property resources. *Property* refers to the rights and obligations of individuals or groups to use the resource base (Bromley, 1991; Hanna, Folke, and Mäler, 1996). It is a bundle of entitlements defining owner's rights, duties, and responsibilities for the use of the resource, or a claim to a benefit or income stream (Bromley, 1992). *Common-property (common-pool) resources* are defined as a class of resources for which exclusion is difficult and joint use involves subtractability (Berkes, 1989; Feeny *et al.*, 1990).

Local, indigenous or traditional knowledge refers to ecological understanding built, not by experts, but by people who live and use the resources of a place (Warren, Slikkerveer, and Brokensha, 1995). *Local knowledge* may be used as a generic term referring to knowledge generated through observations of the local environment in any society, and may be a mix of practical and scientific knowledge (Olsson and Folke, 2001). *Indigenous knowledge* (IK) is used to mean local knowledge held by indigenous peoples, or local knowledge unique to a given culture or society (Warren *et al.*, 1995). In this volume, we use *traditional ecological knowledge* (TEK) more specifically to refer to 'a cumulative body of knowledge, practice and belief, evolving by adaptive processes and handed down through generations by cultural transmission, about the relationship of living beings (including humans) with one another and with their environment' (Berkes, 1999: 8). The word *traditional* signifies historical and cultural continuity, but at the same time we recognize that societies are in a dynamic process of change, constantly redefining what is considered 'traditional.'

TEK started attracting attention through the documentation of a tremendously rich body of environmental knowledge among a diversity of groups outside the mainstream Western world (Johannes, 1981; Colding and Folke, 1997; Berkes *et al.*, 1998, Berkes, Colding, and Folke, 2000; Folke and Colding, 2001). The relationship between TEK and science is controversial, but these two kinds of knowledge should not be thought of as opposites. Rather, it is more useful to emphasize the potential complementarities of the two (e.g., Berkes, 1999; Riedlinger and Berkes, 2001). We deal with local/traditional knowledge for diversity and conceptual pluralism to expand the range of information and approaches for improving resource management.

Each of the six areas summarized here is a 'bridge' spanning different combinations of natural science and social science thinking. Environmental ethics, political ecology, and environmental history help emphasize that all of the examples in this volume have a cultural, historical, political, and ethical context, as seen in several of the chapters. Various chapters build on and contribute to the literature of ecological economics, common property and TEK. The search for resource management alternatives often includes the ecological economics notions of economic systems-within-ecosystems, natural capital, and intergenerational equity. The questions of the control of property rights, the nature of institutions, and their cross-scale interactions are key considerations in many of the chapters. Complexity draws attention to the fact that local and traditional knowledge and management systems should be seen as *adaptive responses* in a place-based context and a rich source of lessons for social–ecological adaptations.

1.4 Social–ecological resilience

Holling (1973) introduced the resilience concept into the ecological literature as a way to understand nonlinear dynamics, such as the processes by which ecosystems maintain themselves in the face of perturbations and change (Gunderson, 2000). As defined by the Resilience Alliance (2002), and as used in this volume, it has three defining characteristics:

- the amount of change the system can undergo and still retain the same controls on function and structure, or still be in the same state, within the same domain of attraction;
- the degree to which the system is capable of self-organization; and
- the ability to build and increase the capacity for learning and adaptation.

To illustrate the first characteristic, consider the case of insectivorous birds and insect outbreaks in the boreal forests of Canada (Holling, 1988). The assemblage of migratory insectivorous bird populations is one of the controlling factors of forest renewal produced by budworm population cycles. The existence of these birds contributes to the resilience of the boreal forest. Mathematical simulations based on long-term studies indicate that the total bird population would have to be reduced by about 75 percent before the system might flip out of the current domain of attraction and into a different one (Holling, 1988).

As the populations of these birds are reduced because of overwintering habitat loss or other factors, the resilience of the boreal forest is also reduced. As a system loses its resilience, it can flip into a different state when subjected to even small perturbations (Levin *et al.*, 1998). Loss of resilience can be modeled or viewed as having a system moved to a new stability domain and being captured by a different attractor. Examples include the transformation of productive grasslands in subtropical Africa into thorny shrublands as a consequence of poor cattle management practices (Perrings and Walker, 1995). It is important to note that the actual point of change cannot easily be predicted. There are threshold effects; the changes are relatively sudden – not necessarily gradual or smooth. Recovery can be costly or nearly impossible (Mäler, 2000), and such flips can be irreversible (Levin, 1999a).

Thus, resilience is concerned with the magnitude of disturbance that can be absorbed or buffered without the system undergoing fundamental changes in its functional characteristics. The issue of disturbance is important. Not only are there natural disturbances, such as forest fires and insect outbreaks, but many human activities, such as resource use and pollution, which also create disturbances. Ecosystem responses to resource use, and the reciprocal response of people to changes in ecosystems, constitute coupled, dynamic systems that exhibit adaptive behavior (Gunderson *et al.*, 1995). This recognition brings into focus the second and third defining characteristics of resilience, those regarding self-organization and learning. It underscores the importance of considering linked social–ecological systems, rather than ecosystems or social systems in isolation (Berkes and Folke, 1998).

Resilience is an important element of how societies adapt to externally imposed change, such as global environmental change. The adaptive capacity of all levels of society is constrained by the resilience of their institutions and the natural systems on which they depend. The greater their resilience, the greater is their ability to absorb shocks and perturbations and adapt to change. Conversely, the less resilient the system, the greater is the vulnerability of institutions and societies to cope and adapt to change (Adger, 2000). Social–ecological resilience is determined in part by the livelihood security of an individual or group. Such security involves, according to Sen (1999), the questions of entitlements and access to resources, the distribution of which is a key element of environmental justice.

The concept of resilience is a promising tool for analyzing adaptive change towards sustainability because it provides a way for analyzing how to maintain stability in the face of change. A resilient social–ecological system, which can buffer a great deal of change or disturbance, is synonymous with ecological, economic, and social sustainability. One with low resilience has limited sustainability; it may not survive for a long time without flipping into another domain of attraction. Here, it should be noted, resilience is not being defined as returning to an equilibrium. This is because we are using a view of ecosystems in which there is no one equilibrium but rather, as a consequence of complexity, multiple states or domains of attraction and multiple equilibria. Thus, ecological stability as a concept is not very useful, and resilience cannot be defined as bouncing back to equilibrium – there is no equilibrium to bounce back to.

In operationalizing this view of resilience, managing for sustainability in socio-economic systems means not pushing the system to its limits but maintaining diversity and variability, leaving some slack and flexibility, and not trying to optimize some parts of the system but maintaining redundancy. It also means learning how to maintain and enhance adaptability, and understanding when and where it is possible to intervene in management. These 'soft' management approaches are necessary because 'hard' management approaches involving quantitative targets for resource production etc. often do not work. Linear models on which 'hard' management depends tend to be incomplete or even misleading in the management of the ecosystems of the world. Equilibrium-based predictive models do not perform well with complex social–ecological systems.

To illustrate policy implications of complexity, Wilson (2000) pointed out with respect to ocean fisheries that the current linear models of resource production (as in single-species management) have to be replaced with a view of ocean ecosystems as multiscale and hierarchical, and the current predominantly top-down institutions with a cross-scale institutional design that matches the hierarchical scale of marine ecosystems. 'These suggested changes in scientific perspective and institutional design will not necessarily solve scientific uncertainties. But they will replace those uncertainties in an institutional context which encourages learning and stewardship' (Wilson, 2000).

Gunderson and Holling (2002) embarked on the volume *Panarchy* with the idea that sustainable futures were inherently unpredictable, rejecting the idea that sustainability can be planned in a rational fashion. In the absence of a linear, mechanical universe that would have permitted simple, rational measures, they argued that the best bet for sustainability involves what we have referred to as the second and third characteristics of resilience – capability for self-organization and capacity for learning and adaptation. Gunderson and Holling provide a synthesis of existing theory for sustainability, complexity, and resilience, and attempt to develop novel extensions of that integration, identifying gaps in knowledge. Several of their conclusions are of significance for the present volume. They find that key unknowns lie in the development of theories to address self-organization at various scales, and to address adaptive change in social–ecological systems.

Another cluster of challenges is in the area of institutions: how do we design institutions and incentive structures that sustain and enhance sources of self-organization and resilience? How can we formulate patterns of emergence of social control mechanisms dealing with environmental problems? How can we create policies to increase the speed of emergence and increase the efficiency of learning? A third cluster of gaps in knowledge concerns the dynamics of disturbance, crisis, response to change, and renewal: how do we facilitate constructive change? Protect and preserve accumulated experience? Build and sustain the capacity of people, economies, and nature for dealing with change?

Gunderson and Holling note that the last decade of the twentieth century saw a cascade of regional and global transformations, biophysical, economic, and political. Such 'gales of change,' they observe, signal periods when the *backloop* of the adaptive renewal cycle dominates, the part of the cycle dealing with disturbance, crisis, response to change, and renewal. To understand the significance of the backloop, we need to review Holling's concept of adaptive renewal cycle.

1.5 Adaptive renewal cycle: emphasis on the backloop

Chapters of the present volume deal with cyclic change as an essential characteristic of all social and ecological systems. Our starting point is the pervasive idea that social systems and ecological systems are dynamic. More specifically, Holling (1986) has argued that ecosystems go through regular cycles of organization, collapse, and renewal. For example, a forest goes through the stages of growth and maturity, followed by a disturbance, such as fire, which releases the nutrients on the way to a new cycle of growth. A business cycle may consist of a company starting up and growing. The company will eventually decline and go out of business, while its parts and the accumulated experience may combine with other sources and reorganize into a new business. Empires start as small states, growing large and eventually collapsing, but giving rise to new nation states and leaving behind organizational legacies in the process. Cyclic change, including birth–death cycles and seasonal cycles, is so ubiquitous in the world that the importance of cycles has been embedded in many traditions of ancient wisdom, including Hinduism and American Indian religions. However, the less wise may see but not recognize the cycle. What may appear as a linear change

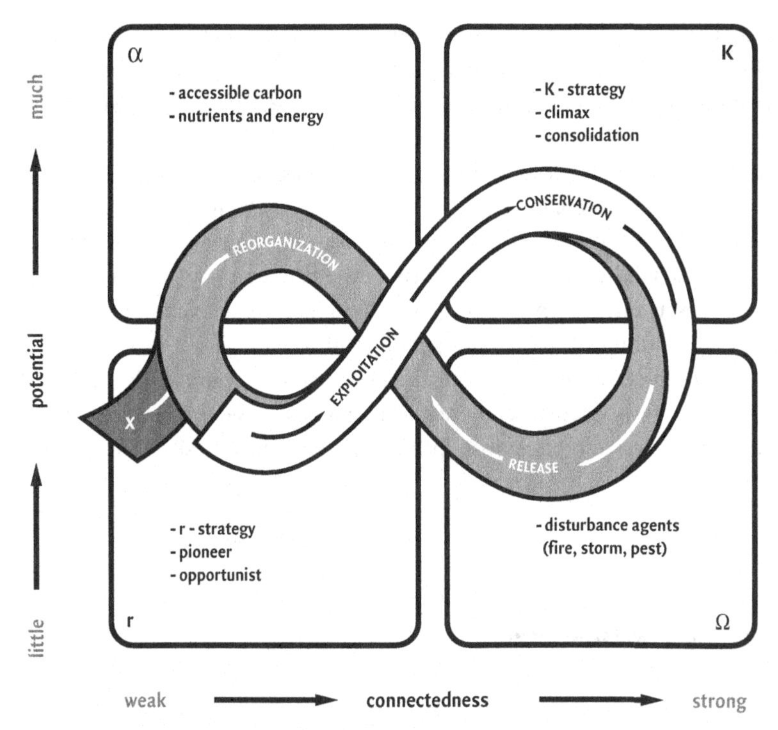

Figure 1.2 The adaptive renewal cycle. A heuristic model of the four system stages and the flow of events among them. The cycle reflects changes in two properties: (1) y-axis: the potential that is inherent in the accumulated resources and structures; (2) x-axis: the degree of connectedness among controlling variables. The exit (marked with an X) from the cycle indicated at the left of the figure suggests, in a stylized way, the stage where the potential can leak away and where a shift is most likely into a less productive and organized system. The shaded part of the cycle is termed the 'backloop' (Holling, 1986, 2001) and concerns the release and reorganization phases.

(e.g., growth) at one temporal scale may in fact be part of a cycle when viewed from a higher-order temporal scale.

Holling's *adaptive renewal cycle* is an attempt to capture some of the commonalities in various kinds of cyclic change (Fig. 1.2). The heuristic model probably does not capture the unique characteristics of different kinds of cycles and the possibilities of divergent responses. But it does provide the insight, for example, that forest succession should be seen, not as a unidirectional process (with climax as endpoint), but as one phase of a cycle in which a forest grows, dies, and is renewed. The cycle in Figure 1.2 consists of four phases: *exploitation, conservation, release*, and *reorganization.*

In a resilient forest ecosystem, these four stages repeat themselves again and again. The first two phases, exploitation (the establishment of pioneering species) and conservation (the consolidation of nutrients and biomass), lead to a climax, in the terminology of classical ecology. But this climax system *invites* environmental disturbances such as fire, insect pest outbreak or disease, and is more susceptible to these disturbances than non-climax forests. When surprise occurs, the accumulated capital is suddenly released, producing other kinds of opportunity, termed creative destruction. Release, which is a very rapid stage, is followed by reorganization in which, for example, nutrients released from the trees by fire will be fixed in other parts of the ecosystem as the renewal of the forest starts again. It is in the reorganization phase that novelty and innovation may occur (Holling, 1986; Holling *et al.*, 1995).

As a complex system, the forest ecosystem is hierarchically scaled. The term *panarchy* is used to capture the dynamics of adaptive cycles that are nested within one another across space and time scales, as shown in Figure 1.3

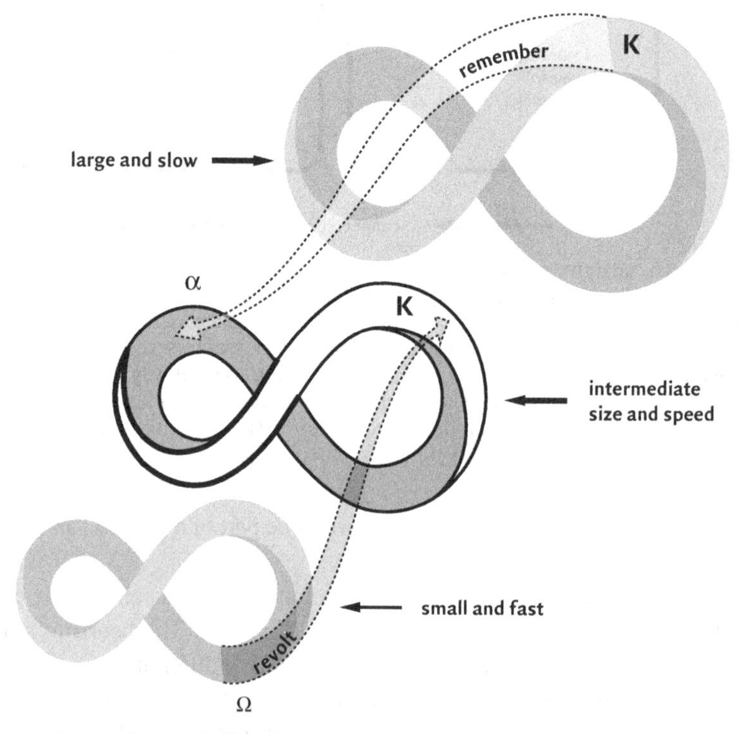

Figure 1.3 Adaptive renewal cycles nested across scales: panarchy. The 'revolt' connection between scales can cause a critical change in one cycle to cascade up to a stage in a larger and slower one. The 'remember' connection facilitates renewal and reorganization by drawing on the memory that has been accumulated and stored in a larger, slower cycle. The 'revolt' and 'remember' connections are exemplified in several of the chapters of the volume and discussed in Chapter 14 in relation to crisis and social–ecological memory. Adapted from Gunderson and Holling (2002).

(Gunderson and Holling, 2002; Holling, 2001). For example, the smallest and the fastest of the three nested 'reclining figure eights' may refer to a tree crown, the intermediate one to a forest patch, and the largest and the slowest to a forest stand. Each level may go through its own cycle of growth, maturation, destruction, and renewal. For institutions, those three speeds might consist of operational rules, collective choice rules, and constitutional rules (Ostrom, 1990). For knowledge systems, the corresponding three scales might be local knowledge, management institutions, and worldview (Folke, Berkes, and Colding, 1998a).

There are many possible connections between phases at one level and phases at another level. The two connections in Figure 1.3 labeled 'revolt' and 'remember' seem to be particularly significant in the context of building resilience. An ecological example of revolt is a small ground fire that spreads to the crown of a tree, then to a patch in the forest, and then to a whole stand of trees. Each step in that cascade of events moves the transformation to a larger and slower level. A societal example may be the transformation of regional organizations by a local activist group.

'Remember' is a cross-scale connection important in times of change, renewal, and reorganization. For example, following a fire in a forested ecosystem, the reorganization phase draws upon the seed bank, physical structures, and surviving species that had accumulated during the previous cycle of growth of the forest, plus those from the outside. Thus, renewal and reorganization are framed by the memory of the system. Each level operates at its own pace, protected by slower, larger levels but invigorated by faster, smaller cycles. The panarchy is therefore both creative and conservative (Holling, 2001) through the dynamic balance between change and memory, and between disturbance and diversity. All living systems, ecological as well as social, exhibit properties of the adaptive cycle, and are nested across scales (Gunderson and Holling, 2002). Several of the chapters provide examples, and the point will be developed further in the synthesis chapter.

Many theories on the management of natural resources and ecosystems have focused on the exploitation and conservation phases of the renewal cycle in order to make management more efficient. This emphasis can be seen in resource management, geared for economic production, that commonly seeks to reduce natural variation in target resources because fluctuations impose problems for the industry that depend on those resources (Holling and Meffe, 1996). Controlling variation, as in the form of natural disturbances, is key in many conventional management systems. This control can be achieved in a number of ways, for example by increased financial investments in harvesting technologies and through energy inputs, such as insecticides, pesticides, and irrigation, as in conventional agriculture. The system is assumed to be stable as long as change can be controlled.

Such measures seek to maintain the system in a configuration of 'optimality,' in the conservation domain characterized by high levels of stored capital. In the forest case, for example, a great deal of planning goes into shortening the growth and succession stages so that the forest reaches the conservation phase, with a high standing crop or biomass of trees. Using a command-and-control approach, managers then try to keep the forest in that state of optimality. Such management may be effective in the short term, but over time, it may reduce resilience in management systems and in the ecosystem itself by making them more vulnerable to disturbances and surprises that cannot be anticipated in advance (Baskerville, 1995; Holling and Meffe, 1996).

Compared to this single-minded interest in the exploitation and conservation phases of the renewal cycle, conventional resource management has largely ignored the release and reorganization phases (Fig. 1.2). Yet, these two *backloop* phases are just as important as the other two (exploitation and conservation phases) in the overall cycle (Folke *et al.*, 1998a). Furthermore, they are of great interest in their own right for a number of reasons.

Crises have a constructive role to play in resource management by triggering the opportunity for renewal, in systems capable of learning and adapting (Gunderson *et al.*, 1995). In economics, Schumpeter (1950) coined the term *creative destruction* to describe the window of opportunity for novelty and creation that was generated by the failures of existing industrial plants with their old technologies. *Novelty*, or the ability to innovate, is an essential element of adaptability and hence of resilience. Of fundamental importance for self-organization is *memory* – memory that allows a system the ability to reorganize after a disturbance. *Memory* is the accumulated experience and history of the system, and it provides the sources for self-organization and resilience. It has both ecological and social components.

Ecological memory is the composition and distribution of organisms and their interactions in space and time, and includes the life-history experience with environmental fluctuations (Nyström and Folke, 2001). Ecological memory includes the species and patterns that persist in a particular area after a disturbance event, together with support areas and the links that connect the disturbed area to the sources of species assemblages that allow reorganization of the system. We return to this concept in more detail in the final chapter.

Social memory refers to the long-term communal understanding of the dynamics of environmental change and the transmission of the pertinent experience, as used, for example, in the context of climate change (McIntosh, 2000: 24). It captures the experience of change and successful adaptations.

Social memory is the arena in which captured experience with change and successful adaptations, embedded in a deeper level of values, is actualized through community debate and decision-making processes into appropriate strategies for dealing with ongoing change (McIntosh, 2000). Memory is an important component of resilience and reductions in social–ecological memory increase the probability for shifts in stability domains (indicated by X on the left side of Fig. 1.2).

There is evidence that some social–ecological systems build resilience through the experience of disturbance, provided that there is memory in the system in the form of both ecological and social sources for reorganization (Berkes and Folke, 2002). This suggests that disturbances may be important for a social–ecological system to 'exercise' its problem-solving skills, and to innovate and adapt. We return to the concepts of social and ecological memory in the concluding chapter to develop conceptual models for integrated social–ecological memory and its role in self-organization.

1.6 The approach and content of the book

Chapters of the present volume emphasize the need to focus on the release and reorganization phases. The change processes captured in these two phases are significant in understanding the dynamics of building adaptive capacity towards sustainability and the well-being of society. Case studies basically deal with disturbance, crisis, and response to change, and their dynamics. The book addresses the resilience of linked social and ecological systems undergoing change, arguing that management systems that fail to address the release and reorganization phases may lose adaptive capacity and become ecologically and economically brittle. The term *navigating* in the title of the book is meant to capture the dynamic process of building adaptive capacity towards sustainability.

We seek to contribute to the search for new approaches, with visions of smaller-scale, more environmentally sound and more democratic and nested resource management systems that are self-organizing, adaptive, and resilient. Our challenge is that the management of such systems is 'made especially difficult by the fact that the putative controllers (humans) are essential parts of the system and, hence, essential parts of the problem,' as Levin (1999b) put it. *Linking Social and Ecological Systems* included a rich set of cases of alternative approaches, and suggested that many of these systems were based on local ecological knowledge and local institutions of the resource users themselves (Folke *et al.*, 1998a). Some of the cases were characterized by decentralized, pluralistic approaches, as had been noted in previous studies of common property institutions; some were based on combinations of local and scientific knowledge; and some on historically accumulated and culturally transmitted knowledge.

The major objective of the *Linking* project was to create a transdisciplinary framework through which we could evaluate management practices based on local ecological knowledge and understanding, and the social mechanisms behind them. Having addressed that objective, we turn in this volume to understanding the *dynamics* of ecosystem–institution linkages, with the more explicit objective of examining ways of *building resilience* to enhance the capacity to deal with change and surprise. Resilience increases the likelihood of avoiding shifts to undesirable stability domains, and provides flexibility and opportunity. Avoiding undesirable stability domains will be a major issue in an increasingly human-dominated world (Scheffer *et al.*, 2001; Jackson *et al.*, 2001).

The challenge is to analyze critical linkages in social–ecological systems, and to generate insights into how to interpret, respond to, and manage feedbacks from complex systems (Folke, Berkes, and Colding, 1998a). Also, we need to find ways to match the dynamics of ecosystems and the dynamics of social systems across scales towards social–ecological resilience (Folke *et al.*, 1998b).

Figure 1.4 defines the area of interest of the volume. On the left-hand side is the natural system, which may consist of nested ecosystems (e.g., a regional ecosystem containing the drainage basin of a river, which in turn consists of a number of constituent watershed ecosystems). On the right-hand side is a set of management practices in use. These practices are embedded in institutions, and the institutions themselves may be a nested set. The linkage between the ecosystem and management practice is provided by ecological knowledge and understanding.

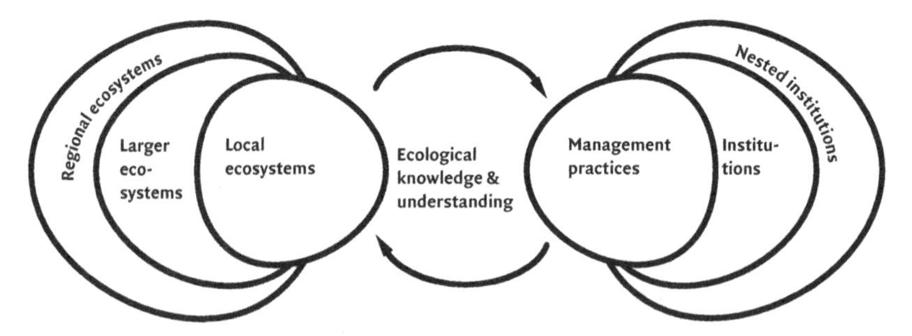

Figure 1.4 A conceptual framework for the analysis of linked social–ecological systems. The focus of the volume is on the dynamics of links among the ecosystem, knowledge (as reflected in management practice), and institutions and how to navigate these dynamics for resilience and adaptive capacity.

As sketched in Figure 1.4, the main focus of the volume is the dynamics of links between the ecosystem, knowledge as reflected in management practice, and institutions. The social–ecological system in Figure 1.4 is an open system. There are a number of influences that impinge on it, including factors such as population growth and urbanization, technology change, communication, effects of markets, international trade and globalization pressures, but the primary focus of the volume is ecosystem–knowledge–institution linkages.

The objectives of the volume (to investigate how human societies deal with change in social–ecological systems, and how capacity can be built to adapt to change and, in turn, to shape change for sustainability) are addressed through the investigation of these interrelationships of ecosystems and social systems. Dynamics of the system are addressed by focusing on four interrelated elements of change and resilience: (1) *disturbance*, which is an essential force in social and ecological change; (2) *diversity*, both social and ecological, which provides the sources for adaptive responses; (3) ecological *knowledge*, which informs institutions and management practice; and (4) *self-organization*, which uses the memory of the system for the renewal process.

Resource and environmental management that suppresses disturbance and diversity will be unsustainable (Gunderson *et al.*, 1995; Gunderson and Holling, 2002). Both disturbance and diversity are essential for building capacity to respond to change. As explored in the chapters of this volume, social–ecological resilience appears to be related to living with disturbance, nurturing diversity, combining sources of knowledge, and creating opportunity for self-organization. The interrelationships among these four elements of change are a recurrent theme in the volume, to which we return in the synthesis chapter.

The book has 13 chapters following this introduction chapter. Part I includes three chapters by Gunderson, Davidson-Hunt and Berkes, and Low *et al.*, which cover the concepts and theory behind the book. Gunderson provides a background emerging from explorations of the resilience idea, and discusses making sense of crisis and surprise, that is, when social systems and natural systems behave in unforeseen ways. He suggests ways of 'surfing' such social–ecological crises, and explores some of the strategies to manage for resilience. Davidson-Hunt and Berkes situate the notion of resilience in the literature that investigates the interface between social systems and natural systems. Evaluating a range of pertinent theories, they analyze the impact and significance of resilience thinking. The chapter by Low and colleagues deals with functional diversity in natural systems and institutional diversity. In linking diversity, resilience, and redundance, they draw attention to the phenomenon of redundance in complex and dynamic social–ecological systems, and its role during the phase of crisis and change, and note the similarities that may be observed in several kinds of resource systems.

Case studies are at the heart of the volume because we seek examples situated in particular places where practice informs theory, and grounded in particular cultural traditions. Part II includes case-study chapters by Carlsson, and Tengö and Hammer, as well as a synthesis of three cases by Colding *et al.* The chapters deal with a diversity of resource types and a diversity of social–ecological practices for managing ecosystem dynamics, illustrating the use of the resilience idea to study how human societies deal with change and build adaptive capacity. In each of the three cases, the emphasis is on disturbance and the ability of a social–ecological system to deal with crises, reorganize itself, and redevelop. To accomplish this, Carlsson investigates changes over a decade-to-century time scale in a case involving a Swedish forest region, demonstrating that diverse and well-organized property rights systems with local monitoring and mechanisms for risk spreading are parts of a strategy for social–ecological resilience. The chapter on agro-pastoralist communities in Northern Tanzania by Tengö and Hammer reveals a bundle of management practices and institutional arrangements for resilience building, and seeks to generate principles for adaptive strategies to deal with change. Colding *et al.* focus on major disturbances like cyclones, floods, and droughts, and show how three distinct groups (from Polynesia, Bangladesh, and Africa) have developed ecological knowledge and adapted to living with disturbance through a bundle of management practices stored in the social memory.

Part III also consists of case studies, with chapters focusing on aspects of knowledge. The chapter by Gadgil *et al.* uses cases from India, Sweden, and northern Canada to analyze the diverse ways in which the knowledge of local resource users can complement scientific knowledge to manage complex system dynamics. The chapter shows that knowledge needs to be embedded in institutions and social organization in an ongoing learning process to be effective. Dynamics of the learning process are also addressed in the chapter by Blann and colleagues. Dealing with adaptive management, Blann *et al.* employ a set of cases from Minnesota to show how government agencies and local groups interact in feedback learning to solve resource management problems in creative ways. Kendrick's chapter is about social and institutional learning and focuses on caribou co-management examples from northern Canada involving aboriginal groups. She explores the evolution of conceptual diversity in caribou resource management as a feedback process of social learning, and the building of trust and respect in a cross-cultural setting.

Part IV consists of chapters with cross-cutting themes that address the dynamics of nested institutions in relation to resource management in complex systems. Seixas and Berkes analyze a Brazilian coastal lagoon fishery that shows several collapse-and-recovery cycles over a 30-year period, and a variety of measures to deal with one problem after another. Their case study demonstrates how resource crisis triggers institutional renewal for ecosystem management at different organization levels and time scales. The case study of Alcorn *et al.* deals with an indigenous peoples' social movement in Indonesia. The movement creates a collective identity for the Dayak people by helping communities build solidarity to face loggers, map their territories, and renew traditional *adat* laws. The chapter reflects on the institutional response of moving from a lower (i.e., local) to a higher (national and international) level of organization in the face of external drivers. Trosper's case study on policy transformations provides insights into the social and political dynamics of the debate between competing interests and competing objectives in the management of national forests of the USA. He illustrates how different response strategies can either erode or enhance social–ecological resilience. He also shows the significance of developing approaches for sustainable ecosystem management to be available when space opens up for reorganization.

The cases and examples are chosen from a diversity of geographic areas, cultures, and resource types to help provide a robust analysis of resilience, change, and adaptive capacity. The chapters are designed to explore different but overlapping aspects of the questions of how human societies deal with change in social–ecological systems, and how resilience, or the capacity to adapt to and shape change, may be nurtured and enhanced. The concluding chapter by the editors presents a synthesis of new insights into the dynamics of linked social–ecological systems for resource and ecosystem management drawing on the case studies. The authors of the volume include both academics and practitioners who come from a diversity of backgrounds, and the authorship represents several social science and natural science disciplines.

References

Adger, W. N. 2000. Social and ecological resilience: are they related? *Progress in Human Geography* 24: 347–64.

Allen, T.F.H. and Starr, T.B. 1982. *Hierarchy: Perspectives for Ecological Complexity.* Chicago: University of Chicago Press.

Anderson, E.N. 1996. *Ecologies of the Heart. Emotion, Belief, and the Environment.* New York: Oxford University Press.

Arrow, K., Bolin, B., Costanza, R., *et al.* 1995. Economic growth, carrying capacity and the environment. *Science* 268: 520–1.

Balee, W., ed. 1998. *Advances in Historical Ecology.* New York: Columbia University Press.

Barbier, E.B., Burgess, J., and Folke, C. 1994. *Paradise Lost? The Ecological Economics of Biodiversity.* London: Earthscan.

Baskerville, G. 1995. The forestry problem: adaptive lurches of renewal. In *Barriers and Bridges to the Renewal of Ecosystems and Institutions*, pp. 37–102, ed. L.H.

Gunderson, C.S. Holling, and S.S Light. New York: Columbia University Press.

Bateson, G. 1979. *Mind and Nature. A Necessary Unity*. New York: Dutton.

Berkes, F., ed. 1989. *Common Property Resources. Ecology and Community-Based Sustainable Development*. London: Belhaven.

Berkes, F. 1999. *Sacred Ecology: Traditional Ecological Knowledge and Resource Management*. Philadelphia and London: Taylor and Francis.

Berkes, F. 2001. Religious traditions and biodiversity. *Encyclopedia of Biodiversity* 5: 109–20.

Berkes, F., Colding, J., and Folke C. 2000. Rediscovery of traditional ecological knowledge as adaptive management. *Ecological Applications* 10: 1251–62.

Berkes, F. and Folke, C. 1994. Investing in cultural capital for a sustainable use of natural capital. In *Investing in Natural Capital*, pp. 128–49, ed. A.M. Jansson, M. Hammer, C. Folke, and R. Costanza. Washington DC: Island Press.

Berkes, F. and Folke, C., eds. 1998. *Linking Social and Ecological Systems. Management Practices and Social Mechanisms for Building Resilience*. Cambridge: Cambridge University Press.

Berkes, F. and Folke, C. 2002. Back to the future: ecosystem dynamics and local knowledge. In *Panarchy: Understanding Transformations in Systems of Humans and Nature*, pp. 121–46, ed. L.H. Gunderson and C.S. Holling. Washington DC: Island Press.

Berkes, F., Kislalioglu, M., Folke, C., and Gadgil, M. 1998. Exploring the basic ecological unit: ecosystem-like concepts in traditional societies. *Ecosystems* 1: 409–15.

Berkes, F., Mahon, R., McConney, P., Pollnac, R.C., and Pomeroy, R.S. 2001. *Managing Small-Scale Fisheries: Alternative Directions and Methods*. Ottawa: International Development Research Centre.

Blaikie, P. 1985. *The Political Economy of Soil Erosion in Developing Countries*. Harlow, UK: Longman.

Blaikie, P. and Jeanrenaud, S. 1996. Biodiversity and human welfare. Discussion Paper No. 72. Geneva: United Nations Research Institute for Social Development (UNRISD).

Bromley, D.W. 1991. *Environment and Economy: Property Rights and Public Policy*. Oxford: Basil Blackwell.

Bromley, D.W., ed. 1992. *Making the Commons Work: Theory, Practice, and Policy*. San Francisco: Institute for Contemporary Studies.

Callicott, J.B. 1994. *Earth's Insights. A Survey of Ecological Ethics from the Mediterranean Basin to the Australian Outback*. Berkeley: University of California Press.

Capra, F. 1996. *The Web of Life*. New York: Anchor Books.

Clark, W.C. and Munn, R.E., eds. 1986. *Sustainable Development of the Biosphere*. Cambridge: Cambridge University Press.

Colding, J. and Folke, C. 1997. The relations among threatened species, their protection, and taboos. *Conservation Ecology* 1 (1): 6. [online] URL:http://www.consecol.org/vol1/iss1/art6

Costanza, R., ed. 1991. *Ecological Economics: the Science and Management of Sustainability*. New York: Columbia University Press.

Costanza, R., Norton, B.G., and Haskell, B.D., eds. 1992. *Ecosystem Health: New Goals for Environmental Management*. Washington DC: Island Press.

Costanza, R., Waigner, L., Folke, C., and Mäler, K.-G. 1993. Modeling complex ecological economic systems: towards an evolutionary dynamic understanding of people and nature. *BioScience* 43: 545–55.

Cronon, W. 1983. *Changes in the Land: Indians, Colonists, and the Ecology of New England.* New York: Hill and Wang.
Feeny, D., Berkes, F., McCay, B.J., and Acheson, J.M. 1990. The tragedy of the commons: twenty-two years later. *Human Ecology* 18: 1–19.
Folke, C., Berkes, F., and Colding, J. 1998a. Ecological practices and social mechanisms for building resilience and sustainability. In *Linking Social and Ecological Systems*, pp. 414–36, ed. F. Berkes and C. Folke. Cambridge: Cambridge University Press.
Folke, C. and Colding, J. 2001. Traditional conservation practices. *Encyclopedia of Biodiversity* 5: 681–93.
Folke, C., Pritchard, L. Jr, Berkes, F., Colding, J., and Svedin, U. 1998b. The problem of fit between ecosystems and institutions. IHDP Working Paper 2. Bonn: International Human Dimensions Programme on Global Environmental Change.
Gadgil, M. and Guha, R. 1992. *This Fissured Land. An Ecological History of India.* Delhi: Oxford University Press.
Greenberg, J.B. and Park, T.K. 1994. Political ecology. *Journal of Political Ecology* 1: 1–12.
Gunderson, L.H. 1999. Resilience, flexibility and adaptive management: antidotes for spurious certitude? *Conservation Ecology* 3: 7. [online] URL: *http://www.consecol.org/vol3/iss1/art7*
Gunderson, L.H. 2000. Ecological resilience – in theory and application. *Annual Review of Ecology and Systematics* 31: 425–39.
Gunderson, L.H. and Holling, C.S., eds. 2002. *Panarchy: Understanding Transformations in Systems of Humans and Nature.* Washington, DC: Island Press.
Gunderson, L.H., Holling, C.S., and Light, S., eds. 1995. *Barriers and Bridges to the Renewal of Ecosystems and Institutions.* New York: Columbia University Press.
Hanna, S., Folke, C., and Mäler, K.-G., eds. 1996. *Rights to Nature.* Washington DC: Island Press.
Holland, J.H. 1995. *Hidden Order: How Adaptation Builds Complexity.* Reading, MA: Addison-Wesley.
Holling, C.S. 1973. Resilience and stability of ecological systems. *Annual Review of Ecology and Systematics* 4: 1–23.
Holling, C.S., ed. 1978. *Adaptive Environmental Assessment and Management.* London: Wiley.
Holling, C.S. 1986. The resilience of terrestrial ecosystems: local surprise and global change. In *Sustainable Development of the Biosphere*, pp. 292–317, ed. W.C. Clark and R.E. Munn. Cambridge: Cambridge University Press.
Holling, C.S. 1988. Temperate forest insect outbreaks, tropical deforestation and migratory birds. *Memoirs of the Entomological Society of Canada* 146: 21–32.
Holling, C.S. 2001. Understanding the complexity of economic, ecological, and social systems. *Ecosystems* 4(5): 390–405.
Holling, C.S., Berkes, F., and Folke, C. 1998. Science, sustainability, and resource management. In *Linking Social and Ecological Systems*, pp. 342–62, ed. F. Berkes and C. Folke. Cambridge: Cambridge University Press.
Holling, C.S. and Meffe, G.K. 1996. Command and control and the pathology of natural resource management. *Conservation Biology* 10: 328–37.
Holling, C.S., Schindler, D.W., Walker, B.W., and Roughgarden, J. 1995. Biodiversity in the functioning of ecosystems: an ecological synthesis. In *Biodiversity Loss*, pp. 44–83, ed. C.A. Perrings, K.-G. Mäler, C. Folke, C.S. Holling, and B.-O. Jansson. Cambridge: Cambridge University Press.

Jackson, J.B.C., Kirby, M.X., Berher, W.H., *et al.* 2001. Historical overfishing and the recent collapse of coastal ecosystems. *Science* 293: 629–38.

Jansson, A.M., Hammer, M., Folke, C., and Costanza, R., eds. 1994. *Investing in Natural Capital: the Ecological Economics Approach to Sustainability.* Washington, DC: Island Press.

Jasanoff, S., Colwell, R., Dresselhaus, S., *et al.* 1997. Conversations with the community: AAAS at the millennium. *Science* 278: 2066–7.

Jentoft, S., McCay, B.J., and Wilson, D.C. 1998. Social theory and fisheries co-management. *Marine Policy* 22: 423–36.

Johannes, R.E. 1981. *Words of the Lagoon. Fishing and Marine Lore in the Palau District of Micronesia.* Berkeley: University of California Press.

Kates, R.W., Clark, W.C., Corell, R., *et al.* 2001. Sustainability science. *Science* 292: 641–2. Statement of the Friibergh Workshop on Sustainability Science. Available from the Internet. URL: http://sustsci.harvard.edu/keydocs/friibergh.htm

Kauffman, S. 1993. *The Origins of Order.* New York: Oxford University Press.

Larkin, P.A. 1977. An epitaph for the concept of maximum sustained yield. *Transactions of the American Fisheries Society* 106: 1–11.

Levin, S.A. 1999a. *Fragile Dominion: Complexity and the Commons.* Reading, MA: Perseus Books.

Levin, S.A. 1999b. Towards a science of ecological management. *Conservation Ecology* 3 (2): 6. [online] URL:http://www.consecol.org/vol3/iss2/art6

Levin, S.A., Barrett, S., Aniyar, S., *et al.* 1998. Resilience in natural and socioeconomic systems. *Environment and Development Economics* 3: 225–36.

Ludwig, D., Hilborn, R., and Walters, C. 1993. Uncertainty, resource exploitation and conservation: lessons from history. *Science* 260: 17, 36.

Lugo, A. 1995. Management of tropical biodiversity. *Ecological Applications* 5: 956–61.

Mäler, K.-G. 2000. Development, ecological resources and their management: a study of complex dynamic systems. *European Economic Review* 44: 645–65.

McCay, B.J. and Acheson, J.M., eds. 1987. *The Question of the Commons. The Culture and Ecology of Communal Resources.* Tucson: University of Arizona Press.

McIntosh, R.J. 2000. Climate, history and human action. In *The Way the Wind Blows: Climate, History and Human Action*, pp. 1–42, ed. R.J. McIntosh, J.A. Tainter, and S.K. McIntosh. New York: Columbia University Press.

McIntosh, R.J., Tainter, J.A., and McIntosh, S.K., eds. 2000. *The Way the Wind Blows: Climate, History and Human Action.* New York: Columbia University Press.

Naess, A. 1989. *Ecology, Community and Lifestyle: Outline of an Ecosophy*, translated and edited by D. Rothenberg. Cambridge: Cambridge University Press.

Norgaard, R.B. 1994. *Development Betrayed: the End of Progress and a Coevolutionary Revisioning of the Future.* New York: Routledge.

North, D.C. 1994. Economic performance through time. *American Economic Review* 84: 359–68.

Nyström, M. and Folke, C. 2001. Spatial resilience of coral reefs. *Ecosystems* 4: 406–17.

Olsson, P. and Folke, C. 2001. Local ecological knowledge and institutional dynamics for ecosystem management: a study of Lake Racken watershed, Sweden. *Ecosystems* 4: 85–104.

Ostrom, E. 1990. *Governing the Commons: the Evolution of Institutions for Collective Action.* Cambridge: Cambridge University Press.

Ostrom, E. 1992. *Crafting Institutions for Self-Governing Irrigation Systems.* San Francisco: Institute for Contemporary Studies Press.

Ostrom, E., Burger, J., Field, C.B., Norgaard, R.B., and Policansky, D. 1999. Revisiting the commons: local lessons, global challenges. *Science* 284: 278–82.

Perrings, C.A., Mäler, K.G., Folke, C., Holling, C.S., and Jansson, B.-O., eds. 1995. *Biodiversity Loss: Ecological and Economic Issues*. Cambridge: Cambridge University Press.

Perrings, C.A. and Walker, B.W. 1995. Biodiversity loss and the economics of discontinuous change in semi-arid grasslands. In *Biodiversity Loss*, pp. 190–210, ed. C.A. Perrings, K.-G. Mäler, C. Folke, C.S. Holling, and B.-O. Jansson. Cambridge: Cambridge University Press.

Resilience Alliance 2002. www.resalliance.org/programdescription

Riedlinger, D. and Berkes, F. 2001. Contributions of traditional knowledge to understanding climate change in the Canadian Arctic. *Polar Record* 37: 315–28.

Santa Fe Institute 2002. www.santafe.edu

Scheffer, M., Carpenter, S., Foley, J., Folke, C., and Walker, B. 2001. Catastrophic shifts in ecosystems. *Nature* 413: 591–6.

Schumpeter, J.A. 1950. *Capitalism, Socialism and Democracy*. New York: Harper and Row.

Sen, A.K. 1999. *Development as Freedom*. Oxford: Oxford University Press.

Turner, B.L., Clark, W.C., Kates, R.W., Richards, J.F., Mathews, J.T., and Meyer, W.B., eds. 1990. *The Earth as Transformed by Human Action. Global and Regional Changes in the Biosphere over the Past 300 Years.* Cambridge: Cambridge University Press.

van der Leeuw, S.E. 2000. Land degradation as a socionatural process. In *The Way the Wind Blows: Climate, History and Human Action*, pp. 190–210, ed. R.J. McIntosh, J.A. Tainter, and S.K. McIntosh. New York: Columbia University Press.

von Bertalanffy, L. 1968. *General Systems Theory*. New York: George Brazilier.

Walters, C.J. 1986. *Adaptive Management of Renewable Resources*. New York: McGraw-Hill.

Warren, D.M., Slikkerveer, L.J., and Brokensha, D., eds. 1995. *The Cultural Dimension of Development. Indigenous Knowledge Systems*. London: Intermediate Technology Publications.

Wilson, J. 2000. Scientific uncertainty and institutional scale: ocean fisheries. Papers of the International Association for the Study of Common Property (IASCP), Bloomington, Indiana, June 2000. *www.indiana.edu/*~iascp2000.htm

Worster, D., ed. 1988. *The Ends of the Earth. Perspectives on Modern Environmental History.* Cambridge: Cambridge University Press.

Part I

Perspectives on resilience

Introduction

A number of volumes have stressed the practical difficulties in attempting to manage ecosystems. The multiple scales of variables, cross-scale connections, and nonlinear interactions generate complex dynamics. Systems of people and nature go through dynamic phases of development, described in resilience theory through the heuristic model of the adaptive renewal cycle. Relatively long periods with little change alternate with short periods of collapse and reorganization in this cycle. During these periods of renewal, resilience can be enhanced or lost, depending on such factors as diversity, redundancy, and memory in the system.

Conventional resource and environmental management is ill-equipped to deal with the challenges of these complexities. Textbook management largely ignores the scale issue, and cross-scale and nonlinear interactions. Adaptive renewal cycles have not normally been part of management thinking, and little attention is paid to the crucial short periods of collapse and reorganization. Diversity has received a great deal of attention from the point of view of the conservation of biological diversity. However, the recognition of functional diversity in adaptive renewal cycles and of its role in the long-term maintenance of ecosystems is relatively recent. Social and cultural diversity, in the form of diversity of knowledge for renewal and reorganization, is also a relatively unexplored area. Redundancy, as distinct from diversity, is important in its own right. Like diversity, redundancy has both ecological and social components, as in institutional redundancy.

These areas require a closer look than that provided in the introduction chapter. Hence, Part I identifies some of these key ideas and perspectives, as a means of providing the groundwork for the chapters to come. Chapter 2 discusses the adaptive interaction between social resilience and ecological crises, especially in cases of surprise, that is, when social–ecological systems behave in unforeseen ways. Chapter 3 reviews and assesses the impact of resilience thinking on the way social–ecological systems have been conceptualized. Chapter 4 provides a boldly speculative perspective that cuts across disciplinary boundaries to explore the implications of redundancy in its various forms.

2

Adaptive dancing: interactions between social resilience and ecological crises

LANCE H. GUNDERSON

2.1 Introduction

Systems of people and nature co-evolve in an adaptive dance (Walters, 1986). Resource systems change as people seek ecosystem services, such as the harvest of stocks, manipulation of key structuring processes, removal of geophysical assets or abation of pollutant concentrations. Meanwhile, as humans are becoming more dependent on these ecosystem services, the ecosystems become more vulnerable to unexpected events. This process that signals a loss of ecological resilience has been described as a pathology of resource development (Holling, 1995).

Complex resource systems are not easily tractable or understood, much less predictable. Nonlinear interactions among multiple variables, scale invariant processes, emergent properties from self-organization and other factors all contribute to unpredictability. Yet, even with these inherent difficulties, we continue attempts at making sense for management and other purposes. Due to a growing empirical base of observation, emergent patterns of these systems, including periods of stability and instability, as well as unexpected behavior due to internal and external changes have been revealed (Gunderson, Holling, and Light, 1995; Berkes and Folke, 1998; Johnson *et al.*, 1999).

This paper builds on earlier work (Holling, 1978; Walters, 1986, 1997; Gunderson *et al.*, 1995; Gunderson, 1999a) to explore these unexpected behaviors in managed ecological systems – perceived as surprises and crises. To begin with, the conceptual basis for understanding these nonlinearities, ecological properties of resilience and adaptive capacity, and analogous properties in institutions are presented. The next section describes a set of different types of surprises, followed by a discussion of how people respond to those different types of surprises. The chapter ends with some tentative propositions on how one might move beyond sense-making and begin to manage for resilience.

2.2 Resilience in ecological and social systems

Complex resource systems link ecological components and social components (including economic systems, institutions, and organizations). Institutions are described as the set of norms, rules that people use to organize activities (Ostrom, 1990). Social systems are comprised of three types of structures (signification, domination, and legitimation) that enable power and resources distributions, patterns of authority in addition to norms, rules, routines and procedures (Giddens, 1987). At the heart of these components and their interaction are the properties of resilience and renewal. Resilience provides these complex systems with the ability to persist in the face of shocks and disturbances. Maintaining a capacity for renewal in a dynamic environment provides an ecological buffer that protects the system from the failure of management actions that are taken based upon incomplete understanding, and therefore allows managers to affordably learn and change.

Resilience of a system has been defined in two very different ways in the ecological literature, each reflecting different aspects of stability (Fig. 2.1). Holling (1973) first emphasized these different aspects of stability to draw attention to the tensions between efficiency and persistence, between constancy and change, and between predictability and unpredictability. The more common definition considers that ecological systems exist close to a stable steady-state. In this context, resilience is described as a return time to a steady-state following a perturbation (Pimm, 1984; O'Neill *et al.*, 1986). This definition has been described as engineering resilience (Holling, 1996) and carries an assumption of a single, global equilibrium (Fig. 2.1A).

The second definition emphasizes conditions far from any stable steady-state, where instabilities can flip a system into another regime of behavior, i.e., to another stability domain (Holling, 1973). In this case resilience is defined as the magnitude of disturbance that can be absorbed before the system redefines its structure by changing the variables and processes that control behavior. This is termed ecological resilience (Walker *et al.*, 1969), as depicted in Figure 2.1B. Those who emphasize the stability domain definition of resilience (i.e., ecological resilience), on the other hand, come from traditions of applied mathematics and applied resource ecology at the scale of ecosystems, e.g., of the dynamics and management of freshwater systems (Fiering, 1982), of forests (Clark, Jones, and Holling, 1979), of fisheries (Walters, 1986), of semi-arid grasslands (Walker *et al.*, 1969), and of interacting populations in nature (Dublin, Sinclair, and McGlade, 1990; Sinclair, Olsen, and Redhead, 1990).

Recent advancements suggest that a third category is needed to describe ecological change. In the above-mentioned definitions of resilience, both are based

A. Engineering resilience (r).

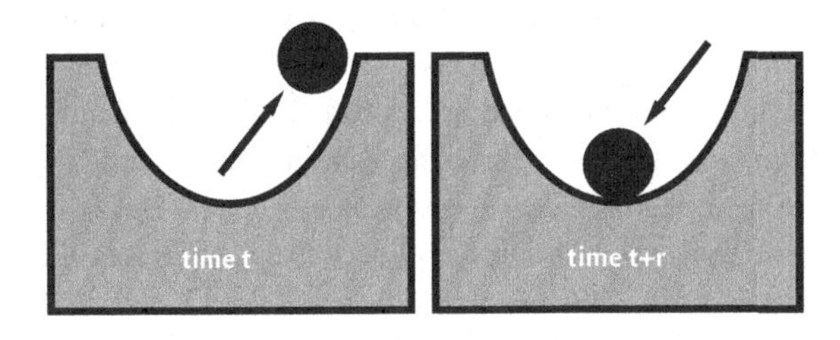

B. Ecological resilience (R).

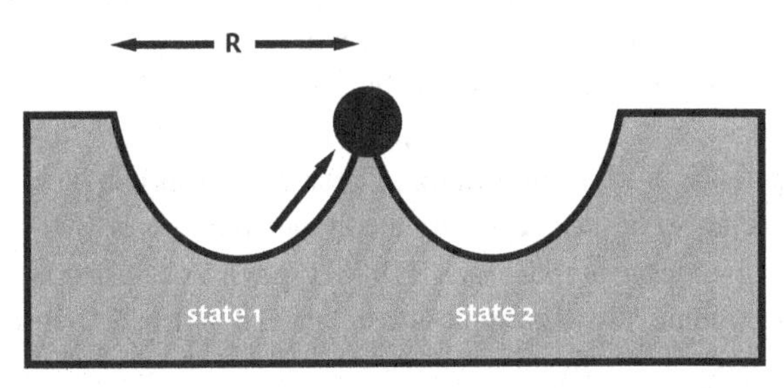

Figure 2.1 Alternative definitions of resilience as represented by a ball and cup model. Cups represent the stability domains of the system, the ball represents the system state, and single arrows represent disturbances to the system. (A) Engineering resilience can be depicted by a global equilibrium (ball resting at the bottom of a cup). When the system is disturbed (ball moves up the side of the cup), resilience is defined as the amount of time (r) for the system to return to the equilibrium state. (B) Ecological resilience is defined as the amount of disturbance that the system can absorb without changing state (stable state 1 or 2), and is measured as the width of the stability domain (R).

on the notion of a system with a stationary stability domain. The structures and processes that produce stability are assumed not to change over time or space, and hence are tractable. In the return-time concept, a single stability domain is implicit, whereas in the ecological resilience concept, multiple steady-states are possible. Yet the kinds of ecological processes that create these stability basins are slowly changing variables: mud in lakes (Carpenter, Ludwig, and Brock, 1999b); species composition in semi-arid rangelands (Walker *et al.*, 1969); soil nutrient concentration in wetlands (Davis, 1994); or spatial connectivity of old trees in spruce budworm forests (Ludwig, Jones, and Holling, 1978). Hence, another term is needed to describe the capacity of a system to adapt to these slower dynamics – described as adaptive capacity (Peterson, Allen, and Holling, 1998; Gunderson, 2000).

When ecosystems are observed to shift in behavior, structure or functions, it is usually signaled as a resource crisis. The literature is replete with examples: sudden blooms of toxic algae in freshwater lakes (Carpenter, Brock, and Hanson, 1999a), emergence of shrubs in semi-arid grasslands (Walker *et al.*, 1969), shifts in species dominance in freshwater wetlands (Davis, 1994). A weak typology of different types of surprising ecosystem behaviors is described in the next section.

2.2.1 Ecological surprise

In these co-evolving systems of humans and natures, surprises are the rule, not the exception. An ecological surprise is defined as a qualitative disagreement between ecosystem behavior and *a priori* expectations – an environmental cognitive dissonance. Brooks (1986) provides a useful typology of surprises in describing the interaction between technology and society, and defines three types; (1) unexpected discrete events, (2) discontinuities in long-term trends, and (3) emergence of new information. These categories can be broadened and placed in the context of the previously mentioned theories of change in ecological systems, by redescribing Brooks' (1986) types as local surprise, cross-scale surprise, and true novelty. Examples and elaboration are described in the next paragraphs.

Local surprises can often be addressed by recognizing broader-scale processes. Unexpected discrete events can be part of broader-scale fluctuations or variation of which there is little or no local knowledge. An ecological example of this is the cycle of flood and drought over the southeastern USA, which is part of global atmospheric and oceanic coupling known as El Niño/Southern Oscillation (ENSO). In these cases, the ignorance of broader and longer-term processes and human limits on perception both contribute to the local surprise.

The next class of surprise deals with abrupt, nonlinear or discontinuous behavior of a system that, after analysis, can be attributed to an interaction between key variables that operate at distinctly different scale ranges. That is, the surprise is due to a faster variable interacting with slower variables. In ecological systems, examples include spatially contagious processes, such as forest fires, which only occur when there is an interaction amongst a trigger such as a spark, dry fuel, and sufficient fuel for the fire to carry. In this example, ignition frequency (such as lightning strike rate) is a 'fast' variable and the spatial distribution and amount of fuel characterize more slowly changing variables

(Holling, 1986). In these dynamics lay the interactions for qualitative shifts in stability domains of resource systems, such as Walker *et al.*'s (1969) analysis of subtropical savanna grazing systems, or in the dynamics between stable states of lake systems (Scheffer 1998; Carpenter *et al.*, 1999b).

The final type of surprise is genuine novelty – that is, something truly unique and new or not previously experienced by humans (or at least outside the breadth of captured experience for a culture in a new situation). These types of surprise can generate change, the consequences of which are inherently unpredictable. Most examples of new technologies (such as the personal computer in the late 1970s) fall into this category (Tenner, 1996). In resource systems, invasions by exotic species are this type of surprise. The invasion of alien trees such as *Myrica faya* in Hawaii (D'Antonio and Vitousek, 1992) or *Melaleuca quinquenervia* in Florida (Myers, 1983) alters key ecosystem processes such as nutrient cycling, water relations, and fire patterns. Perhaps the greatest surprises ahead of us are the unforeseen planetary impacts of humans, such as facilitated species movement, nutrient cycling changes, changes in land-use, and the creation of new substances.

2.2.2 Ecological crises

Ecological crises are a special type of surprise. That is, a surprise becomes a crisis when it reveals an unambiguous failure of policy. As elsewhere (Gunderson *et al.*, 1995), the term *policy* is used in this context to describe the rules, norms, behavior, and infrastructure of management action (Ostrom, 1990).

Not all ecological surprises lead to crises of policy. In situations where flexibility in policy exists, variations in external events can be easily managed without a policy change. Examples of this would include adaptive responses of fire management agencies to the outbreak of fires that were associated with ENSO fluctuations. In 1998, fire management officials in Florida were able to manage severe drought and fire conditions by calling on a large pool of firefighting resources to deal with unexpected fire outbreaks. In other surprises, such as the massive die-off of seagrass in Florida Bay in the late 1980s (Robblee *et al.*, 1991), little or no change in policy can occur, even though they were viewed as ecological crises. The point here is that some surprises can be managed, without leading to a policy change, whereas others result in shifts in policy. There is further discussion of this dichotomy later in the chapter.

On the other hand, not all crises stem from ecological surprises. Shifts in ecological policy can be due to broader social reforms or changes in the way people view or understand resource issues. Light, Gunderson, and Holling (1995) indicated a major reformation of policy in Everglades water management in the early

1980s due to a shift in public perception of water quality concerns, when little or no ecological change had been documented. Costanza and Greer (1995) suggest that policy reformations of water quality monitoring and action in Chesapeake Bay were rooted in a changing sense of stewardship and responsibility by key people (activists, legislators, and scientists) who lived on the bay.

Rather than try to understand the role of crises and surprises in a broad set of conditions, this chapter focuses on how people respond to those ecological crises that signal the erosion of resilience and a shift of stability domains. In a review of large-scale ecosystems (Gunderson and Pritchard, 2002), we found that at least three different pathways can lead to the loss of resilience and inevitable surprise. The first pathway involves the addition of key substances into the ecosystem. Examples include the addition of low levels of phosphorus into the Everglades wetlands (Davis, 1994) or lakes (Carpenter and Cottingham, 1997). The second pathway involves removal of key resources or sources of resilience. Examples of these include the removal of soil in tropical forests (Lugo *et al.*, 2002), or removal of drought-tolerant plant species via overgrazing (Walker, 1995). The third pathway involves the manipulation of keystone ecological processes by human intervention, be they self-organized spatial patterns of forests with fire (Peterson *et al.*, 1998) or budworm (Ludwig *et al.*, 1978) or of key trophic relationships in coral reefs (McClanahan, Done, and Polunin, 2002). In each of these groupings, the differences in mechanisms that led to a loss of resilience created different circumstances for understanding and action during and after the ecological crises. How humans respond to these crises and make sense of the unexpected shift in stable states of nature is the topic of the next section.

2.3 Responding to crises

As Thompson (1983) states, we have no escape from having to 'manage the unmanageable.' Given that humans will continue to cope with systems that are partly knowable and partly unknowable, the ways in which people begin to make sense and develop dynamic responses are linked to the types of surprises and crises. The relationship between different types of uncertainty is key: how people choose to deal with uncertainty appears to either increase or decrease the resilience of an ecosystem. It is the ecological resilience that allows managers a margin of failure. There is growing evidence that acknowledgement and confrontation of uncertainty add resilience to managed systems (Gunderson, 1999a). But the subtleties and nuances of how that uncertainty is managed are complicated and not very well understood.

People involved in the practice of resource management are all linked by the need for understanding. Yet in these complex resource issues, uncertainty is pervasive. Partitioning that uncertainty is an initial step for an approach that involves confronting and the hope of winnowing. Lee (1993) recognized that resource uncertainties can be separated into categories of social uncertainty (agreement on social objectives, norms) and technical uncertainty (understanding and explaining the mechanisms associated with a resource issue). Resolving the technical uncertainties of resource issues has generally been the domain of a technical and expert community, as described in the following section. Researchers such as Williams and Matheny (1995) and Pritchard and Sanderson (2002) suggest that the social uncertainty category of Lee (1993) is generally addressed by two groups: a political community and a stakeholder community. The second section that follows describes how social systems (including stakeholders and political components) manage uncertainty.

A key unresolved issue, however, is how these three communities interact (or not) around notions of ecosystem management and resilience (Pritchard and Sanderson, 1999, 2002). Often, what is most critical is the management of uncertainty among the three communities (technical, stakeholder, and political). A brief discussion of the interaction among technical, stakeholder, and political communities is presented in the third section that follows on integrating activities.

2.3.1 Technical and expert community

In these complicated systems, a great deal of uncertainty has been assigned to the technical and expert community for resolution. The experience and practice during this century have been to turn to scientists, the heart and soul of technology and technologic solutions, as the fountains of understanding. But there has been a growing sense that traditional scientific approaches are not working, and indeed make the problem worse (Ludwig, Hilborn, and Walters, 1993). Two reasons why rigid scientific and technological approaches fail are because they tend to focus on the wrong types of uncertainty and on narrow types of scientific practice. Many formal techniques of assessment and policy analysis presume a system near equilibrium, with a constancy of relationships, and that uncertainties arise not from errors in tools or models, but from lack of appropriate information on which to base the models. Walters (1997) outlines the challenges faced by the scientific and technical community in attempting to assess and resolve uncertainties of resource issues, including lack of data across appropriate scales and difficult cross-scale modeling problems, among others.

A conflict also arises between two views of science that contributes to perpetuate uncertainty rather than winnow it. One mode of science focuses on parts of the system and deals with experiments that narrow uncertainty to the point of acceptance by peers; it is conservative and unambiguous by being incomplete and fragmentary. The other view is integrative and holistic, searching for simple structures and relationships that explain much of nature's complexity. The latter view provides the underpinnings for an approach to dealing with resource issues called adaptive environmental assessment and management (Holling, 1978; Walters, 1986), which assumes that surprises are inevitable, that knowledge will always be incomplete, and that human interaction with ecosystems will always be evolving.

Another common pathology in environmental assessments involves the level of complexity in analysis and explanation. Many of these assessments engage scientists and technical experts to compile existing information in two forms: one is a set of facts or observations about the status of the resource and the second is in generating a set of plausible explanations about what has generated the ecological surprise. Accounts of recent examples indicate that the assessments generally reveal a paucity of reliable data. Sapp (1999) describes the generation of large research and monitoring activities around the world following the coral reef crisis in the 1960s associated with population outbreaks of the crown-of-thorns starfish. Similar results occurred following the seagrass die-off in Florida Bay in the late 1980s (Robblee *et al.*, 1991). Whereas a common response is to increase monitoring activities in order to learn more about the ecosystem, a more interesting set of activities deals with how the technical community attempts to sort through competing explanations or hypotheses that explain the crisis.

Alternative sets of hypotheses can always be generated to explain an ecological crisis. Using the examples just mentioned, at least four hypotheses were originally proposed for the crown-of-thorns outbreak: increased dredging, warmer sea temperatures, typhoons, and nuclear fallout (Sapp, 1999). In the Florida Bay example, at least seven hypotheses were proposed (Robblee *et al.*, 1991; Gunderson and Walters, 1999), including hypersalinity, excess nutrients, diseases, loss of grazers, lack of hurricanes. In each of these cases, the construct involved a single variable relationship. That is, one factor was supposed to have exceeded a level to which the ecological system was pre-adapted. Psychologists argue that in part it is due to cognitive limits (Dörner, 1996). In both examples (and in others), what generated a more successful explanation was the development of frameworks (models) that allowed two things: a rigorous comparison between data and hypotheses, and models that contained a minimum complexity. This debate still goes on, but there is evidence from these examples and others that a small set (more than three and less than five) of structuring variables can be used to explain much of these dynamic patterns and interactions (Holling, Gunderson, and Peterson, 2002). This was certainly the case in the crown-of-thorns outbreaks (Bradbury, 1990) and the Florida Bay seagrass die-off (Gunderson and Walters, 1999).

In successful assessments, the models and data are used to exclude or invalidate hypotheses, leaving the remaining ones to be put at risk through management actions. Yet that too can be problematic, especially when different actions are required to sort among hypotheses (Walters, Gunderson, and Holling, 1992; Gunderson, 1999b). For example, in Florida Bay, the alternative explanations each suggested a different management strategy and action to attempt invalidation, along with varying social and political trade-offs. If the salinity hypotheses were true, then actions aimed at controlling nutrient inputs would be a costly misappropriation of resources.

Assessments and understandings can also be linked to the nature of the surprise and the mechanisms associated with the loss of ecological resilience. Understanding an ecological crisis that arose from a local surprise is perhaps the most tractable. Longer-term data sets and broader spatial perspectives both allow for linking the crisis to variables that are distant in space and time. Examples of these include the increased understanding of the role of ENSO (Glantz, Katz, and Nicholls, 1991) in driving local crises. Another example is that the recruitment patterns of key fish stocks in the Baltic are linked to weather systems in the north Atlantic (Jansson and Velner, 1995). Even the cross-scale surprises are becoming more tractable not just through increased data collection, but also through more sophisticated modeling. Many of the crises and surprises noted elsewhere (Gunderson *et al.*, 1995, Gunderson and Walters, 1999; Sapp, 1999) were resolved to the point at which actions could be initiated by this iteration between models and data. It is the true novelties that will continue to be the most intractable of the surprises.

Many authors argue that the failures of agency-based resource management and the abilities of those agencies to respond to ecological crises are in part due to technical limitations (Walters, 1997). Yet that community is often embedded in and operates within multiple sets of other human-dominated systems. A brief discussion of those communities and how they deal with different types of uncertainty is the subject of the next paragraphs.

2.3.2 Social systems

A unique property of social systems in response to uncertainty is the generation of novelty. Novelty is key to dealing with surprises or crises. Humans are unique in that they create novelty that transforms the future, and often it is the ecosystem crisis that spawns brief periods of creativity.

Often, new types and arrangements of management institutions are created after resource crises. These institutions can be formal, government-based agencies, such as the State of Florida Water Management Districts which were created following the drought of 1971 in the Everglades (Light *et al.*, 1995). Other formal agencies can span gaps in existing governance, such as the interstate compact that created the Northwest Power Planning Council in the Columbia River Basin (Lee, 1993). Often, scientific-based epistemic communities arise, such as the Baltic Management Commission in the Baltic Sea (Jansson and Velner, 1995), or a similar group in the Mediterranean (Haas, 1990).

The epistemic communities often include a broad spectrum of views, and provide a forum well suited to address surprises. Thus, one interpretation of these arrangements is that these institutions are set up to resolve different types of uncertainties. They provide a venue in which some technical and social uncertainties can be resolved (Lee, 1993; Pritchard and Sanderson, 1999; Pritchard, Folke, and Gunderson, 2000).

Yet there are many situations in which the institutions constantly struggle with resolving those uncertainties. That conservatism or institutional inertia can be described as an inability to re-invent themselves and adapt to changing conditions. Nor do agencies appear capable of generating either novel solutions to or policies for chronic resource issues. Indeed, one of the few mechanisms for agency change is the advent of an ecological crisis, as has been argued elsewhere (Gunderson *et al.*, 1995). But there are many other reasons why social systems are so conservative and inflexible regarding policy changes.

One reason why management institutions have such high moments of inertia is that they utilize (directly or indirectly) ambiguities and uncertainties of resource issues to maintain a *status quo*. With a pragmatic focus on policy implementation, most agencies seem to have a two-fold strategy that is aimed at reinforcing the *status quo*: prove that extant policies are wrong, and do not act until one is confident of what to do next. Many agencies focus on implementation, without realizing that narrow implementation schemes often subvert policy intent, or realizing that implementation is an organic process that changes over time and reveals the failure of policy, not its success (Gunderson *et al.*, 1995).

Another conserving strategy is seen by vested interests that have political and social sway over agencies. Whereas science uses uncertainty to drive the engine of inquiry, vested interest groups use and foster uncertainty to maintain a *status quo* policy. There are many examples – take the actions of sugar farmers in the Everglades following claims that nutrient run-off was changing the structure and function of pristine areas in the Everglades. Prominent scientists were hired to generate alternative hypotheses (other than those that involved phosphorus), which for a while stalemated any movement towards resolving the crisis. Similar results of dis-information campaigns are chronicled for health, climate change, and biodiversity issues (Ehrlich and Ehrlich, 1996). Vested interests are not the only groups that generate or defend pet hypotheses. Agency scientists often generate policy recommendations that are politically correct in the sense of gaining what they view as a favorable policy. Take the Florida Bay example mentioned earlier. Most agency scientists gravitated to the freshwater run-off–salinity hypotheses as the reason for the large-scale seagrass crisis. It held political sway to the point where extra water was indeed delivered to Everglades National Park and Bay, with the counterproductive result of regeneration being delayed rather than accelerated (Walters, 1997). These examples further highlight the point that science is a highly social process, with lots of tacit and implicit factors influencing and shaping an 'objective' process.

2.3.3 Integrating activities

Currently, there is a lot of activity aimed at more integration of agencies, stakeholders, and citizens with regard to resource issues (Westley, 1995; Pritchard and Sanderson, 2002). Community-based resource management, citizen science, amongst others, are common buzz words and approaches. Examples include federal task forces or sustainability commissions that attempt to invite and engage all interested parties to help resolve the chronic resource issues. There are some examples, e.g., Florida Everglades (Gunderson, 1999b) and the Mediterranean Sea (Haas, 1990), of these meshing groups helping to link formerly disparate communities of agencies, stakeholders, and citizens around the issue of ecosystem restoration and resolution of chronic issues. These groups often fill 'structural holes' in problem domains, i.e., where explicit or implicit partitioning of responsibilities still leaves some uncertainties not addressed. These meshing functions appear to be a robust solution to filling obvious and critical gaps to resolve these complex issues.

Yet, in terms of resolving chronic issues or breaking through the types of inertia previously mentioned, most of the seemingly successful integrating activities appear to arise from temporary groups. The history of the Everglades provides an example. Following the drought of 1971, the Governor of the state of Florida called a symposium of the experts on how to deal with water supply issues. The convention created the design of water management districts in the state (Light *et al.*, 1995). The convention was then dissolved, passing the charge to resolve water supply and flood issues to a formal institution. The Everglades also provide examples in which formalizing these types of organizations was unsuccessful. Numerous technical committees have been formed to deal with resolving uncertainties that required integration across scientific disciplines or even between science and political arenas. Yet history indicates that these formal temporary groups tend to resolve little, if anything, of an issue that they were established to address. So this inconsistency raises a paradox: how to sustain a necessary role for integration across agency boundaries and across scientific disciplines, when most of the seemingly successful groups are temporary. This paradox is addressed in the next section, with some suggestions on how to create more sustaining types of institutions and inter-organizational activities.

2.4 Surfing ecological crises

2.4.1 Learning-based institutions

Perhaps it is time to rethink the paradigms or foundations of resource management institutions, and to place more emphasis on the development of sustaining foundations for dealing with complex resource issues. Learning is a long-term proposition, which requires a ballast against short-term politics and objectives. Another shift will probably require a change in the focus of actions away from management by objectives and determination of optimum policies, towards new ways to define, understand, and manage these systems in an ever-changing world. That focus should not be solely on the variables of the moment (water levels, population numbers) and their correlative rates, but rather on more enduring system properties such as resilience, adaptive capacity, and renewal capability. This framework involves both the human components of the system (operations, rules, policies, and laws) and the biophysical components of the landscape and its ecosystems. The shift of focus to learning basis is likely to require flexible linkages with a broader set of actors or network. Another way of saying this bluntly is: until management institutions are able and willing to embrace uncertainty and systematically learn from their actions, adaptive management will not continue in its original context, but rather will be redefined in a weak context of 'flexibility in decision-making' (Gunderson, 1999b).

In order to meet these challenges for learning-based institutions, there is a need to develop new theory and expand old theories addressing issues of scale. In times of uncertainty, there is nothing more practical than theory (Holling, 1995). Walters (1997) cites the cross-scale problem as a severe obstacle in most assessment/modeling activities. Development of new theories is needed to help address ecosystem and natural resource dynamics across space and time scales. Over the last 40 years, time and space have been separated for analytical purposes. Most field-scale ecologic investigations either freeze space and experiment over time or freeze time and look at spatial patterns (witness the explosion and ubiquity of Geographic Information Systems (GIS) technology in resource management agencies). Perhaps there are practical reasons for this pattern, but also it can be explained in part because of underlying theoretical frameworks. There is a growing sense that this separable-dimension framework will result in different outputs from assessments, suggesting the need for integration.

There is a growing trend towards addressing the issues and theories of cross-scale interactions in ecology (Levin, 1992). These frameworks include the types of resilience arguments mentioned earlier, with multiple stability domains in various systems, and the controls among such domains (Gunderson *et al.*, 1997), and patterns of cross-scale discontinuities – where textural discontinuities in cross-scale structures create templates or signatures of similar lumpiness in animal community structure (Holling, 1992).

2.4.2 Can we manage for resilience?

In addition to developing better theories and trained professionals, perhaps we should attempt to develop new paradigms or schema that underlie resource management approaches. One place to start is with the notions of resilience and adaptive capacity. These theoretical concepts identify at least two strategies: those that people employ in order to manage for resilience in a resource system and properties that contribute to flexibility in human organizations. In order to add resilience to managed systems, a number of strategies are employed: increase the buffering capacity of the system, manage for processes at multiple scales, and nurture sources of innovation and renewal, as elaborated in the following paragraphs.

Most activities for buffering tend to address the engineering type of resilience, that is, mitigating the effects of unwanted variation in the system in order to facilitate a return time to a desired equilibrium. In many agricultural systems, resistance to change is dealt with by a combination of barriers to outside forces (tariffs, fences, etc.) and internal adjustments such as cost control mechanisms (Conway, 1993). Water resource systems can be designed for resilience by increasing the buffering capacity or robustness through redundancy of structures (and flexibility of operations) rather than fewer, larger structures and rigid operational schemes (Fiering, 1982). Berkes and Folke (1998) suggest that traditional approaches (which they define as traditional ecological knowledge) buffer managed systems by not allowing unpredictable or large perturbations to threaten ecosystem structure and function by allowing smaller-scale perturbations to enter the system. One such example is the Cree fishers' use of a mixed-size mesh net to harvest multiple age classes, thereby preserving an age-class structure that mimics a natural population (Berkes, 1995). This stable age structure helps buffer widely varying reproductive success.

Resource systems that have been sustained over long time periods increase resilience by managing processes at multiple scales. Returning to the example of the Cree in northern Canada, Berkes (1995) argues that multiple spatial domains are part of their fishing practices and multiple temporal domains in their hunting. While fishing, the Cree monitor catch per unit effort. When they notice the rate dropping, they immediately move to alternative fishing sites; over longer time frames, they rotate fishing effort to more remote sites.

Another example is in the Everglades water management system, where, in the mid 1970s, water deliveries to Everglades Park were based upon a seasonally variable, but annually constant, volume of water. This system was changed in the mid 1980s to a statistical formulation that incorporated interannual variation into the volumetric calculation (Light and Dineen, 1994).

Berkes and Folke (1998) argue that local communities and institutions co-evolve by trial and error at time scales in tune with the key sets of processes that structure ecosystems within which the groups are embedded. Many of the crises chronicled in Gunderson *et al.* (1995) were created by an inherent focus on one scale for management, and reformations of learning recognized the multiple scales by which the ecosystem was functioning.

Another way in which people manage for resilience in resource systems is by concentrating on sources of innovation and renewal. Many forms of catastrophic insurance provide this function, by creating a fiscal reservoir that can be tapped, should structures need to be replaced. Another mechanism that explicitly plans for renewal in resource systems is the scheme of market-based property rights systems developed for Australia. Young and McCay (1995) argue that adding flexibility and renewable structure to property rights regimes will increase resilience. They indicate that market-based property right schemes (licenses, leases, quotas or permits) should have built-in sunset (termination) to the scheme, with stable arrangements (entitlements, obligations) in the interim years. These principles complement Ostrom's (1990) findings of successful institutions allowing for stakeholders to participate in changing rules that affect them.

Finally, the ability of institutions to renew themselves following crises or to generate new and novel solutions to resource problems appears to be a pragmatic adaptation. A few key ingredients appear necessary to facilitate the movement of systems out of crisis through a reformation. In the review of management histories in Western systems (Gunderson *et al.*, 1995), these included functions of learning, and engagement and the ability to tap into deeper understanding and trust. Lee (1993) calls this 'social learning;' it is a process that combines adaptive management within a framework of collective choice.

Weick (1995) describes a number of sources of organizational resilience including improvisation, virtual role systems, the attitude of wisdom, and norms of respectful interaction. Other authors (Berkes and Folke, 1998) describe this as cultural capital, comprised of the institutions, traditional knowledge, common property systems, which are the mechanisms by which people link to their environment. It is such linkages and connectivity across time and among people that help navigate transitions through periods of uncertainty to provide social resilience.

2.5 Summary and conclusions

Walters (1997), Johnson *et al.* (1999), Gunderson (1999b), and the cases in this volume have stressed the practical difficulties that humans face in attempting to manage ecosystems. The multiple scales of variables, cross-scale, and nonlinear interactions generate the multi-stable behaviors in ecosystem dynamics. The surprises generated by this multi-stable behavior create a range of problems for management (Carpenter *et al.*, 1999a). All of these compound the difficulties of managing, let alone managing adaptively.

Adaptive management in its early form focused on confronting the uncertainty of resource dynamics through actions designed for learning (Holling, 1978; Walters, 1986). This has evolved from a process of testing a single hypothesis about the system to sorting among multiple hypotheses, each of which may have different social and management implications (Gunderson, 1999b). Other layers of complexity arise from having adequate monitoring or data to put these hypotheses at risk (Walters, 1997). Ludwig (1995) considers harvesting strategies under increasing layers of uncertainty, and shows that increasing uncertainty generally leads to increasing caution in harvesting and a strengthened precautionary principle.

The challenges posed have a technical dimension and a social dimension. The technical challenge has two parts as well. The first is to develop a framework that will allow for a process of formulating testable hypotheses, and the second is how to choose among multiple hypotheses. The models of complex adaptive systems that appear in Carpenter *et al.* (1999a) are useful frameworks for the problem of formulating hypotheses. These have a long history of use in the process of adaptive assessment (Walters, 1997). The process of constructing these types of models is much more important than the model itself (Walters, 1986).

The technical challenge of sorting among competing hypotheses is problematic, although Walters (1986) and Hilborn and Mangel (1997) give quantitative guidance. The second challenge is the social arena. The types of organizational complexity raised by Westley (2002) and political pathologies (Pritchard and Sanderson, 2002) generate barriers for the adoption of adaptive management in Western bureaucratic agencies. Adaptive management has been socially challenged through practices such as self-serving interests of management agencies, career concerns and greed among scientific experts, and dis-information campaigns by opposing sides who exploit the uncertainty of resource issues to maintain the *status quo*.

Uncertainty pervades resource issues. That uncertainty is manifest as ecological surprises and crises. In a typology, different types of surprises have been noted: local, cross-scale surprises and true novelty. Each connotes a different type of adaptive response. Ecological crises occur when it is revealed that extant policy fails. Surprises and crises are linked to ecological properties of resilience and adaptive capacity, whereas dynamic responses are related to institutional adaptability and flexibility. This tension is in contrast to the conservative properties of resource management institutions – institutions that in many cases were created to resolve key uncertainties. Those uncertainties have technical components, political components, and stakeholder-citizen components. Few arenas exist that seem to successfully embrace these different types of uncertainties. Positive, adaptive responses appear to involve novelty – tools, models, and theory that focus on both understanding the ecological dimensions of a crisis and on the institutions that focus on creativity and learning.

Acknowledgements

This is a contribution from the Resilience Network, funded by a grant from the John D. and Catherine T. MacArthur Foundation. I thank Buzz Holling, Rusty Pritchard, Steve Carpenter, Brian Walker, Buz Brock and other members of the Resilience Network for their educating discussions.

References

Berkes, F. 1995. Indigenous knowledge and resource management systems: a native Canadian case study from James Bay. In *Property Rights in a Social and Ecological Context*, pp. 99–109, ed. S. Hanna and M. Munasinghe. Washington DC: Beijer International Institute and World Bank.

Berkes, F. and Folke, C., eds. 1998. *Linking Social and Ecological Systems: Management Practices and Social Mechanisms for Building Resilience.* Cambridge: Cambridge University Press.

Bradbury, R.H. 1990. *Acanthaster and the Coral Reef: a Theoretical Perspective.* Proceedings of a workshop held at the Australian Institute of Marine Science, Townsville, Aug. 6–7, 1988. Berlin, New York: Springer-Verlag.

Brooks, H. 1986. The typology of surprises in technology, institutions and development. In *Sustainable Development of the Biosphere*, pp. 325–47, ed. W.C. Clark and R.E. Munn. Laxenburg, Austria: IIASA.

Carpenter, S.R., Brock, W., and Hanson, P. 1999a. Ecological and social dynamics in simple models of ecosystem management. *Conservation Ecology* 3(2): 4. [online] URL: http://www.consecol.org/vol3/iss2/art4

Carpenter, S.R. and Cottingham, K.L. 1997. Resilience and restoration of lakes. *Conservation Ecology* 1: 2. URL: http://www.consecol.org/vol1/iss1/art2

Carpenter, S.R., Ludwig, D., and Brock, W. 1999b. Management of lakes subject to potentially irreversible change. *Ecological Applications* 9(3): 751–71.

Clark, W.C., Jones D.D., and Holling, C.S. 1979. Lessons for ecological policy design: a case study of ecosystem management. *Ecological Modelling* 7: 2–53.

Conway G.R. 1993. Sustainable agriculture: the trade-offs with productivity, stability and equitability. In *Economics and Ecology: New Frontiers and Sustainable Development,* pp. 88–101, ed. E.B. Barbier. London: Chapman and Hall.

Costanza, R. and Greer, J. 1995. Chesapeake Bay. In *Barriers and Bridges to the Renewal of Ecosystems and Institutions*, pp. 169–213, ed. L.H. Gunderson, C.S. Holling, and S.S. Light. New York: Columbia University Press.

D'Antonio, C.M. and Vitousek, P.M. 1992. Biological invasions by exotic grasses, the grass/fire cycle and global change. *Annual Review of Ecology and Systematics* 23: 63–87.

Davis, S.M. 1994. Phosphorus inputs and vegetation sensitivity. In *The Everglades: the Ecosystem and its Restoration*, pp. 357–78, ed. S. Davis and J. Ogden. Florida: St Lucie Press.

Dörner, D. 1996. *Logic of Failure: Why Things Go Wrong and What We Can Do to Make Them Right*, translated by R. and R. Kimber. New York: Metropolitan Books.

Dublin, H.T., Sinclair, A.R.E., and McGlade, J. 1990. Elephants and fire as causes of multiple stable states in the Serengeti-mara woodlands. *Journal of Animal Ecology* 59: 1147–64.

Ehrlich, P. and Ehrlich, A. 1996. *The Betrayal of Science and Reason.* Washington DC: Island Press.

Fiering, M. 1982. Alternative indices of resilience. *Water Resources Research* 18: 33–9.

Giddens, A. 1987. *Social Theory and Modern Sociology.* Cambridge: Polity Press.

Glantz, M.H., Katz, R.W., and Nicholls, N. 1991. *Teleconnections Linking Worldwide Climate Anomalies: Scientific Basis and Societal Impact.* Cambridge: Cambridge University Press.

Gunderson, L.H. 1999a. Resilient management: comments on ecological and social dynamics in simple models of ecosystem management by S.R. Carpenter, W.A. Brock, and P. Hanson. *Conservation Ecology*, 3(2): 7 [online] URL: http://www.consecol.org/vol3/iss2/art7

Gunderson, L.H. 1999b. Resilience, flexibility and adaptive management: antidotes for spurious certitude. *Conservation Ecology* 3(1): 7. [online] URL: http://www.consecol.org/vol3/iss1/art7

Gunderson, L.H. 2000. Resilience in theory and practice. *Annual Review of Ecology and Systematics* 31: 425–36.

Gunderson, L.H. and Pritchard, L. Jr 2002. *Resilience and the Behavior of Large Scale Ecosystems*. Washington DC: Island Press.

Gunderson L.H., Holling, C.S., and Light, S.S. 1995. *Barriers and Bridges to the Renewal of Ecosystems and Institutions*. New York: Columbia University Press.

Gunderson, L.H., Holling, C.S., Pritchard, L., and Peterson, G. 1997. Resilience in ecosystems, institutions, and societies. *Beijer Discussion Papers No. 95*. Stockholm: Beijer International Institute of Ecological Economics, The Royal Swedish Academy of Sciences.

Gunderson, L. and Walters, C.J. 1999. Florida Bay; alternative or alternating states? Unpublished manuscript, Department of Zoology, University of Florida.

Haas, P. 1990. *Saving the Mediterranean: the Politics of International Environmental Cooperation*. New York: Columbia University Press.

Hilborn, R. and Mangel, M. 1997. *The Ecological Detective*. Princeton, NJ: Princeton University Press.

Holling, C.S. 1973. Resilience and stability of ecological systems. *Annual Review of Ecology and Systematics* 4: 1–24.

Holling, C.S., ed. 1978. *Adaptive Environmental Assessment and Management*. New York: John Wiley and Sons.

Holling, C.S. 1986. The resilience of terrestrial ecosystems: local surprise and global change. In *Sustainable Development of the Biosphere*, pp. 292–317, ed. W.C. Clark and R.E. Munn. Cambridge: Cambridge University Press.

Holling, C.S. 1992. Cross-scale morphology, geometry and dynamics of ecosystems. *Ecological Monographs* 62(4): 447–502.

Holling, C.S. 1995. What barriers? What bridges? In *Barriers and Bridges to the Renewal of Ecosystems and Institutions*, pp. 3–34, ed. L.H. Gunderson, C.S. Holling, and S.S. Light. New York: Columbia University Press.

Holling, C.S. 1996. Engineering resilience versus ecological resilience. In *Engineering Within Ecological Constraints*, pp. 32–43, ed. P.C. Schulze. Washington DC: National Academy Press.

Holling, C.S., Gunderson, L.H., and Peterson, G. 2002. Panarchies and sustainability. In *Panarchy, Understanding Transformations in Systems of Humans and Nature*, pp. 63–102, ed. L.H. Gunderson and C.S. Holling. Washington DC: Island Press.

Jansson, B.O. and Velner, H. 1995. The Baltic, sea of surprises. In *Barriers and Bridges for the Renewal of Ecosystems and Institutions*, pp. 292–372, ed. L. Gunderson, C.S. Holling, and S.S. Light. New York: Columbia University Press.

Johnson, N.K., Swanson, F., Herring M., and Greene S., eds. 1999. *Bioregional Assessments: Science at the Crossroads of Management and Policy*. Washington DC: Island Press.

Lee, K. 1993. *Compass and the Gyroscope*. Washington DC: Island Press.

Levin, S.A. 1992. The problem of pattern and scale in ecology. *Ecology*. 73: 1943–67.

Light, S.S. and Dineen, J.W. 1994. Water control in the Everglades. In *The Everglades: the Ecosystem and its Restoration*, pp. 47–84, ed. S. Davis and J. Ogden. Delray Beach, FL: St Lucie Press.

Light, S.S., Gunderson, L.H., and Holling, C.S. 1995. Everglades, the evolution of management in a turbulent ecosystem. In *Barriers and Bridges to the Renewal of Ecosystems and Institutions*, pp. 103–68, ed. L.H. Gunderson, C.S. Holling, and S.S. Light. New York: Columbia University Press.

Ludwig, D. 1995. A theory of sustainable harvesting. *Siam Journal of Applied Mathematics* 55: 564–75.

Ludwig, D., Hilborn, R., and Walters, C. 1993. Uncertainty, resource exploitation, and conservation: lessons from history. *Science* 260: 17, 36.

Ludwig, D., Jones, D.D., and Holling, C.S. 1978. Qualitative analysis of insect outbreak systems: the spruce budworm and forest. *Journal of Animal Ecology* 47: 315–32.
Lugo, A., Scatena, F.N., Silver, W., *et al.* 2002. An exploration of resilience mechanisms in subtropical wet and subtropical dry forests in Puerto Rico. In *Resilience and the Behavior of Large Scale Systems*, ed. L.H. Gunderson and L. Pritchard Jr. Washington, DC: Island Press. (In press.)
McClanahan, T., Done, T., and Polunin, N.C. 2002. Resiliency of coral reefs. In *Resilience and the Behavior of Large Scale Systems*, ed. L.H. Gunderson and L. Pritchard Jr. Washington, DC: Island Press. (In press.)
Myers, R.L. 1983. Site susceptibility to invasion by the exotic tree *Melaleuca quinquenervia* in South Florida. *Journal of Applied Ecology* 20: 645–58.
O'Neill, R.V., DeAngelis, D.L., Waide J.B., and Allen T.F.H. 1986. *A Hierarchical Concept of Ecosystems*. Princeton: Princeton University Press.
Ostrom, E. 1990. *Governing the Commons: the Evolution of Institutions for Collective Action*. Cambridge: Cambridge University Press.
Peterson, G.D., Allen, C.R., and Holling C.S. 1998. Ecological resilience, biodiversity and scale. *Ecosystems* 1: 6–18.
Pimm, S.L. 1984. The complexity and stability of ecosystems. *Nature* 307: 321–6.
Pritchard, L. Jr, Folke, C., and Gunderson, L. 2000. Valuation of ecosystem services in institutional context. *Ecosystems* 3: 36–40.
Pritchard, L. Jr and Sanderson, S. 1999. The power of politics in agent-based models of ecosystem management: comment on 'Ecological and social dynamics in simple models of ecosystem management' by S.R. Carpenter, W.A. Brock and P. Hanson. *Conservation Ecology* 3(2): 9. [online] URL: http://www.consecol.org/vol3/iss2/art9
Pritchard, L. Jr and Sanderson, S. 2002. The dynamics of political discourse in seeking sustainability. In *Panarchy, Understanding Transformations in Systems of Humans and Nature*, pp. 147–69, ed. L.H. Gunderson and C.S. Holling. Washington DC: Island Press.
Robblee, M.B., Barber, T.R., Carlson, J.P.R., *et al.* 1991. Mass mortality of the tropical seagrass *Thalassia testudinum* in Florida Bay (USA). *Marine Ecology Progress Series* 71: 297–9.
Sapp, J. 1999. *What is Natural? Coral Reef Crisis*. New York: Oxford University Press.
Scheffer, M. 1998. *The Ecology of Shallow Lakes*. London: Chapman and Hall.
Sinclair, A.R.E., Olsen P.D., and Redhead, T.D. 1990. Can predators regulate small mammal populations? Evidence from house mouse outbreaks in Australia. *Oikos* 59: 382–92.
Tenner, E. 1996. *Why Things Bite Back: Technology and the Revenge of Unintended Consequences*. New York: Knopf.
Thompson, M. 1983. A cultural bias for comparison. In *Risk Analysis and Decison Processes: the Siting of Liquid Energy Facilities in Four Countries*, pp. 232–62, ed. H.C. Kunreuther and J. Linnerooth. Berlin: Springer.
Walker, B. 1995. Conserving biological diversity through ecosystem resilience. *Conservation Biology* 9 (4): 747–52.
Walker, B.H., Ludwig, D., Holling, C.S., and Peterman, R.M. 1969. Stability of semi-arid savanna grazing systems. *Ecology* 69: 473–98.
Walters, C.J. 1986. *Adaptive Management of Renewable Resources*. New York: Macmillan.
Walters, C.J. 1997. Challenges in adaptive management of riparian and coastal ecosystems. *Conservation Ecology* 1(2):1 [on line] URL:http://www.consecol.org/vol1/iss2/art1

Walters, C., Gunderson, L., and Holling, C.S. 1992. Experimental policies for water management in the Everglades. *Ecological Applications* 2(2): 189–202.
Weick, K. 1995. *Sensemaking in Organizations.* Thousand Oaks, CA: Sage.
Westley, F. 1995. Governing design: the management of social systems and ecological management. In *Barriers and Bridges to the Renewal of Ecosystems and Institutions*, pp. 391–427, ed. L.H. Gunderson, C.S. Holling, and S.S. Light. New York: Columbia University Press.
Westley, F. 2002. The devil is in the dynamic. In *Panarchy, Understanding Transformations in Systems of Humans and Nature*, pp. 333–60, ed. L.H. Gunderson and C.S. Holling. Washington DC: Island Press.
Williams, B.A. and Matheny, A.R. 1995. *Democracy, Dialogue, and Environmental Disputes: the Contested Languages of Social Regulation*. New Haven: Yale University Press.
Young, M. and McCay, B.J. 1995. Building equity, stewardship and resilience into market-based property rights systems. In *Property Rights and the Environment*, pp. 87–102, ed. S. Hanna and M. Munasinghe. Washington DC: Beijer International Institute and World Bank.

3

Nature and society through the lens of resilience: toward a human-in-ecosystem perspective

IAIN J. DAVIDSON-HUNT AND FIKRET BERKES

3.1 Introduction

There is a long history in several disciplines of trying to understand the relationship between ecological and social systems. The issue is often glossed as the nature/culture and environment/society dichotomies. Glacken (1967) has provided an extensive and wide-ranging survey of the ways in which the relationship between nature and society have been conceptualized within Western thought up to the eighteenth century. With the Age of Enlightenment, humans were extracted from the environment. The separation of nature and society became a foundational principle of Western thought and provided the organizational structure for academic departments. Since that time, Western thought has oscillated between positions in which nature and society were treated as distinct entities, and one in which articulations between the two were examined.

One of the early attempts to provide a model of natural system–society articulation was the one constructed by Karl Marx in the nineteenth century (Ingold, 1980; Wolf, 1982; Harvey, 1996). The discussion of the relationship between nature and society continued during the twentieth century in many different disciplines. There has been the human ecology of Park (1936), the cultural ecology of Julian Steward (1955), the ecological anthropology of Gregory Bateson (1973, 1979), Netting (1974, 1986, 1993), Vayda and McCay (1975), the ideas of Carl Sauer (1956) and other human geographers, the environmental history of William Cronon (1983, 1995) and Donald Worster (1977, 1988), the ethnoecology of Conklin (1957) and others (Toledo, 1992; Nazarea, 1999) and the emerging political ecology of Greenberg and Park (1994), Peet and Watts (1996) and others. The literature pertaining to the nature and society relationship spans many disciplines and has consolidated in the last two decades or so into half a dozen subdisciplines, as reviewed in Chapter 1.

Since the 1970s, an emerging body of literature has emphasized that ecological systems are characterized by nonlinear processes and multiple equilibria instead of stability: surprises (perceived reality departing qualitatively from expectation, in the sense of Holling, 1986), threshold effects, and system flips. Implications of this perspective are being explored for social systems and social–ecological systems as well (Vayda and McCay, 1975; Zimmerer, 1994; Abel, 1998; Zimmerer and Young, 1998; Biersack, 1999; Kottak, 1999; Scoones, 1999). Following Chapter 1, institutions are a key link between social systems and ecosystems. To the extent that much of the environment is used as a shared resource, common property institutions are an important consideration in this body of literature. Because resources and human uses occur across a variety of scales, both spatial and temporal, linking social and ecological systems is a cross-scale problem (Holling, Berkes, and Folke, 1998).

Taken as a whole, this body of thinking questions the utility of Cartesian models which maintain the separation between nature and society. Such a position reverses Descartes' *cogito ergo sum* (I think therefore I am) to *sum ergo cogito* (I am therefore I think), which may be called a human-in-ecosystem, or 'dwelling,' perspective (Descola and Pálsson, 1996; Ingold, 2000). One line of thought which has been woven throughout this broader literature is the idea of resilience as reformulated by the ecologist C.S. Holling (1973). Holling's resilience is utilized as a way to think about the relationship between nature and society and about the boundary between the two.

This chapter extends the work of Holling and Sanderson (1996), Berkes and Folke (1998) and others to explore the contribution of the concept of resilience for understanding social–ecological system linkages. Chapter 1 discusses several main streams of integrative approaches; this chapter narrows the discussion. The main area of emphasis is the exploration of how the concept of resilience has influenced paradigms of nature and society; the antecedents of these paradigms; and the way ahead for operationalizing the use of the resilience concept in human-in-ecosystem applications. The chapter begins with a broad overview of different ways in which Western thought has conceptualized the relationship between nature and society (Section 3.2). It looks briefly at the contending theories of environmental determinism and historical possibilism and the way in which cultural ecology was posited as a resolution of this dichotomy. The chapter then traces the sources of a systems approach for understanding nature–society interactions (Section 3.3). Ecological anthropology and other systems approaches represent a movement toward a consideration of the mutual influences of ecological and social processes, instead of treating social and ecological systems as linked but separate domains.

The chapter then turns to an examination of how the concept of resilience has been utilized within the social science literature to develop an ecological, or relational, understanding of humans in the environment (Section 3.4). This section is based upon a review of the Social Science Citation Index (SSCI) for those papers which cite Holling (1973), Holling and Goldberg (1971), and Vayda and McCay (1975). We do not claim that this review captures all resilience references that can be found across the spectrum of social science disciplines, but only those that are cited in SSCI and that have a direct linkage to the three seminal papers.

In this literature, the emergence of concepts such as generalist and specialist strategies, uncertainty and surprise, and adaptation and centralization is developed, expressing the mutual influences of social and ecological processes. In Section 3.5, we turn away from an examination of how the resilience concept has been utilized in the literature, to examine how emerging theory has begun to shift the practice of applied conservation and development. This section emphasizes the continuity between theory and practice and builds upon propositions such as sustainability science and adaptive management to show that theory and practice have become closely linked and place specific (Chapter 1). Section 3.6 concludes the chapter by considering some of the syntheses and questions emerging from resilience research and practice. This intersection between resilience research and practice provides insights for moving toward a human-in-ecosystem perspective.

3.2 Environmental determinism, possibilism, and cultural ecology

Cultural ecology and ecological anthropology emerged from the tension between two contending 'grand' theories which have been termed 'anthropogeography' and 'historical possibilism' (Geertz, 1963; Moran, 1979). Anthropogeography is a variant of environmental determinism which has its origins in human geography. It proposes that the environment is the causal agent for the behavior of social systems. According to environmental determinism, 'a "temperate" or "balanced" climate, ethnocentrically defined, was responsible for the virtuous qualities of the area's inhabitants. As a result, they were destined to rule and control the "lesser" domains where populations were more lethargic, less courageous, and less intelligent' (Moran, 1979: 24). Environmental determinism was often used by many societies to explain the relationship between their own society and other people. The early Greek, Roman, and Arab empires, along with the dominant European countries of the eighteenth and nineteenth centuries, all utilized environmentally deterministic theories in this manner (Moran, 1979).

Anthropogeography was a particular manifestation of environmental determinism, developed by Friedrich Ratzel in the late 1800s, which suggested that the interrelation of groups with their habitats produced specific kinds of cultural traits. Moran (1979) has suggested that Ratzel's main thesis included the following propositions: habitat was primary in bringing about cultural diversity; similarities between cultural groups were explained as occurring due to the diffusion of traits by migrating groups; and human cultural evolution emerged out of the territorial competition between migrating groups. Ratzel's theory began a trend that viewed human beings as limited by their habitat in their range of responses, and human culture as shaped by environmental conditions. According to this view, environment was postulated to cause change in human societies, thereby accounting for human evolution through a process of trait selection by the environment.

Historical possibilism was developed by Franz Boas in the early 1900s as an alternative theory to explain the interactions between humans and nature. Historical possibilism suggested that 'nature circumscribes the possibilities for humans, but historical and cultural factors explain what possibility is actually chosen' (Moran, 1979: 34). In Boas's view, humans choose what they want to use in nature, and it is those cultural decisions, not nature itself, which influence the trajectory of human societies and cultural change. In order to refute environmental determinism, historical possibilism constructed the concept of culture as the basis of human adaptation. Boas emphasized inductive studies that focused on the empirical cultural traits of different human groups as a means to counterbalance the deductive theories of environmental determinism.

Through studies carried out in the early and mid-twentieth century, anthropologists, geographers and other social scientists were able to demonstrate that many different cultural traits were found in areas sharing similar biophysical environments. It was culture, and not geography, which led to differences between human groups. Traits shared by groups were explained by diffusion from one culture area to another. How far a cultural trait had diffused from its origin was believed to demonstrate the antiquity of a trait. Although this approach corrected for environmentally deterministic theories, it led to another problem because of its emphasis on the idea of culture. Culture became a 'superorganic' entity that subordinated individual humans to its patterns (Wolf, 1982; Moran, 1990). The causal agent shifted from the environment to culture; change emerged from historical and cultural forces while the environment acted as the setting in which these forces were played out.

The cultural ecology of Julian Steward emerged out of the debate between these two opposing theories as another perspective on the relationship between nature and society. One of Steward's objectives was to move social theory back toward a consideration of the evolutionary or adaptive relationship between human society and nature (Steward, 1955). To return to the question of how societies change, the purpose of cultural ecology was to 'develop a methodology for determining regularities of form, function and process which recur cross-culturally among societies found in different cultural areas' (Steward, 1955: 3).

Whereas other writers sought to formulate cultural development in terms of supposed universal stages, Steward's objective was to seek causes of cultural change. Cross-cultural comparisons were made through an examination of the 'cultural core,' which was defined as the 'recurrent constellations of basic features . . . which have similar functional interrelationships resulting from local ecological adaptations and similar levels of sociocultural integration' (Steward, 1955: 6). The basis of Steward's evolutionary theory was 'multilinear evolution,' which he defined as 'the methodological position [which] assumes that certain basic types of culture may develop in similar ways under similar conditions but that few concrete aspects of culture will appear among all groups of mankind in regular sequence' (Steward, 1955: 4).

Steward's cultural ecology, by focusing on empirical features rather than on deductive and universal theories, was an important reframing of understanding the relationship between cultural change and the environment. First, he pushed the focus toward the relationship between the environment and cultural features, and how adaptation or change emerged out of the relationship over time. Second, he emphasized empirical cases by which similarities could be found across cultures so that theories of process could be built. Third, he recognized that human perception of the environment played a role in nature–society relationships and adaptation.

3.3 Ecological anthropology and the rise of systems approaches

Geertz (1963), in his book *Agricultural Involution*, provided a new challenge to the model of cultural ecology. He suggested that an ecological approach should utilize an ecosystem model, whereby humans were one component of an ecological system. As Geertz put it, 'the ecological approach attempts to achieve a more exact specification of the relations between selected human activities, biological transactions, and physical processes by including them within a single analytical system, an ecosystem' (Geertz, 1963: 3). This mode of analysis trains attention on the pervasive properties of systems *qua* systems (system structure, system equilibrium, system change) rather than on the point-to-point relationships between paired variables of the 'culture' and 'nature' variety (Geertz, 1963). One significant aspect of Geertz's (1963) approach is that he suggested a unified systems model which would include both biological and social entities and processes, an idea which would not be fully explored again until the 1990s. His main criticisms of Steward's (1955) cultural ecology model were that it still perpetuated the nature/society dichotomy, and reduced the number of variables which might be considered in understanding cultural change to those emerging from the adaptive relationship between nature and society.

Geertz showed in *Agricultural Involution* that changes in Indonesian society were not attributable to ecological processes but emerged from political, commercial, and intellectual developments. The features of society do not change only as a result of changes in the cultural core, as those features adapt to the environment, but may also change for reasons which are unrelated to subsistence technology. Geertz's (1963) ecosystem approach attempted to put humans into a unified system, while insisting that the system account for social and political structures, functions, and processes along with the biological.

The application of the systems approach and use of ecological concepts moved the field into what was termed the new ecological anthropology (Vayda and McCay, 1975; Moran, 1979, 1990). While Geertz (1963) recognized that social, political, and biological variables should be included in a systems approach, it proved difficult to operationalize such an ecosystem approach. Rather, ecological anthropology turned toward the study of human adaptation by utilizing the principles of biological ecology (Vayda and Rappaport, 1968).

A number of different approaches were utilized to study human adaptation within an ecosystem framework. One of these is the use of energy flows and of cybernetics, or information flows, in the study of rituals (Rappaport, 1967). The systems approach was not without its critics. Some of the most glaring problems included the teleological fallacy which overemphasized organizational versus individual goal-seeking behavior (Alland and McCay, 1973); an overemphasis on the role of energy (Moran, 1979, 1990); the assumption of equilibrium and functional behavior at the expense of historic change (Vayda and McCay, 1975); lack of consideration of the role of the individual (Keesing, 1976; Borofsky, 1994); and lack of attention to boundary and scale (Moran, 1979, 1990).

Many of these criticisms emerged out of the renewed emphasis on the primacy of the individual in the theory of biological evolution (Alland and McCay, 1973; Richerson, 1977). However, the challenge for both the systems and the evolutionary approaches has been to account for Geertz's (1963) early observation that changes in societies cannot solely be explained by the adaptation of either cultures or individuals. The environment/society dichotomy, and the location of causality for societal adaptation within this dichotomy, has remained a continuing tension within ecological anthropology.

Another perspective on adaptation and change comes from a consideration of landscape and history. History as a social process has implications for social–ecological system interactions because history helps explain why landscapes look the way they do. Ecologists' attention to history challenged the notions of climax and equilibrium by revealing cycles and multiple equilibria (Holling *et al.*, 1995). Likewise, historians' attention to ecology revealed landscapes altered by human action (e.g., Cronon, 1983), leading to the questioning of notions of 'wilderness' and pristine environments (Balee, 1998). Hence, not only are resources socially and politically constructed (Harvey, 1974), many landscapes of the world are also socially and politically constructed. The political ecology perspective has taken up the challenge; it 'expands ecological concepts to respond to this inclusion of cultural and political activity within an analysis of ecosystems that are significantly but not always entirely socially constructed' (Greenberg and Park, 1994: 1).

These new developments and the interaction of ecological anthropology, historical ecology (environmental history), and political ecology have brought new perspectives to the analysis of the old culture/individual dichotomy. The main area of interest has been to explore different paradigms of nature and society in order to build a new model which allows humans to understand themselves as an integral part of the environment (Descola and Pálsson, 1996; Escobar, 1999; Ingold, 2000). A human-in-ecosystem model is an important first step, as models are the means by which humans translate perceptions into information, knowledge, and institutions. Such models will exhibit similarities to many of those constructed by non-Western societies (e.g., see Chapter 12). A human-in-ecosystem model will require addressing the environment/society dichotomy, incorporating evolutionary and historical processes into the model, and creating concepts that are sensitive to both (Holling *et al.*, 1998). A key concept that was developed in the ecological literature and utilized in the analysis of the linkages in social–ecological systems is resilience.

3.4 Resilience for nature–society linkages

While the concepts of single versus multiple equilibrium systems, stability, change, resistance, and resilience have been developed in the ecological literature, they are concepts that are not unfamiliar to social scientists. In 1975 Andrew Vayda and Bonnie McCay drew from the work of Holling and Goldberg (1971) and Holling (1973) to suggest that resilience may be a more useful concept to understand human adaptation than stability and resistance. Vayda and McCay (1975: 298) stated, 'ecological systems that have survived are "those that have evolved tactics to keep the domain of stability, or resilience, broad enough to absorb the consequences of change".' The consequence for social systems is that resilience means '... remaining flexible enough to change in response to whatever hazards or perturbations come along' (Vayda and McCay, 1975: 299). Resilience, in this sense, is co-terminous with flexibility, and stresses the ability of individuals, households or groups to respond to disturbances and survive (McCay, 1981; Lamson, 1986). The concept of resilience appeared in the ecological literature at a time when culture was held to be the force by which humans confronted nature and shaped it to their purposes, as well as the superorganic entity which constrained individual human behavior (Anderson, 1973; Moran, 1979, 1990; Wolf, 1982). In a similar vein, Rappaport (1967) had published his study that suggested the use of ritual acted to translate complex ecological processes into binary switches. This process allowed humans to make the appropriate decision, in Rappaport's analysis, to maintain an ecological equilibrium. The assumption behind this functionalist approach was that culture was an equilibrium-based system in which the equilibrium was near the carrying capacity of the environment (Vayda and McCay, 1975; Moran, 1990).

A culture, in this perspective, was analogous to a climax forest. The role of system processes was to maintain a society at a specific balance, and to move it back toward the equilibrium point after a disturbance. However, the carrying capacity of an environment is not fixed, but varies with factors such as institutions and technologies. Rejecting equilibrium-based concepts calls for a more dynamic understanding of the processes that link nature and society (Abel, 1998). A static understanding of the relationship between equilibrium and disturbance has also been a major focus of population/community ecology. Ecologists studying recovery after disturbance often use the term resilience to mean the return time to the same equilibrium that is predefined on the basis of the carrying capacity of the environment. Holling *et al.* (1995) refer to that approach as 'engineering resilience,' an interpretation quite different from a dynamic understanding of ecological or ecosystem resilience, which is the focus of this chapter.

Utilizing Holling and Goldberg (1971) and Holling (1973), Vayda and McCay (1975) challenged Rappaport's concept of culture as an equilibrium-based system. What Rappaport saw as a stability-maintaining mechanism could be interpreted instead as a resilience-building mechanism. To Vayda and McCay, resilience, as a property of social systems, allowed individuals and societies to change in the face of environmental challenges such as hazards. The resilience concept required 'investigating possible relationships between such characteristics of hazards as their magnitude, duration, and novelty, and the temporal and other properties of people's responses; abandoning an equilibrium centered view and asking instead about change in relation to homeostasis; and studying how hazards are responded to not only by groups but also by individuals' (Vayda and McCay, 1975: 302).

The concept of resilience helped to move ecological anthropology toward a dynamic, ecological perspective that investigated processes of change and equilibrium and disequilibrium, through an examination of the relationships among the environment, individuals, and groups. Subsequent literature used the concept of resilience to explore three related themes: (1) generalist and specialist strategies; (2) uncertainty and surprise; and (3) adaptation ability and degree of centralization.

3.4.1 Generalist and specialist strategies

The Africanist literature on hunter–gatherers and pastoralists parallels the above-cited work but differs from it in that it uses Holling's ideas about resilient and stable systems in a comparative fashion. The basic premise of this optimal foraging strategy thesis is that in locations where resources are unpredictable (referred to as 'resilient systems') a generalist strategy is pursued, whereas in areas with predictable resources ('stable systems') a specialist strategy emerges. Yellen (1977) utilizes Holling's view as part of what he calls the stability–time hypothesis. 'Holling (1973) draws the useful distinction between stability and resilience. Stable systems are those which tend to return quickly to equilibrium after a temporary disturbance and can be best described with equilibrium models. In resilient systems, there may be no single point of equilibrium. Individual components may be subject to rapid, unpredictable change; however, basic relationships between components or populations remain the same' (Yellen, 1977: 264).

Yellen (1977: 270) goes on to suggest that, in reference to hunters and gatherers who live in desert environments characterized by unpredictability, 'in relatively severe, variable environments of low predictability, populations exhibit resilience and the ability to persist over time.' One of the ways in which hunters and gatherers have adapted to these unpredictable resource fluctuations is through flexible forms of social organization in which the composition of groups can change readily:

> In an environment subject to severe and unpredictable change, it is obviously advantageous for a population to be able to alter its distribution rapidly in order to put the most people in places where the most resources are available. Among the !Kung, for example, the ability for rapid movement in a number of possible directions is provided primarily through kin ties and the obligations generated by them . . . Other social patterns such as the belief that people with the same name have special obligations toward each

other, and their system or inheritance in which individuals may have rights in scattered places, also *increase the number of possible residence locales and permit rapid changes in residence, if need be.*

(Yellen, 1977: 270–1, emphasis added).

Brooks, Gelburd, and Yellen (1984) developed this idea into a comparative approach by contrasting African hunter–gatherers ('generalists') with African pastoralists ('specialists'). The 'generalist' strategy was characterized by the following features of social organization: '... small social groups, absence of rights to property, emphasis on bilateral kin relations, and egalitarian social structure with minimal formal political and legal structures' (Fratkin, 1986: 270). In contrast, the 'specialist' strategy was characterized by '... longer occupation of settlements, cooperative herding, corporate ownership and lineal inheritance, increased value of children's labor, formal political and legal structures, the emergence of material accumulation and differential wealth, and increased birth rates' (Fratkin, 1986: 270). The study contrasted two pastoralist societies of northern Kenya, the Ariaal, who live in an unpredictable environment, and the Rendille, who reside in a more predictable environment.

The Ariaal differ from the cattle-keeping Rendille in their production of two major types of livestock, camels and cattle, in the marginal and variable environment of the Ndoto Mountain–Kaisut Desert interface. Their subsistence strategy is more generalist than the Rendille, allowing them to emphasize either cattle, camels, or small stock production when conditions drastically change, such as in the nineteenth century during the Rinderpest epidemic that destroyed Ariaal and Samburu cattle. The Rendille, wholly dependent on their camels and small stock in the constant but limited resources of the desert, have a specialized subsistence strategy that is less resilient to sharp changes...'

(Fratkin, 1986: 283, 284).

The relationship between resilience and social organization is also explored by Dyson-Hudson, Meekers, and Dyson-Hudson (1998), who suggest that the Turkana follow a generalist strategy to cope with environmental diversity and stochasticity. The characteristic structure of social organization is a residential and production unit called a camp. Camps are composed of temporary shelters and corrals, a nuclear family, and a herd. The herd moves many times a year and may subdivide and undertake independent movements. In order to undertake this activity, a herd owner '... needs several skilled and responsible herder managers with a detailed knowledge of livestock and of the South Turkana environment' (Dyson-Hudson *et al.*, 1998: 22). An individual production unit with fluctuating herd size and a loose system of social organization, such as the Turkana, also requires a network of kin and friends who act as supporters in disputes and provide insurance during bad years.

Dyson-Hudson and colleagues suggest that one of the mechanisms by which resilience is built into the Turkana social system is through marriage rules. It is the marriage rules which 'prevent the society from fragmenting into tiny, independently operating human/livestock units, which would lack the adequate resources to provide the diversity of labor and management skills to cope with a harsh, severely fluctuating environment' (Dyson-Hudson *et al.*, 1998: 42).

3.4.2 Uncertainty and surprise

Gunderson, Holling, and Light (1995) have developed a typology of response to surprise. Type I surprises lead to '... adaptations to risk that are amenable to economic rationality on an individual level, including risk-reducing strategies and risk spreading or risk pooling across independent individuals' (Folke *et al.*, 1998). Type II surprises, panarchy surprises, are those that occur at a regional or global scale due to the interlinkages amongst scales. Adaptations to type II surprises require the coordinated effort of many individuals through existing or readily formed institutions (Folke *et al.*, 1998). The final type of surprises are those which by definition are unpredictable. Type III surprises, or true novelty, result from phenomena that are outside the range of human experience or memory. Folke *et al.* (1998) suggest that latent mechanisms for reorganization, learning, and renewal may provide a society with the ability to adapt successfully following type III surprises. Resilience emerges from the institutional inventory of a society to deal with these different types of surprises.

The response of individuals and societies to uncertainty and surprise has also been dealt with by a number of other authors who likewise utilize the concept of resilience to explain how societies deal with natural resource uncertainties. For example, Winterhalder (1983), writing about Cree-Ojibwa moose hunting in Northern Ontario, pointed out the importance of resilience in adjusting hunting strategies in a patchy and uncertain environment. One author who has used the resilience idea in some detail is McCay, who discussed the relationship between resource uncertainty and fisherfolk in Fogo Island, Newfoundland.

> Variability in the size of year-classes is inherent in Atlantic cod populations ... Temporal and spatial variations in wind, currents, and other factors generate changes in cod migratory behavior and availability ... Unpredictable variability in cod is compounded by the effects of storms, high winds, and Arctic ice on the ability of fishermen to get out to the grounds and use their gear. The result is a highly uncertain and fluctuating resource. Codfish production figures on Fogo Island vary as much as two-fold from year to year ... In addition, in any given year cod may be abundant on some fishing grounds but scarce on others.
>
> *(McCay, 1978: 405).*

Fisherfolk perception of resource cycles across time (seasons) and space provides them with the means to formulate goals and strategies (coping mechanisms), which over time develop into adaptive strategies (in the sense of Bennett, 1969). Fisherfolk perceive the environment as the textured seascape of patterns and rhythms in which they dwell and make their livelihood. Technology, knowledge, and social organization are forged within the patterns and rhythms of the seascape and help the fisherfolk make a livelihood from a fish population that fluctuates widely over time and space. Resilience is reflected in the technology, knowledge, and social organization. The particularities of such a livelihood can be seen in the Fogo Island fishing strategies.

Cod fishing crews ... maintained two or three traps, placed in widely spaced 'berths', or fishing spots. If fish were scarce in one berth, they might be abundant in another. Similarly, crews used or kept on hand a wide variety of fishing gear, and often two or more different kinds of boats, which permitted a rapid switch from one technology to another. If cod failed to come into shallow inshore waters where the 'trap' fishery took place, the crew might set gill-nets or use hook-and-line gear in deeper near-shore waters. In addition, family firms maintained the capital equipment and recruitment process necessary for engaging in a 'fall fishery,' which they relied upon as insurance against the failure of the more intensive and normally more productive 'summer fishery.' Seasonally available salmon and lobster also provided buffers against the failure of cod. Moreover, a long tradition of being 'fisherman–farmers,' and more recently 'jacks of all trade,' provides yet another means of coping. When fishing was poor, lumbering, construction work, subsistence farming, and the use of government transfer payments (unemployment insurance since 1959 and welfare assistance) provided alternative or supplemental sources of income.

(McCay, 1978: 405–6).

Resource uncertainty and unpredictability (type I and II surprises) need to be distinguished from truly novel or unexpected change (type III surprises). McCay (1978) suggests that there are two possible responses when a resource cycle departs from the expected cycle: diversification and intensification. Occupational pluralism, or diversification, referring to a 'general "spreading of the risk" and expanding alternative modes of coping with environmental problems,' is relevant to both general uncertainty and surprise (McCay, 1978: 410). During an unexpected resource event, the first response is to diversify into minimal, less costly, and more reversible alternatives. This is a 'wait and see' or a 'weather the storm' strategy, whereby an individual or group undertakes alternative activities to see if the resource cycle returns to the expected pattern: '... minimal responses to perturbation may be valuable in providing a built-in time lag for evaluating the magnitude, duration, and other characteristics of problems, as well as the effectiveness of solutions. They thereby minimize the chance that costly and irreversible responses are activated for what might turn out to be trivial or transient problems' (McCay, 1978: 415–16).

If the expected pattern does not reemerge, an 'intensification' strategy may be adopted, whereby people will make an 'increased commitment to an investment in one or another mode of resource procurement [which are] "deeper", more costly, and less reversible' (McCay, 1978: 410). If these new strategies are adaptive, they may provide a long-term solution that restores flexibility to social actors and units. Diversification is the appropriate strategy if the ecological system has remained within the domain of stability, whereas intensification would be the necessary strategy if the properties of an ecological system change dramatically and permanently, a 'flip' in Holling's (1986) terminology.

In some cases, it is difficult to distinguish between a cycle of long periodicity and a true surprise. Resource cycles of 50 to 100 years may not be within the memory of a group of people and their institutions. If surprise is a result of long-term cycles, it may be possible to return to the previous livelihood cycle at some future point. There is some evidence that traditional societies may be able to deal with certain classes of long cycles such as once-in-a-generation tropical hurricanes (Lees and Bates, 1990) and caribou population cycles that may be on the scale of a century (Berkes, 1999). The mechanisms that provide resilience seem to be those that help start up the reorganization phase of the adaptive renewal cycle, such as the knowledge held by elders and the use of oral histories (Berkes and Folke, 2002).

Whether elders or oral history can help in the case of real surprises and 'large, infrequent disturbances' (LIDs, Turner and Dale, 1998) is not clear. Also, system flips from one stability domain to another, as in the potential case of climate change (Holling, 1986), create a unique set of adaptation and resilience problems. Folke, Berkes, and Colding (1998) have pointed out that one way traditional societies and other groups seem to have dealt with surprises is to create small disturbances that would help forestall much larger disturbances and surprises. They point out several cases in *Linking Social and Ecological Systems* that indicate that disturbance management is an adaptive response of many groups. Nurturing sources of ecosystem disturbance and renewal maintains the capacity of an ecosystem to absorb perturbations, thus preventing flips.

3.4.3 Adaptation ability and degree of centralization

Another class of uses of the stability/resilience idea deals with the question of why some societies (groups, companies) may have failed to adapt and 'went out of the evolutionary game' (McGovern, 1980; Lamson, 1986; Hurst, 1995;

King, 1995, 1997). One of the commonalities to these approaches is that flexibility of social organization allows societies to adapt to resource cycles and surprises more effectively than do rigid hierarchies. This is the essential message of McGovern's examination of the Norse colony in Greenland during AD 985–1500:

> There is little doubt that the Greenlandic Norse economy, as established in the little climatic optimum, faced serious if not fatal challenges in the 14th century. With full inner-fjord resource space, heavy investment in ceremonial architecture, and strong linkages to distant and increasingly disinterested European markets, Norse society of ca A.D. 1300 showed a *dangerous lack of resilience (in the sense of Holling, 1973)* in the face of waning extractive efficiency, fluctuating resources, and Inuit competition.
>
> *(McGovern, 1980: 270, emphasis added).*

McGovern (1980: 272) identified three characteristics which may reduce resilience in societies: (1) treating innovation as inherently dangerous to elites; (2) controlling social ideology, in this case through the medieval church, to punish deviance and reinforce orthodoxy; and (3) centralizing decision-making powers. In his view, the suppression of innovation, adherence to orthodoxy, and centralized decision-making in the face of environmental change impair adaptive response. In the case of the Norse colony, the movement to utilize coastal resources by poor farmers threatened the land-holding elite of the colony, and was prevented through social controls. The later work of McGovern and colleagues continued to use the resilience concept to deal with the question of environmental degradation in North Atlantic offshore islands colonized by medieval Scandinavians:

> Modern climatic data indicate that it is the resilience and stress-resistance of pasture communities that would be most altered as Norse farmers sailed north and west – not gross species composition or initial resistance to grazing pressure. As modern experience suggests, it is not easy to judge pasture resilience until damaging overgrazing has already occurred. Stocking levels appropriate to wind-sheltered areas may be disastrous on nearby exposed slopes, as small holes in the groundcover are rapidly widened, soon turning into a swiftly advancing erosion front that is difficult to halt.
>
> *(McGovern et al., 1988: 125).*

Much of the literature on the question of adaptation ability and centralization deals with institutions. King (1995) reviewed the common features of four societies (referred to as 'surprise-avoiding communities') that seem to have survived for a long time without reducing the natural environment's ability to support life, and concluded that the critical point was the communal use of resources. Common property resource management institutions in these four societies allowed the natural spatial and temporal variability in the environment 'to a degree almost unimaginable today' (King, 1995: 976). Because resources were held in common, King argued, human-made boundaries did not place an artificial grid on the landscape to impede natural flows and cycles.

Although common property is no guarantee of prudent ecological practice, one of the ways in which common property institutions are supportive of resilience is through locally adapted practices based on ecological knowledge and understanding (Folke *et al.*, 1998). It has been documented by many cases in *Linking Social and Ecological Systems*, as well as elsewhere, that local-level institutions learn and develop the capability to respond to environmental feedbacks faster than do centralized agencies. Being 'on the ground,' they are physically closer to the resources, there is no separation of the user from the manager, and there is more learning-by-doing in accumulating a base of practical ecological knowledge (Berkes and Folke, 1998).

Another way of looking at this issue is that large, centralized resource management agencies are susceptible to making large mistakes (Gunderson *et al.*, 1995). It is less risky for managers and users alike to make smaller mistakes and to learn from those smaller mistakes. Local-level common property institutions help decentralize environmental decision-making and diffuse the risk. Also, to reflect on McGovern (1980), they produce opportunities for innovation – innovation that is important for the integrity of adaptive renewal cycles. One of the most striking aspects of common property institutions is their diversity. This is in sharp contrast to conventional resource management that has reduced, since the middle of the twentieth century, the diversity in and experimentation of the ways in which environment and resources are managed (Berkes and Folke, 1998).

The concepts of resilience and the centralization of power become increasingly important for the intentional management of resources. For example, people are able to intentionally change the temporal and spatial characteristics of terrestrial resource cycles through technologies such as fire (Johnson, 1999; Lewis and Ferguson, 1999). By changing the range of oscillation of the resource cycle and the spatial characteristics of the system, people attempt to gain stability by reducing the size of the stability domain. The result is a reduction in resilience, but an increase in short-term stability. Such human interventions within an ecosystem are intentional acts that exist along a continuum from burning a berry patch to producing berries in an industrial monoculture. The trade-off is between resilient but fluctuating resource cycles on the one hand, and stable resource production which is increasingly vulnerable to surprise on the other (Finlayson and McCay, 1998). As production becomes more stable, in the short term, it can attract the investment of human-made capital (Berkes and Folke, 1994) to maintain or increase stability. For example, there is a substantive difference in the investment of human-made capital by a harvester of wild berries versus the industrial production of such berries. As people and industries specialize and invest in a resource, there is also a centralization of resource management power at a higher scale. The priority of resource management becomes the maintenance of stability and not resilience (Holling, 1995). McCay has summarized this process in the following manner:

> With centralization of power and control, there is a greater likelihood that inappropriate responses or errors in the scale of response will occur ... or that centralization will itself worsen the initiating environmental problem ... In addition, it becomes difficult for individuals and local communities to maintain their own flexibility, or ability to respond effectively to an uncertain and changing environment, because of their increased dependency, and the specialization attendant upon political development. It is also politically difficult for them to regain responsibility over the management of their local environments when the nature of their environmental problems is such that a lower level of regulation might be more appropriate. One reason for this is the public policy comes to serve the special purposes of certain powerful groups or individuals through the process of usurpation ... Accordingly, it may not be in their interests to allow regulations which have come to attain values other than environmental control to be changed in favor of local communities.
>
> *(McCay, 1981: 372).*

Stable commodity production systems and centralized resource management may be efficient and desirable, but it also necessary to recognize that they may increase society's long-term vulnerability to uncertainty and surprise. There is the risk that centralized resource management may create more surprises by altering ecosystem dynamics, such as disturbance regimes, and the functioning of slow structuring variables that are critical for resilience.

3.5 Operationalizing resilience: dwelling in the eco-commons

Exploring the environment/society dichotomy has required a stroll through a vast literature spanning several disciplines. In this section, we turn to a consideration of some recent practical applications of a human-in-ecosystem approach. These conceptualizations essentially use an ecological perspective that attempts to move beyond the individual/culture and nature/culture oppositions through a focus on *processes*. As such, they are consistent with the resilience concept and provide ways of operationalizing resilience.

Dwelling is a perspective that begins with the premise of the integrated concept of humans-in-nature (Berkes and Folke, 1998). It is the practical and perceptual engagement of humans with others of the dwelt-in-ecosystem. Knowledge of the environment, in this perspective, is '... not of a formal, authorized kind, transmissible in contexts outside those of its practical application. On the contrary, it is based in feeling, consisting in the skills, sensitivities and orientations that have developed through long experience of conducting one's life in a particular environment' (Ingold, 2000: 25). An ecological approach is necessary to understand how skills for living within an ecosystem are built, as skills are not properties of individuals but of '... the total field of relations constituted by the presence of the organism-person, indissolubly body and mind, in a richly structured environment' (Ingold, 2000: 353). Learning, or enskilling, is a process which may be described as the 'education of attention' as elders create structured contexts through which the novice can build his or her own perceptual skills in relation to the total environment, biophysical and social (Ingold, 2000).

What are the practical applications of a dwelling perspective that takes into account an ecological perspective of knowledge, skill, and learning? One way to examine this question is to look at the practice of selected initiatives which have applied human-in-ecosystem theoretical approaches through 'on the ground' projects. Three projects are chosen for discussion, each one reflecting a dwelling or humans-in-ecosystem perspective: (1) the Sense of Place (SoP) project directed by Gary Nabhan of the Arizona Sonora Desert Museum in Tucson, Arizona; (2) the People's Biodiversity Registers (PBR) program in India led by Madhav Gadgil and others; and (3) the Kagiwiosa-Manomin (KM) project of the Wabigoon First Nation in Canada initiated by Joe Pitchenese and Andrew Chapeskie. All three projects share the twin emphasis on the importance of access and use of resources and the process of knowing–learning–remembering the ecosystem through livelihood activities. This can be seen through a brief summary of the three projects and their foundational statements.

The *Sense of Place* project has worked in collaboration with local organizations to explore the unique natural and cultural resources, place names and vocabularies, songs, foods, and other traditions of the Sonora watershed. The project works with both indigenous and non-indigenous communities in the southwestern USA and northwestern Mexico to build a strong sense of identity, heritage, and relationship to the surrounding terrain.

> When a culture remembers and incorporates particular springs, sacred mountains, fields, buildings, marketplaces and ceremonial grounds as 'places of the heart', these places are less likely to be unnecessarily exposed to external threats that diminish them. We believe that strong community institutions such as museums, libraries, historical societies, cultural centers and gardens can help nurture and support the unique features of a place and its peoples, acculturate newcomers and slow detrimental change.
>
> *(Nabhan, 2001: 1).*

The *People's Biodiversity Registers* program was initiated by the Foundation for Revitalization of Local Health Traditions, with the initial purpose of documenting community-based knowledge of medicinal plants and their uses for 52 communities spread across India. The scope of the project has since been expanded to examine all elements of biodiversity by recording the knowledge and perceptions of lay people, primarily rural and forest-dwelling communities, of living organisms and their ecological setting across an expanded geographic region.

All knowledge and wisdom ultimately flow from practices. But their organization differs amongst the different streams of knowledge. Folk knowledge is maintained, transmitted, augmented almost entirely in the course of applying it in practice; it lacks a formal, institutionalized process of handling it . . . [Folk knowledge and wisdom] must therefore be supported in two ways; through creating more formal institutions for their maintenance, and most importantly, by creating new contexts for their continued practice.

(Gadgil et al., *2000: 1307).*

Kagiwiosa-Manomin was an initiative started by Joe Pitchenese of Wabigoon First Nation to establish Ojibway tenure for manomin ('wild rice,' *Zizania aquatica* L.), an Ojibway harvesters' cooperative, a manomin processing facility based on historic Ojibway processing methods, and organic and bulk markets in North America and Europe. The project emerged from the realization that the retention, transmission, and adaptation of knowledge about manomin were linked to the practice of manomin harvesting, the Ojibway identity in northwestern Ontario, Canada and the need for supplemental income.

The Anishinaabeg of the Wabigoon Lake Ojibway Nation are continuing to struggle to retain an indigenous management regime pertaining to the growing and harvesting of wild rice. Indeed they have begun to reclaim control over the processing of the product through the establishment of a wild rice processing business in the community which is seeking to work within the customary rule-making framework of the community itself which regulates the utilization of the resource.

(Chapeskie, 1986: 131–2).

The three projects exemplify human-in-ecosystem perspectives applied in different contexts, with each project emphasizing different approaches. One of the main differences among the three projects is the degree to which the participants derive their livelihood from a direct use of the products of their local ecosystem. Table 3.1 summarizes an attempt to capture the core components of a human-in-ecosystem approach. Seven themes emerge out of the grouping of these core components: (1) use of spatially bounded management units; (2) relational networks; (3) embeddedness; (4) knowing–learning–remembering; (5) cultural identity and sense of place; (6) institution building; and (7) livelihood activities.

Table 3.1 *Core components of a dwelling or human-in-ecosystem perspective*

Component/initiative	Sense of Place (SoP)	People's Biodiversity Registers (PBR)	Kagiwiosa-Manomin (KM)
1a Spatial boundaries – organizing principle for ecosystem-based projects	XXX	X	X
1b Emphasis on cross-scales, as in nested watersheds	XXX	XX	X
1c Importance of the local-scale, larger-scale efforts bringing together local efforts	XXX	XX	XX
2a Emphasis on understanding relationships between people and other species	XXX	XX	XX
2b Emphasis on understanding relationships among people	XXX	X	XXX
2c Emphasis on communication, relational networks, and cross-scale institutions	XXX	XX	XX
3a Actor behavior (individuals, organizations) embedded in social structures and cultural processes	XXX	XXX	XXX
4a Knowledge transmission (spatial diffusion) among groups	XXX	XX	XX
4b Knowledge transmission (temporal diffusion) between generations	XXX	X	XX
4c People's participation in documentation and mapping of local ecosystem	XXX	XXX	XX
5a Cultural identity and sense of place	XXX	X	XXX
6a Institution building	XXX	XX	XXX
6b Emphasis on commons institutions	XX	XX	XXX
7a Livelihood activities in the ecosystem	XXX	XXX	XXX

Numbering: primary numbers refer the reader to the subheadings of Section 3.5. For example, numbers 6a and 6b refer to two specific ideas which emerge from Section 3.5.6.
XXX = strongly emphasized in the case study; XX = emphasized in the case study; X = mentioned, or suggested, but not emphasized in the case study.

3.5.1 Spatially bounded management units

The SoP project emphasizes the importance of recognizing nested ecological and institutional units. However, it also raises the problem that the boundaries or scale of an ecological unit do not always match social or institutional units. Institutions representing different classes or ethnic groups may be found within one ecological unit. This situation becomes increasingly apparent as the ecological unit is scaled up. The solution pursued by the SoP project regarding this problem of fit between ecological and institutional units was to work at multiple scales (Folke *et al.*, 1998). The project worked intensively with ethnic or other local communities while also encouraging events that brought people together within a regional watershed. Over time it is hoped that these 'communities of interest' may form into larger representative institutions which can cope with larger-scale ecological units. The use of watershed units is common in traditional ecosystem-like concepts (Berkes *et al.*, 1998), as well as in large-scale management systems (Gunderson *et al.*, 1995).

3.5.2 Relational networks

Understanding the networks of relationships among people, and among people and other species was seen, to varying degrees, as an important element by all three projects. The existence, or creation, of networks was seen as the basis of communication among the inhabitants of an ecosystem. Such communication was seen to provide a system of feedbacks among the inhabitants and allow for the appropriate adjustments in behavior. Events as simple as annual picnics which brought together different communities, sharing food, crafts, stories, and song, school field trips in which elder members of the community could share knowledge with youth about places in the landscape, or workshops where people could learn how to make things from local species, were all seen to contribute to building communication and feedback loops within an ecosystem (Anderson, 1996).

3.5.3 Embeddedness and behavior

All three initiatives concurred that the behavior of individuals and institutions was embedded within social structures and cultural values. Behavior toward other humans and other species cannot be explained solely through an analysis based on aggregating the economizing or rational choice preferences of individuals. This is especially evident in the SoP project that has supported creative writing and other artistic endeavors to express the relationship among people and among people and other species of the Sonora watershed. The assumption is that the values that embed economizing preferences and rational choices will influence the contours of such preferences and choices (McCay and Jentoft, 1998).

3.5.4 Knowing–learning–remembering

All three projects reflect an emphasis on the importance of supporting, or re-engaging, people in the processes of knowing, learning, and remembering an ecosystem through practical activities. Such activities allow people to build their own perceptions of an ecosystem and to share their perceptions within and between generations. The dwellers of an ecosystem were brought, or initiated, into these three projects as active participants in the documentation, creation, and communication of knowledge, institutions, technologies, and values. Such strategies are effective with contemporary, heterogeneous societies as well as with traditional or relatively homogeneous societies such as those in the three projects. For example, much of the success of 'citizens' science' projects in Minnesota is based on knowing–learning loops created by the active involvement of local citizens groups (see Chapter 9; Light, 1999).

3.5.5 Cultural identity and sense of place

Cultural identity and sense of place were explicitly mentioned by the SoP and KM initiatives. Identity and sense of place are complex concepts but were considered to be linked to the practical activities of people, people's perceptions of an ecosystem, and the relational networks that people build within an ecosystem. The projects supported activities which allowed for the creation and strengthening of both individual and collective identities. Individual and collective identities and senses of place appear to be emergent properties of particular ecosystems and relational networks as built through on-going practical experiences and communication. This has been explored in detail with indigenous peoples who stress that their cultural identity emerges, in part, from the land they occupy (Basso, 1996; Berkes, 1999). A number of studies indicate that cultural identity and sense of place can be important for a wide spectrum of societies, for example in Pakistan (Butz, 1996), England (Butz and Eyles, 1997), the USA (Nabhan, 1997), and Sweden (Olsson and Folke, 2001).

3.5.6 Institution building

The PBR and the KM projects both place an emphasis on institutions at the scale of collective property rights and specific livelihood activities. Larger-scale common property institutions are considered to be the mechanism by which people continue adapting livelihoods dependent upon the goods of local ecosystems (Alcorn and Toledo, 1998). Access allows people to carry out the practical activities that create the contexts for knowing, learning, and remembering. It is this practical engagement with resource harvesting that allows harvesters to collectively codify and reformulate the specific rules of resource harvesting at the scale where resource harvesting occurs. Dwelling requires access to resources through some form of property rights, while allowing people the flexibility to generate specific institutions out of their practical engagements with the ecosystem. However, the case of KM in particular emphasizes that such an approach should recognize that periods of stability and change are inherent to common property institutions at both local and higher scales. Institutions should not be frozen in time, the traditional should not be traditionalized, as Ingold (2000) puts it, but should be generated out of the processes of knowing–learning–remembering.

3.5.7 Livelihood activities

All three of the initiatives recognize that practical activities that create linkages among humans and other constituents of an ecosystem are the foundation upon which a human-in-ecosystem approach is built. The PBR and KM cases work with people who already have a strong linkage to the ecosystem through their livelihood activities. Practical activities that link people to other humans and other constituents of the ecosystem are already in place. The creation of organizations that promote horizontal communication networks within ecosystems is considered to be a major goal (see Chapter 10). The main challenge for these initiatives is whether people who already dwell in the eco-commons can continue to do so while their livelihoods become increasingly integrated into global markets and cultural processes (Hernández Castillo and Nigh, 1998). The SoP initiative works with a variety of people with diverse relationships to other inhabitants of the ecosystem. These initiatives have tended to create new organizational contexts for practical activities. However, a major challenge in this context has been the disjuncture in the inter-generational learning of practical activities, as instructional time has become dominated by schooling. The SoP project, in particular, has focused on linking practical activities and inter-generational communication within a schooling context.

In sum, the consideration of the three case studies points toward two practical applications. (1) Ensuring that people who are attentive to the land are able to continue making a living in a landscape is an effective way to nurture social–ecological systems. Dwelling in the eco-commons breaks down the opposition between livelihood and nature. (2) For people whose livelihoods are no longer closely connected to the land, we need to create learning contexts to reconnect them to the land. These are situations in which people learn representations about the land by becoming attentive to the land and building their own memories and skills in relationship with the land.

The first application would require public policy that supports a dwelling perspective: the co-evolution of skills, memories, institutions, property rights, organizations, and landscapes as they emerge out of people's livelihoods. The second would require public policy to support organizations which focus less on a museum concept of 'heritage' and more on experiential activities that enskill people to be attentive to humans, other animals, and life processes of the landscape in which they dwell. How would such public policy be developed? What are the major factors and impediments? These considerations take us into the realm of political economy of institutional development, which is outside the scope of this chapter and a major topic in its own right.

3.6 Conclusions

The relationship between nature and society has been examined in Western thought through a number of different models, including environmental determinism, possibilism, and cultural ecology. The use of systems approaches appeared in ecological anthropology in the 1960s and the 1970s. These models improved our understanding of the linkage between social and ecological systems by considering two-way feedback relations between the two. The use of the resilience concept in social sciences first appeared with the critique of some of the equilibrium-based systems models (Vayda and McCay, 1975). A rich literature subsequently emerged, using the resilience approach to develop an understanding of humans-in-ecosystem. There were several areas of emphasis: generalist and specialist strategies as adaptations; uncertainty and surprise; and the issue of centralization, flexibility, and resilience.

This literature shows in some detail that the concept of resilience has provided a significant means to uncover new insights for the understanding of social–ecological system linkages. Although this literature dates back to the mid-1970s, it lacks coherence because it is scattered across a number of disciplines. Many of these fields are interdisciplinary, including ecological anthropology in which the largest single concentration of contributions is found. Others include environmental psychology (Lamson, 1986), human geography (Zimmerer, 1994), business management (Hurst, 1995), development studies (Adger, 2000), and political science/environmental policy (Holling and Sanderson, 1996; Kates and Clark, 1996).

The three practical applications of the human-in-ecosystem approach, or the dwelling approach, to use Ingold's (2000) terminology, show how theory and practice have become closely linked and place specific. Each case uses an ecological perspective that attempts to move beyond the Cartesian nature/culture, environment/society dichotomies through a focus on process. To various degrees, each of the three cases uses ecological boundaries; pays attention to scale; emphasizes ecological relationships in which humans are a part; recognizes embeddedness; actively encourages ecological knowledge generation and transmission; builds a sense of place; builds commons institutions; and emphasizes livelihood activities within the ecosystem. By suggesting that 'nature and culture must be seen as co-created' (Scoones, 1999: 486), the cases illustrate one way in which the human-in-ecosystem approach may be operationalized.

Resilience provides key insights for such social–ecological system integration. It moves the emphasis away from a focus on form to a focus on process; from equilibrium to change; and from static relationships to dynamics. The shift in emphasis is not insignificant. Resilience helps shift the analysis from simple models of cause and effect, to complex systems and nonlinear relationships. It points the way ahead by identifying new research directions: concerning the issues, first, of scale in space and time and, second, of key variables. We conclude by dealing with each in turn.

Resilience focuses attention on the characteristics of the temporal dynamics of a human-in-ecosystem perspective. History reveals that processes are often cyclical in that they lead to periods of stability followed by periods of rapid change or adaptive lurches (Gunderson *et al.*, 1995). Processes pertain not only to different spatial scales but also to different temporal scales. A maple tree sheds and grows its leaves yearly, while a maple forest may give way to a birch forest over a 500-year period. Cyclical processes are occurring in both cases, but change is often difficult to recognize when time has 'slowed down.' Time may also slow down at different points of a cycle. In the case of institutions, for example, periods of stability may last hundreds of years, whereas periods of instability occur over a time scale of years or decades. This shift in rates of change tends to divert our attention away from processes and create the illusion that causality is attributable to stable forms such as long-enduring institutions and climax landscapes.

A resilience emphasis raises the question of clusters of processes that occupy different spatial and temporal scales. Holling (1992) observed that nature is 'lumpy,' with the key variables clustering around different spatial and temporal scales. Such observations suggest a powerful tool to simplify complexity by looking for evidence of clustering and discontinuities in spatial and temporal scales, and by analyzing structuring processes that produce these patterns (Gunderson and Holling, 2002). Such an approach has been providing insights for ecosystem analysis, and it is clearly relevant for understanding social–ecological systems as well.

For example, common property institutions specify the rights of a community to a resource. In some cases, these institutions appear stable over the time scale of a century. However, these same property rights regimes may allow a community to negotiate and change rules that pertain to the harvest of a specific resource over a much shorter period, a season or a year. Are there identifiable slow (e.g., cultural values) and fast (e.g., operational rules) variables in such cases which would improve our ability to understand commons management? Regarding spatial scale, the persistence of common property institutions pertaining to a small geographic area may depend upon the existence of larger-scale institutions (e.g., Alcorn and Toledo, 1998). In turn, institutions at various scales may be embedded in globalizing processes, some of which may be characterized by rapid rates of change.

Gunderson and Holling (2002) have postulated that slower variables structure a complex system by serving as potential bifurcation points for the faster variables. If this idea is applicable to social–ecological systems in general, the identification of the key variables at various scales becomes important for understanding the system, not only from an academic point of view, but also for managing change. Thus, the use of the notion of resilience – emphasizing system survival, self-organization and adaptive change – provides leads for the identification of key variables at various scales. The list in Table 3.1, the core components of a dwelling perspective, may suggest future research directions for the identification of these variables in integrated systems of nature and society.

References

Abel, T. 1998. Complex adaptive systems, evolutionism, and ecology within anthropology: interdisciplinary research for understanding cultural and ecological dynamics. *Georgia Journal of Ecological Anthropology* 2: 6–29.

Adger, W.N. 2000. Social and ecological resilience: are they related? *Progress in Human Geography* 24: 347–64.

Alcorn, J.B. and Toledo, V.M. 1998. Resilient resource management in Mexico's forest ecosystems: the contribution of property rights. In *Linking Social and Ecological Systems: Management Practices and Social Mechanisms for Building Resilience*, pp. 216–49, ed. F. Berkes and C. Folke. Cambridge: Cambridge University Press.

Alland, A. Jr and McCay, B. 1973. The concept of adaptation in biological and cultural evolution. In *Handbook of Social and Cultural Anthropology*, pp. 143–77, ed. J.J. Honigmann. Chicago: Rand McNally and Co.

Anderson, E.N. 1996. *Ecologies of the Heart: Emotion, Belief and the Environment*. Oxford: Oxford University Press.

Anderson, J.N. 1973. Ecological anthropology and anthropological ecology. In *Handbook of Social and Cultural Anthropology,* pp. 179–239, ed. J.J. Honigmann. Chicago: Rand McNally and Co.

Balee, W., ed. 1998. *Advances in Historical Ecology*. New York: Columbia University Press.

Basso, K.H. 1996. Wisdom sits in places: notes on a Western Apache Landscape. In *Senses of Place*, pp. 53–89, ed. S. Feld and K.H. Basso. Santa Fe: School of American Research Press.

Bateson, G. 1973. *Steps to an Ecology of Mind*. London: Granada.

Bateson, G. 1979. *Mind and Nature: a Necessary Unity*. New York: Dutton.

Bennett, J. 1969. *Northern Plainsman: Adaptive Strategy and Agrarian Life*. Arlington Heights, IL: AHM Publishing Co.

Berkes, F. 1999. *Sacred Ecology: Traditional Ecological Knowledge and Resource Management*. Philadelphia and London: Taylor & Francis.

Berkes, F. and Folke, C. 1994. Investing in cultural capital for a sustainable use of natural capital. In *Investing in Natural Capital: the Ecological Economics Approach to Sustainability*, pp. 128–49, ed. A.M. Jansson, M. Hammer, C. Folke, and R. Costanza. Washington DC: Island Press.

Berkes, F. and Folke, C., eds. 1998. *Linking Social and Ecological Systems: Management Practices and Social Mechanisms for Building Resilience*. Cambridge: Cambridge University Press.

Berkes, F. and Folke, C. 2002. Back to the future: ecosystem dynamics and local knowledge. In *Panarchy: Understanding Transformations in Systems of Humans and Nature*, pp. 121–46, ed. L.H. Gunderson and C.S. Holling. Washington DC: Island Press.

Berkes, F., Kislalioglu, M., Folke, C., and Gadgil, M. 1998. Exploring the basic ecological unit: ecosystem-like concepts in traditional societies. *Ecosystems* 1: 409–15.

Biersack, A. 1999. Introduction: from the 'new ecology' to the new ecologies. *American Anthropologist* 101: 5–18.

Borofsky, R. 1994. Introduction. In *Assessing Cultural Anthropology*, pp. 1–19, ed. R. Borofsky. New York: McGraw-Hill.

Brooks, A.S., Gelburd, D.E. and Yellen, J.E. 1984. Food production and cultural change among the Kung San: implication for prehistoric research. In *From Hunters to Farmers,* pp. 293–310, ed. J.D. Clark and S.A. Brandt. Berkeley: University of California Press.

Butz, D. 1996. Sustaining indigenous communities: symbolic and instrumental dimensions of pastoral resource use in Shimshal, northern Pakistan. *The Canadian Geographer* 40: 36–53.

Butz, D. and Eyles, J. 1997. Reconceptualizing senses of place: social relations, ideology and ecology. *Geografiska Annaler* 79: 1–25.

Chapeskie, A. 1986. Indigenous law, state law and the management of natural resources: wild rice and the Wabigoon Lake Ojibway Nation. *Law and Anthropology* 5: 129–66.

Conklin, H.C. 1957. *Hanunoo Agriculture. Report of an Integral System of Shifting Cultivation in the Philippines*. FAO Forestry Development Paper No. 5. Rome: FAO.

Cronon, W. 1983. *Changes in the Land: Indians, Colonists, and the Ecology of New England*. New York: Hill and Wang.

Cronon, W., ed. 1995. *Uncommon Ground: Toward Reinventing Nature*. New York: Norton & Co.

Descola, P. and Pálsson, G., eds. 1996. *Nature and Society: Anthropological Perspectives*. London: Routledge.

Dyson-Hudson, R., Meekers, D., and Dyson-Hudson, N. 1998. Children of the dancing ground, children of the house: costs and benefits of marriage rules (South Turkana, Kenya). *Journal of Anthropological Research* 54: 19–47.

Escobar, A. 1999. After nature: steps to an antiessential political ecology. *Current Anthropology* 40: 1–30.

Finlayson, A.C. and McCay, B.J. 1998. Crossing the threshold of ecosystem resilience: the commercial extinction of northern cod. In *Linking Social and Ecological Systems: Management Practices and Social Mechanisms for Building Resilience*, pp. 311–38, ed. F. Berkes and C. Folke. Cambridge: Cambridge University Press.

Folke, C., Berkes, F., and Colding, J. 1998. Ecological practices and social mechanisms for building resilience and sustainability. In *Linking Social and Ecological Systems: Management Practices and Social Mechanisms for Building Resilience*, pp. 414–36, ed. F. Berkes and C. Folke. Cambridge: Cambridge University Press.

Folke, C., Pritchard, L. Jr, Colding, J., Berkes, F., and Svedin, U. 1998. *The Problem of Fit between Ecosystems and Institutions*. IHDP Working Paper No. 2. Bonn: IHDP.

Fratkin, E. 1986. Stability and resilience in East African pastoralism: the Rendille and the Ariaal of Northern Kenya. *Human Ecology* 14: 269–86.

Gadgil, M., Seshagiri Rao, P.R., Utkarsh, G., Pramod, P., Chhatre, A., and Members of the People's Biodiversity Initiative. 2000. New meanings for old knowledge: the People's Biodiversity Registers program. *Ecological Applications* 10: 1307–17.

Geertz, C. 1963. *Agricultural Involution: the Process of Ecological Change in Indonesia*. Berkeley: University of California Press.

Glacken, C. 1967. *Traces on the Rhodian Shore: Nature and Culture in Western Thought from Ancient Times to the End of the 18th Century*. Berkeley: University of California Press.

Greenberg, J.B. and Park, T.K. 1994. Political Ecology. *Journal of Political Ecology* 1: 1–12. [online] www.library.arizona.edu/ej/jpe/jpeweb.html

Gunderson, L.H. and Holling, C.S., eds. 2002. *Panarchy: Understanding Transformations in Systems of Humans and Nature*. Washington DC: Island Press.

Gunderson, L.H., Holling C.S., and Light, S.S., eds. 1995. *Barriers and Bridges to the Renewal of Ecosystems and Institutions*. New York: Columbia University Press.

Harvey, D. 1974. Population, resources and the ideology of science. *Economic Geography* 50: 256–77.

Harvey, D. 1996. *Justice, Nature and the Geography of Difference*. Oxford: Blackwell Publishers.

Hernández Castillo, R.A. and Nigh, R. 1998. Global processes and local identity among Mayan coffee growers in Chiapas, Mexico. *American Anthropologist* 100: 136–47.

Holling, C.S. 1973. Resilience and stability of ecological systems. *Annual Review of Ecology and Systematics* 4: 1–23.

Holling, C.S. 1986. The resilience of terrestrial ecosystems: local surprise and global change. In *Sustainable Development of the Biosphere,* pp. 292–317, ed. W.C. Clark and R.E. Munn. Cambridge: Cambridge University Press.

Holling, C.S. 1992. Cross-scale morphology, geometry, and dynamics of ecosystems. *Ecological Monographs* 62: 447–502.

Holling, C.S. 1995. What barriers? What bridges? In *Barriers and Bridges to the Renewal of Ecosystems and Institutions*, pp. 3–34, ed. L.H. Gunderson, C.S. Holling, and S.S. Light. New York: Columbia University Press.

Holling, C.S., Berkes, F., and Folke, C. 1998. Science, sustainability and resource management. In *Linking Social and Ecological Systems*, pp. 342–62, ed. F. Berkes and C. Folke. Cambridge: Cambridge University Press.

Holling, C.S. and Goldberg. M.A. 1971. Ecology and planning. *Journal of the American Institute of Planners* 37: 221–30.

Holling, C.S. and Sanderson, S. 1996. Dynamics of (dis)harmony in ecological and social systems. In *Rights to Nature*, pp. 57–85, ed. S. Hanna, C. Folke, and K.-G. Mäler. Washington DC: Island Press.

Holling, C.S., Schindler, D.W., Walker, B.H., and Roughgarden, J. 1995. Biodiversity in the functioning of ecosystems: an ecological synthesis. In *Biodiversity Loss: Economic and Ecological Issues,* pp. 44–83, ed. C.A. Perrings, K.-G. Maler, C. Folke, C.S. Holling, and B.-O. Jansson. Cambridge: Cambridge University Press.

Hurst, D.K. 1995. *Crisis and Renewal: Meeting the Challenge of Organizational Change*. Boston: Harvard Business School Press.

Ingold, T. 1980. *Hunters, Pastoralists and Ranchers: Reindeer Economies and their Transformations*. Cambridge: Cambridge University Press.

Ingold, T. 2000. *The Perception of the Environment: Essays on Livelihood, Dwelling and Skill.* London: Routledge Press.

Johnson, L.M. 1999. Aboriginal burning for vegetation management in Northwest British Columbia. In *Indians, Fire and the Land in the Pacific Northwest*, pp. 238–54, ed. R. Boyd. Corvallis: Oregon State University Press.

Kates, R.W. and Clark, W.C. 1996. Expecting the unexpected. *Environment* March: 6–11, 28–34.

Keesing, R.M. 1976. *Cultural Anthropology: a Contemporary Perspective*. New York: Holt, Rinehart and Winston.

King, A. 1995. Avoiding ecological surprise: lessons from long-standing communities. *Academy of Management Review* 20: 961–85.

King, T.D. 1997. Folk management among Belizean lobster fishermen: success and resilience or decline and depletion? *Human Organization* 56: 418–26.

Kottak, C.P. 1999. The new ecological anthropology. *American Anthropologist* 101: 23–35.

Lamson, C. 1986. Planning for resilient coastal communities: lessons from ecological systems theory. *Coastal Zone Management Journal* 13: 265–79.

Lees, S.H. and Bates, D.G. 1990. The ecology of cumulative change. In *The Ecosystem Approach in Anthropology*, pp. 247–77, ed. E.F. Moran. Ann Arbor: University of Michigan Press.

Lewis, H.T. and Ferguson, T.A. 1999. Yards, corridors, and mosaics: how to burn a boreal forest. In *Indians, Fire and the Land in the Pacific Northwest*, pp. 164–84, ed. R. Boyd. Corvallis: Oregon State University Press.

Light, S., compiler 1999. *Citizens, Science, Watershed Partnerships, and Sustainability in Minnesota: the Citizens' Science Project*. Minneapolis: Minnesota Department of Natural Resources.

McCay, B.J. 1978. Systems ecology, people ecology, and the anthropology of fishing communities. *Human Ecology* 6: 397–422.

McCay, B.J. 1981. Optimal foragers or political actors? Ecological analysis of a New Jersey fishery. *American Ethnologist* 8: 356–82.

McCay, B. and Jentoft, S. 1998. Market or community failure? Critical perspectives on common property research. *Human Organization* 57: 21–9.
McGovern, T.H. 1980. Cows, harp seals, and churchbells: adaptation and extinction in Norse Greenland. *Human Ecology* 8: 245–75.
McGovern, T.H., Bigelow, G.F., Amorosi, T., and Russell, D. 1988. Northern islands, human error, and environmental degradation. *Human Ecology* 16: 225–70.
Moran, E.F. 1979. *Human Adaptability: an Introduction to Ecological Anthropology*. North Scituate: Duxbury Press.
Moran, E.F. 1990. Ecosystem ecology in biology and anthropology: a critical assessment. In *The Ecosystem Approach in Anthropology: From Concept to Practice*, pp. 3–40, ed. E.F. Moran. Ann Arbor: The University of Michigan Press.
Nabhan, G.P. 1997. *Cultures of Habitat: On Nature, Culture and Story*. Washington DC: Counterpoint.
Nabhan, G.P. 2001. The launch of a transboundary, transcultural sense of place project. [online] http://www.desertmuseum.org/place/launch.html
Nazarea, V.D., ed. 1999. *Ethnoecology: Situated Lives/Located Lives*. Tucson: The University of Arizona Press.
Netting, R. McC. 1974. Agrarian ecology. *Annual Review of Anthropology* 3: 21–56.
Netting, R. McC. 1986. *Cultural Ecology*, 2nd edn. Prospect Heights, IL.: Waveland Press.
Netting, R. McC. 1993. *Smallholders, Householders: Farm Families and the Ecology of Intensive Sustainable Agriculture*. Stanford, CA: Stanford University Press.
Olsson, P. and Folke, C. 2001. Local ecological knowledge and institutional dynamics for ecosystem management: a study of Lake Racken watershed, Sweden. *Ecosystems* 4: 1–23.
Park, R. E. 1936. Human ecology. *American Journal of Sociology* 42: 1–15.
Peet, R. and Watts, M., eds. 1996. *Liberation Ecologies: Environment, Development, Social Movements*. London: Routledge.
Rappaport, R.A. 1967. *Pigs for the Ancestors: Ritual in the Ecology of a New Guinea People*. New Haven: Yale University Press.
Richerson, P.J. 1977. Ecology and human ecology: a comparison of theories in the biological and social sciences. *American Ethnologist* 4: 1–27.
Sauer, C.O. 1956. The agency of Man on the earth. In *Man's Role in Changing the Face of the Earth*, Vol. 1, pp. 49–69, ed. W.L. Thomas, Jr. Chicago: University of Chicago Press.
Scoones, I. 1999. New ecology and the social sciences: what prospects for a fruitful engagement? *Annual Review of Anthropology* 28: 479–507.
Steward, J.H. 1955. *Theory of Culture Change: the Methodology of Multilinear Evolution*. Urbana: University of Illinois Press.
Toledo, V.M. 1992. What is ethnoecology? Origins, scope and implications of a rising discipline. *Ethnoecológica* 1(1): 5–21.
Turner, M.G. and Dale, V.H. 1998. Comparing large, infrequent disturbances: what have we learned? *Ecosystems* 1: 493–6.
Vayda, A.P. and McCay, B.J. 1975. New directions in ecology and ecological anthropology. *Annual Review of Anthropology* 4: 293–306.
Vayda, A.P. and Rappaport, R. 1968. Ecology, cultural and noncultural. In *Introduction to Cultural Anthropology*, pp. 477–97. ed. J.A. Clifton. Boston: Houghton-Mifflin.
Winterhalder, B. 1983. The boreal forest, Cree-Ojibwa foraging and adaptive management. In *Resources and Dynamics of the Boreal Zone*, pp. 331–45, ed.

R.W. Wein, R.R. Riewe, and I.R. Methven. Ottawa: Association of Canadian Universities for Northern Studies.

Wolf, E. 1982. *Europe and the People without History.* Berkeley: University of California Press.

Worster, D. 1977. *Nature's Economy: a History of Ecological Ideas.* Cambridge: Cambridge University Press.

Worster, D., ed. 1988. *The Ends of the Earth: Perspectives on Modern Environmental History.* Cambridge: Cambridge University Press.

Yellen, J.E. 1977. Long term hunter–gatherer adaptation to desert environments: a biogeographical perspective. *World Archeology* 8: 262–74.

Zimmerer, K.S. 1994. Human geography and the 'new ecology': the prospect and promise of integration. *Annals of the Association of American Geographers* 84: 108–25.

Zimmerer, K.S. and Young, K.R., eds. 1998. *Nature's Geography: New Lessons for Conservation in Developing Countries.* Madison: The University of Wisconsin Press.

4

Redundancy and diversity: do they influence optimal management?

BOBBI LOW, ELINOR OSTROM, CARL SIMON, AND JAMES WILSON

The command-and-control approach, when extended uncritically to treatment of natural resources, often results in unforeseen and undesirable consequences. A frequent, perhaps universal result of command and control as applied to natural resource management is reduction of the range of natural variation of systems – their structure, function, or both – in an attempt to increase their predictability or stability.

(Holling and Meffe, 1996: 329).

4.1 Introduction

In many fields, there are fashions in favored approaches – what is assumed to be 'best.' A recurrent theme in American academia – particularly among students of public administration, policy analysis, and resource economics – has been to criticize 'redundancy' in government, decrying the number of governments that exist in the USA and the competition that exists among them. Consider education policy: beliefs that large numbers of schools were inefficient and that massive consolidation would be effective led to the reduction of 'redundant' school districts in a massive campaign during the first half of the twentieth century. In 1932, there were almost 130 000 school districts in the USA. This number was halved by 1952 and quartered by 1962, and halved once again by the early 1970s. The massive consolidation of school districts has slowed down during the past two decades. However, today we have around 15 000 school districts in the USA for a population that has almost doubled since the campaign to consolidate schools was initiated (see Ostrom, Bish, and Ostrom, 1988). During the heat of this policy reform, research was almost non-existent on the effect of school size, number of schools in a region, and related issues. Since the 1970s, considerable research on the effects of these variables on school performance has provided contrary evidence to the implicit theory used by policy makers to support the school consolidation movement. A recent study for the National Bureau of Economic Research, for example, finds that having a larger number of schools in a metropolitan area is associated with higher average student performance (as measured by students' educational attainment, local wages, and test scores). These areas were also characterized by lower per-pupil spending (Hoxby, 1994; see also Pritchett and Filmer, 1999). Now, after years of trying to increase size and reduce numbers of schools, policy makers are reconsidering the consequences of these past reforms and recommending new efforts to create more responsive schools through a variety of structural reforms.

Similarly, earlier empirical studies of redundancy in public services found that redundancy did not have the adverse consequences frequently attributed to it and that improvements (rather than reductions) in performance were frequently associated with redundant arrangements.[1] During the 1960s and 1970s, the 'Metropolitan Reform' movement was the dominant way of thinking about urban government (see Hawley and Zimmer, 1970). The existence of many units of government, seen as redundant and inefficient, was thought to cause many problems. Multiple units of government were further viewed as competitive, and providing a means whereby the rich could escape without contributing to the provision of public services needed by the poor and disadvantaged living in the central city (reviewed in Stephens and Wikstrom, 1999; Hawkins and Ihrke, 1999).[2] But fashions change. In the 1990s, scholars called for the elimination of all of the redundant suburbs by creating one city for a metropolitan area (Rusk, 1993), at the same time as other problems were seen as best solved by extensive stakeholder analyses.

Fashions have influenced the management of ecosystems as well. Over time, ecological 'redundancy,' of multiple similar species, was seen as buffering an ecosystem from stresses. Yet, in contrast, because ecosystems were seen as complex and because of actors' ignorance and self-interest, centralized rules were seen as necessary to restrain local actions that were injurious to the interests of the broader society. This reasoning influenced the formation of important national laws (in the USA the Clean Air Act, Clean Water Act, Fisheries Conservation and Management Act) and international protocols (e.g., Kyoto). All of these initiatives were designed to address large-scale ecological problems; all tend toward the creation of 'one-size-fits-all' rules. It is no surprise that such attempts have problems: both scale and conflicting self-interest of actors can be difficult. Nation-states might attempt to subvert international plans so they may continue to pollute their neighbors; states and provinces argue against national rules so that they might pass their environmental costs on to their neighbors, and so on. Although the perceived need for centralization may be softened in clear instances of local heterogeneity (see below), redundancy among local units is, in some circles, viewed as giving rise to and encouraging collective dilemmas.

These examples raise a puzzle for anyone interested in establishing effective governance arrangements. The systems are complex, and it is difficult to imagine how to test for efficiency in appropriate ways. We know little about what *level of redundancy*, from zero to complete, would be optimal. Common-sense arguments have frequently been used as the foundation for policy: some relationships are thought to be self-evident. Action is proposed on the basis of these self-evident truths – but without testing the 'self-evident' hypotheses. When reforms are based on self-evident 'truths' that do not have a solid empirical and theoretical foundation, they can generate counterproductive results – as they have with American schools.

We suggest that it is short-sighted and ineffective to derive policy from untested assumptions. Both the 'redundancy is inefficient' and the countervailing 'local decisions are best' approaches seem too often to be prescriptions, rather than decisions arising from analysis of function. Here we examine redundancy as a widespread attribute of many types of systems: genetic, human engineered, complex adaptive, ecological, and governance systems. We define functionally different kinds of redundancy, and suggest a path for deciding the optimal level of redundancy in particular systems.

4.1.1 Challenged policy assumptions

No single model describes governance of ecological resources, and often there is debate. However, many important contemporary environmental policies rest on critical assumptions we wish to challenge. These include:

- Ecological organization is characterized by high levels of connectivity over large spatial scales. Hence, 'management over the range' (of connectivity) is necessary.
- Local users of natural resources cannot really be trusted to take a long-term perspective and to pay attention to the externalities they cause, because their short-term interests are seldom the same as those of the greater group or the ecosystem itself. Local users, trapped in dilemmas, will overuse or even destroy valuable resources unless they are prevented from doing so by government action.
- It is possible to plan for the efficient and equitable use of resources covering a large region – to design an optimal one-size-fits-all management system – by doing systematic analysis for a region as a whole. This assumes that existing variability within the region is irrelevant to efficient design.
- Organization or order is generated by centralized direction, generally through hierarchical systems of superior–subordinate relationships.
- The presence of a large number of governance regimes is a sign of inefficiency.

Under these assumptions, ideal management would involve a single governmental unit devising rules for managing local resources, and ensuring that these rules are monitored and enforced. Redundancy or diversity of resource–governance units would be inefficient. However, many natural resource regimes are locally self-organized and quite robust and functional, in contrast to the assumptions above (e.g., Ostrom, 1990; Baland and Platteau, 1996). In the Maine lobster fishery, for example, self-organizing groups tend to arise in local fishing areas (Acheson, 1993; Wilson, 1997). They probably number up to a hundred along the coast. Such actors seldom all use the same rules for either organization or resource utilization. Knowledge about how these local systems operate – and sometimes even their existence – often does not exist in a state or regional center, let alone the national capital. We think there are conditions in which redundancy and diversity at a local level enhance performance – as long as there are also overlapping units of government that can: (1) resolve conflicts, (2) aggregate knowledge across diverse units, and (3) insure that when problems occur in smaller units, a larger unit can temporarily step in if needed.

4.1.2 Is it time to reconsider redundancy?

Rather than prescriptions, which often arise from over-generalizing specific results, we seek to define and analyze redundancy across systems (see Low *et al.*, 1999, Costanza *et al.*, 2001). We examine redundancy not as *always costly*, or *absolutely required*, but simply as one attribute of a system, with consequences for the way a system performs – under some conditions improving, under other conditions decreasing, overall system performance. Costs are always associated with redundancy, because building more than one unit involves the use of energy, materials, and time that could be used for other purposes. System performance can, in turn, be measured along multiple dimensions, e.g., capacity to cope with risk and uncertainty; adaptation to exogenous change; error reduction through repetitive learning or through learning from others; matching system responses to local conditions; and ability to reduce the probability of system failure. Whether the benefits (of improved system performance) are worth the costs (of added time, effort, and resources used to build multiple units) of redundancy depends on: (1) the type of problems faced in governing a system, (2) how the particular kind of redundancy copes with these problems, and (3) the cost of the particular type of redundancy. In any system, we should be able to calculate an *optimal level of redundancy.*

Many natural systems – genetic, ecological, physiological, and behavioral – exhibit considerable redundancy, of various types. The existence of apparently profitable redundancy in natural systems suggests that it is time to re-examine the effects of redundancy in policy decisions. Consider, for example: most ecosystems exhibit some biodiversity – much of which constitutes functional redundancy from the perspective of the entire ecosystem (see below). Examine the trophic structure of most healthy ecosystems, and you will find numerous herbivores whose diets are not identical, but similar. This 'redundancy' renders the system relatively robust in the face of exogenous changes. From a manager's perspective, redundancy, and resulting robustness, mean many management decisions are relatively safe, unlikely to precipitate ecological crises. (We note, however, that from a genetic diversity perspective, or that of a conservation biologist, similar species are not, in fact, redundant.)

Here we examine redundancy in several systems. In genetic systems, there has been substantial recent research to determine why genomes appear so redundant. In engineering and information systems there has been a self-conscious effort to understand the importance of redundancy. We examine the role of redundancy in contemporary theory of complex adaptive systems. Redundancy in ecological systems challenges several assumptions: that most resource systems are so thoroughly interconnected that they must be considered as one large system governed by one large, administrative entity. We then examine governance systems, in which it appears that redundancy can buffer the system in the face of decision errors. We conclude that, despite the complexity and superficial diversity, in all these systems: (1) there is some optimum level of redundancy, depending on a variety of conditions (e.g., ecological, technological, institutional, informational); and (2) redundancy arises, is maintained, or disappears in different systems, as a result of its benefits and costs to actors at different levels.

4.2 The meanings of redundancy

'Redundancy' is defined by the *Oxford English Dictionary* (OED) as 'the state or quality of being redundant; superfluity, superabundance;' 'redundant' is further defined as 'excessive, abounding too much.' The word, paradoxically, has substantially different meanings in the fields we survey, yet most, like the definition from the OED, carry a negative connotation. We argue that understanding function would change our view of redundancy as 'superfluity.' This has happened in some fields already. In cybernetics dictionaries, redundancy is defined as one minus the ratio of the actual uncertainty to the maximum uncertainty. That is, redundancy does not increase the amount of information actually transmitted, but is essential to combat noise, to assure reliability, and to maintain communication (e.g., http://pespmc1.vub.ac.be/ASC/REDUNDANCY.html).

In most of our discussions, we refer to redundancy of multiple *units* (building blocks) within some larger *system*. The units may be genes in an individual, individuals in a community, physical parts in an engineered object, jobs in a political or social organization, firms in an industry – any organized subunits (themselves composed of other units or parts), or anything that is itself part of a larger system. The functional form of the units differs and so must the appropriate level of focus. We identify the following kinds of redundancy (Table 4.1).

4.2.1 Redundancy within a level of a multi-level system

1. *Multiple identical in-use copies (of rules or units).* This is common in genetic systems and human-designed systems, for example, and can serve two functions:
   (a) *Encounter rates.* The existence of multiple exits in a commercial airliner means that, in emergencies, any passenger, regardless of seat location, is likely to find an exit quickly. In city police departments, the assignment of numerous foot-patrolmen in local neighborhoods has been considered redundant – but the probable encounter rate (of children needing directions, small break-ins) is increased. One might argue that there are 'too many' gas stations – but the more gas stations, the better the encounter rate for consumers.
   (b) *Dosage–response* curve. If each unit contributes similar strength of response, multiple copies confer increased total strength. The evolution of pesticide resistance in mosquitoes, discussed below, exemplifies this kind of redundancy: identical copies of a resistant allele confer additive resistance. Similarly, local riot control effectiveness is, over some range, a function of the number of police assigned.
2. *Multiple similar in-use copies.* The production of similar – but not identical – antibodies means that, as a pathogen counters the currently most-effective antibody, new variations exist to confer immunity to the host. At a slightly different level, the evolution of gametic sex is a device that produces novel genetic combinations in new offspring: meiosis 'scrambles' genetic material, to produce haploid 'samples' of genetic combinations in egg and sperm, for combining with gametes of another individual. The resulting diversity of genetic combinations in offspring is advantageous in changing environments (e.g., see Williams, 1975; Maynard Smith, 1978). Economic competitors faced with a heterogeneous consumer environment develop a variety of differentiated products to cope successfully with changing and heterogeneous tastes. Chess and checkers strategies proliferate as slight variants on standard ploys.

Table 4.1 *Kinds of redundancy in selected systems*

		Kind of system			
	Proposed advantages	Genetic	Genetic algorithms	Designed physical systems	Governance
Within 'level'	Identical multiple in-use copies Encounter rates	R	R	R	R, L
	Multiple in-use copies Dose response	Ch		Ch, R	R, L
	Nonidentical multiple in-use copies	L	L, Exp, Err	L, Exp	L, Err
	Many rules → one outcome	R, Ch	L	Ch, R	Err
	Spare tires	——	——	R, Ch	Ch, R
	Reduced margin of error in designed systems			Ch, R	R, Ch, Err
Horizontal units at 1 level	Multiple nonidentical units	L, Ch, Exp, R	L	(?)	Ch, Exp
Multi-level	High-level rules general	L, Ch	Ch, R	Ch, R	Ch, R, Rec
	Low-level rules specific				
	Duplication of rules low and high		(?)		R, C, Rec
	Occasionally high replaces low		(?)	R, C	R, C, Rec

L = matching local conditions ('requisite variety') – nonuniform response to nonuniform conditions; Ch = response to exogenous change – large stresses (pesticides–mosquito, acts of God), – nonuniform responses to nonuniform stresses; Exp = experimentation (from L) (and learning from others) – V-notch (Newfoundland) – municipal associations and their communities, – faster exploration of solution space; R = Risk reduction – backup systems within unit; C = Reducing cost of institutional failure – high/low (Nixon grand jury), – low/high (Segregation, Spotted Owl); Err = Error reduction through repetitive learning – repetitive learning; Rec = Recombination to find appropriate scale – joint powers agreement, V-notch (Mass).

3. *Many rules, one outcome.* In genetic coding systems, considerable redundancy exists (e.g., Hurst, 1996; Freeland and Hurst, 1998). For example, there are 16 amino acids – but 3^3 ways of coding for them. This means that a deleterious mutation in one of the coding sites does not result in loss of the (essential) amino acid – a sort of insurance policy to avoid the risk of mistakes in one-to-one coding.
4. *Backup systems not currently in use ('spare tires').* Such devices, as with point 3, function to reduce risk of failure. John Doyle (see below) has estimated that a few hundred systems could run a Boeing 777 if no uncertainties were faced; instead, there are some 150 000 systems.
5. *Redundant strength to reduce margin of error.* An example of this is common, initially costly insurance against potential extreme conditions and unforeseen changes. For example, gaps in bridges are often larger than apparently necessary, to accommodate later expansion and contraction in hot and cold conditions. Daily and Ehrlich (1995: 55) stress that 'Society should no more assume abundant functional redundancy among population and species and exterminate them *ad lib* than a pilot should pop rivets from the wing of an aircraft and sell them based on a similar redundancy assumption.'

4.2.2 Redundancies across multiple levels

6. *General high-level rules, specific low-level rules.* This is a common design in genetic algorithm systems, as well as in many governance systems. A constitution, for example, provides general powers to specific government units. Within these constitutional rules, the units establish public policies that specify general rights and duties for participants in the polity. Participants, in turn, create many operational rules about specific activities that are consistent with public policies and constitutional rules. Thus, there are always at least three levels of rules operating in governance systems.
7. *Duplication of high-level and low-level rules.* The criminal codes of lower jurisdiction frequently duplicate some of the criminal code of a higher-level jurisdiction. If the one level does not prosecute a suspected criminal, the other system is potentially available. (In American jurisprudence, a person does not have to stand trial in both jurisdictions, so the potential double jeopardy is eliminated once proceedings are completed by one level.)
8. *Occasional replacement of rules at one level by rules at another level.* In engineered systems, this reduces the risk of failure or human error (e.g., ABS braking systems). In federal systems, the national government may decide to regulate some area of activity due to previously nonexistent 'spillovers.' The regulation of banking in the USA was largely handled at the state level

until Congress allowed banks to cross state lines – opening up the need for much more regulation at the national level (Polski, 2000).

These examples may seem diverse and unconnected; in fact, there are real homologies. Particular functional kinds of redundancy occur in systems not usually thought of as related – yet the function of redundancy is similar in the different systems. Many of the statements about redundancy in governance systems, for example, have parallels in the immune system's antibodies (see Farmer, Packard, and Perelson, 1986).

Several points are important here.

1. In genetic, ecological, political, and market systems, redundancy is likely to arise from the self-interest of the redundant units (e.g., Ostrom 1987, 1991, 1997). It may continue to exist even when it has no positive effect on the entire system, and will be lost only if it creates such severe costs that the entire system fails. In contrast, in such human-designed phenomena as engineered systems, there is strong selection on the entire system's coherence, and redundancy may be designed in for the sake of functionality of the entire system.
2. Repetition of identical or similar units may, or may not, fit the OED explanation of 'superfluity,' depending on whether they have functional importance.
3. Governance systems show parallels to other systems in many kinds of redundancy (Table 4.1) – but in many cases we have as yet insufficient information about the associated costs and benefits to define optimal levels of redundancy.

4.2.3 Redundancy in genetic systems

Geneticists have puzzled for a long time over the existence of repetitions – redundancies – and apparent nonsense genes in genomes. Organisms have many genes: the smallest known genome in nature is almost twice the size of best estimates for the minimal necessary genome (Maniloff, 1996), and genetic redundancy appears to be common (e.g., Goldstein and Holsinger, 1992; Tautz, 1992; Thomas, 1993; Brookfield, 1997). However, clues exist: for example, 'despite the apparent redundancy in the yeast genome, more than half of all yeast genes contribute detectably to competitive fitness' (Smith *et al.*, 1996: 2073). Although a duplication can arise and become fixed through drift, clearly the rate of fixation results from the relative advantage (or disadvantage) of the duplication for the organism (e.g., as a buffer: Clark, 1994). Nowak *et al.* (1997) modeled four cases that explain the commonness of genetic redundancy; in three of the four cases, redundancy is stable. Wagner (2000) noted that, along with overlapping gene function, 'one or more genes with similar functions' (redundancy) is a principal mechanism protecting an organism's physiological and developmental processes from the deleterious effects of mutation.

Gene duplications arise spontaneously at high rates in bacteria, bacteriophages, insects, and mammals. They are generally viable (Fryxell, 1996), but only a small fraction of all duplicated genes is retained, and an even smaller proportion evolves new functions, because the probability of 'nonfunctionalization' is comparatively high. Nonetheless, in very large populations, there may be a significant probability of a duplicated gene evolving a new function (Walsh, 1995; Nadeau and Sankoff, 1997).

As we examine genomes more closely and learn more, we discover that many duplicate genes and apparent 'nonsense' genes are in fact functional. Clearly duplication may serve the interests of the duplicated unit. It is more interesting to ask when duplication serves the interests of the whole genome – the organism. Clark's (1994) models show clearly that any duplication can only invade when it provides a *direct advantage to the organism*. Invariant repetitions (redundant copies) of gene sequences may occur when some threshold level of a genetic product is important. This type of redundancy is advantageous (and common) when there exists a metabolic need to produce large quantities of specific RNAs or proteins (Ohno, 1970). An increase in the number of genes can occur quite rapidly under selection for increased amounts of a gene product. Some spectacular examples include the evolution of resistance to organophosphorous insecticides in aphids (Field and Devonshire, 1998), mosquitoes, and *Drosophila* (Mouchès *et al.*, 1986; Maroni *et al.*, 1987; Callaghan *et al.*, 1998). Each gene contributes some amount of resistance, and repeated genes mean increased resistance.

Genes that are spatially separated from other genes that work with them are more likely than nearby genes to become duplicated as a form of risk reduction. Because genetic material can 'cross over' in replication, a gene can become further separated from its necessary co-genes (and, after meiosis to form egg or sperm, it may end up in a different sex cell – with loss of function). Loss of a single-copy gene is usually deleterious. There are two solutions: spatial clustering ('supergenes') of genes that work together, or duplication of separated genes.

Frequently, a nonfunctional (silent) pseudogene arises from a duplicate allele. Perhaps because these are typically harmless (duplicates exist), they may be maintained. Many have been documented: the human pseudogene *yh* in the β-globulin family contains numerous defects. Chimpanzees and gorillas, our closest relatives, have the same number of genes and pseudogenes as humans, suggesting that the pseudogene arose before the species diverged. Here is a redundancy that may have no positive benefit, but does not appear to cost the organism. Even when duplicates have no advantage, they may go to fixation, suggesting that costs are low (Clark, 1994). We should be cautious, however. Wagner (1999) calculated mean equilibrium redundancy (which depends on fitness effects of mutations); he noted that while selection will slow the 'decay' of redundancy caused by mutation and genetic drift, some mutations may only be 'neutral' because their effects on gene products are absorbed by the epigenetic system.

Novel function can arise after duplications; when this occurs, the original redundancy disappears. Some complex genes may have arisen this way: ovomucoid gene, $\alpha 2$ allele of haptoglobin, antifreeze glycoprotein genes (Graur and Li, 2000: 259–63, Table 6.1). 'Variant repeats' are copies with small differences (multiple nonidentical copies), as in the knirps and knirps-related genes in *Drosophila* (González-Gaitán *et al.*, 1994). The repeats occasionally come to perform new or different functions (e.g., thrombin and trypsin, lactalbumin and lysozyme). Differentiation typically requires a large number of substitutions, so one would think this sort of duplication-leading-to-new-function would be rare. However, sometimes surprisingly few substitutions after duplication can give rise to novel functions. For example, lactate dehydrogenase can be converted into malate dehydrogenase by replacing just one of 317 amino acids (Graur and Li, 2000: 264). This kind of redundancy allows rather cheap experimentation. We could view the evolution of pleiotropy and divergence in function over time as a trade-off between two countervailing forces: mutation, which tends to add diversity, and selection for robustness and resilience (Wagner, 1998, 2000).

Larger-scale redundancies are more complicated to understand. A well-known deleterious example is Down's syndrome, a type of polysomy (duplication of a complete chromosome) called trisomy 21. Repetition of whole chromosomes seems to be disadvantageous, and this kind of redundancy is rare.

At an even larger scale, polyploidization is the addition of one or more complete sets of chromosomes to the original set. When genetically distinct sets of chromosomes are combined (as is common in plants), the condition is called allopolyploidy. Autopolyploidy (especially autotetraploidy) occurs in many organisms (Nagl, 1990). However, tetraploids seem to have survived rarely: they suffer prolonged division time, increased nucleus volume, increased chromosomal disjunctions, and other difficulties. In these cases, redundancy gives rise not only to slower function, but also to internal dysfunctions. A few cases of fully functional tetraploidy are known, in which the duplication has no effect on the phenotype (e.g., the flowers *Chrysanthemum* and *Rosa*, the leptodactylid frog *Odontophrynus*, and goldfish).

In some plants, polyploidy reduces inhibitions to selfing and hybrid infertility, so that individual plants isolated at the edge of a habitat can reproduce by selfing (e.g., Stebbins, 1974) – an advantage. In those cases in which polyploidy 'works,' an ancient polyploid is no longer distinguishable today from a diploid (Cavalier-Smith, 1985). Thus, the large size of some genomes may reflect assimilated genetic redundancy.

The bottom line is that certain generalities hold for a variety of (otherwise apparently unrelated) cases of genetic redundancy: redundancy typically serves the interests of the duplicated unit; and redundancy may or may not serve the interests of any larger unit in which it is embedded. In genetic systems, the persistence and/or proliferation of replicated subunit depends in part on the relative *efficiency* of large units with replicated subunits, compared to those without. This may involve efficiency of communication and ability to respond to stimuli (see Tautz, 1992; Clark, 1994; Wagner, 1998, 2000).

When will replication serve the interests of both the replicator and the larger group? We can think of several conditions: when conditions differ for subunits ('experimental' nonidentical units); when replication increases the total response possible (as in the development of pesticide resistance in mosquitoes). In contrast, there are cases in which what is ideal from the point of view of the replicated unit may be costly from the viewpoint of the larger unit (e.g., 'outlaw' genes, driving Y chromosomes). In these cases, persistence of the replicated unit will depend on the relative ability of the replicated subunit to protect itself from elimination or consolidation by the whole group.

4.2.4 Redundancy in engineering systems

Genetic systems (and their mimic, genetic algorithms: Holland, 1995) begin with some elements of randomization; then the relative survival and reproduction of the elements result from differential performance. Indeed, in some complex manufacturing problems, this approach has been used with great success (Norman and Bean, 1999). In most engineered systems, however, the intent is to design in optimality from the start.

Redundancy, because it has costs, might seem suboptimal, but many systems must function in a variety of environments. Further, when failure would be very costly (e.g., engine failure in an airplane), the expense of redundant elements may be worthwhile. The 'robust integration of systems of systems' (Carson and Doyle, 1999) can provide reliable performance in changing and uncertain environments. Consider the Internet, the portable compact disk player, VLSI design, the Boeing 777, and the Mars Pathfinder as examples of the robust integration of systems of systems. As John Doyle (1999) noted:

> The Boeing 777 has millions of parts, mostly rivets, but 150,000 distinct subsystems, many of which are themselves highly complex components ... What's important, though, is that the overwhelming proportion of the millions of parts in a modern commercial aircraft or the thousands of genes in biological organisms, is there purely for robustness and uncertainty management. For the 777, some uncertainties are flight timing, weather, routing, other traffic, turbulence in the boundary layer, payload size and location, uncertainty in components due to manufacturing and aging, and so on ... Now imagine an idealized laboratory setting in which uncertainty is greatly reduced or eliminated... For the case of the idealized 777, a working vehicle could probably be built with a few hundred subsystems, rather than 150,000 ... This interplay between complexity and robustness... is both the most essential issue in complex systems, and the least understood... Major success stories, such as... the Boeing 777, have been the result of highly structured and systematic processes, with an almost obsessive attention to robustness.

Computer design follows similar principles. Consider the following, from an Intel Application Note (Intel LXT332 Redundancy Applications; http://developer.intel.com):

> The primary concern in most high speed data networks is reliability. Redundancy is one way to protect and ensure reliability in the event of catastrophic failure. At low data rates, redundancy may not make sense, but as the number of lower data ports are multiplexed to the higher bit streams, it begins to play a major role. Because of this, most major network multiplexers and bandwidth managers use redundancy techniques to ensure data integrity.

In this industry, also, are some of the most developed procedures for analyzing the costs and benefits of redundancy (e.g., Hampson, 1997).

4.2.5 Redundancy in ecological systems

In ecosystems, 'redundancy' raises questions of biodiversity and ecosystem function (e.g., Frank and McNaughton, 1991; Naeem and Li, 1997; Grime, 1998). Here, redundancy is typically of the 'multiple nonidentical copies' sort within ecosystems, or across ecosystems. Ecology has a history of postulating that species diversity enhanced primary productivity, stability, resistance to invaders (e.g., MacArthur, 1955; Margalef, 1969; Frank and McNaughton, 1991; McGrady-Steed, Harris, and Morin 1997; Naeem and Li, 1997), and resilience. The reality is much more complicated (e.g., Tilman, 1996; McGrady-Steed *et al.*, 1997; Symstad *et al.*, 1998; Naeem, Hahn, and Schuurman, 2000; Lehman and Tilman, 2000), and still in dispute (e.g., Finlay, Maberly, and Cooper, 1997; Grime, 1997; Wardle, Bonner, and Nicholson, 1997; Bengtsson, 1998; Hodgson, Thompson, and Wilson, 1998 (and the commentary that follows); Andren and Balandreau, 1999).

Does species, or functional, redundancy influence how ecological systems react to external and internal changes? Scholars today suggest that, within ecosystems, functional redundancy (different species occupying roughly the same niches) can, if co-dependencies are not too developed, often afford resiliency to the ecosystem (e.g., Risser, 1995; Tilman, 1996; Grime, 1998; Naeem *et al.*, 2000; Lehman and Tilman, 2000). Greater temporal stability appears to be afforded by higher productivity at higher diversity, and competitive interactions. The relative importance of each varies at different levels of diversity (Lehman and Tilman, 2000), and few broad generalizations are possible. Scale is very important (e.g., Pankhurst *et al.*, 1996; Groffman and Bohlen, 1999), and empirical results can support either the 'null' or the 'resilience' hypothesis. For example, in rangeland ecosystems, it is the communities with greater species diversity (and thus presumed redundancy) that are most easily and often invaded by new colonizer species (Levine, 2000)!

It used to be fashionable in environmental circles to repeat the mantra 'everything is connected to everything else.' Perhaps this was once a useful caution, even if untrue (Budiansky, 1995: 56–64). But we are better served by real understanding, rather than rhetoric, as Levins (1992) noted:

> All things are indeed connected if we follow chains of causation through their devious twists and turns. But everything is not strongly, directly or significantly connected to everything else. The analogy between an ecosystem and an individual organism simply does not hold up. The relation between, say, the liver and the heart is not the same as the relation between gazelles and gnus. The relative autonomy of linked subsystems is as important in understanding nature as their connectedness – we can in fact change some things without changing others – thus ecosystems are best understood not as harmoniously balanced wholes but as loosely coupled semi-autonomous sub-systems.

In other words, while there are important connections within ecosystems, they tend to be relatively important at local, rather than larger, levels. It is the *un*connectedness, or lack of critical dependence, that is important in analyzing redundancy. Most ecological 'connections' are spatially restricted and weak; further, their directions are unpredictable. If ecological systems were, in fact, unitary and fully connected, succession should indeed be (as early ecologists believed) a unitary phenomenon in any ecosystem, with a single endpoint. Rather, we find that (1) there is a significant element of chance in what species might arrive, (2) natural selection operates, so that (3) what particular species succeed depends on the specific local conditions. Thus, oak-hickory succession is not a singular phenomenon, but a multiply-replicated event, the outcome of which depends on local conditions. Yet oak-hickory successions are also recognizably similar, although the general relationships play out slightly differently depending on local conditions. The result is redundancy (multiple nonidentical in-use copies) of local, loosely connected ecological subunits.

In these important regards, ecosystems are qualitatively different from genetic systems, in which natural selection operates on the complex of genes carried in the organism (at the 'system' level), and engineering and political systems in which design for efficiency is deliberate. Natural selection affects the relative survival and reproduction of organisms most strongly; effects at the ecosystem level are, for the most part, simple epiphenomena. So, while managers may care about redundancy effects for our human ends, we have no evidence that redundant ecosystems always are, or are not, better in any way than simpler systems. This is the source of the complexity in the ecological literature (above).

In ecological systems, our desire for biodiversity as a contributor to stability in the face of fluctuations may rely more on the 'nonidentical' aspect of species redundancy than we have typically considered (references above). Ives, Klug, and Gross (2000), modeling complex communities from modular 'subcommunities' with random characteristics, found that it was not species diversity *per se* that generated stability, but rather the existence of species groupings with different characteristics – that community-level stability arises when species with a diversity of characteristics, sometimes overlapping, exist. Griffiths *et al.* (2000), in an empirical study, fumigated soil communities, progressively reducing soil microbial species diversity. There was no direct relationship between diversity and function: some functions (decomposition) were enhanced by reduced diversity, while others (nitrification) were compromised. Our concepts of species diversity may be enhanced by asking about the possible *kinds* of redundancy (see above) represented by examples of biodiversity.

Spatial redundancy (lesser-connected spatial repetitions) can benefit managers. It is precisely the relative independence of areas or subsystems that allows them to persist as natural areas near inhabited areas. A world in which everything *were* tightly connected to everything else would be a world in which wilderness areas, parks, greenways, and so on, might well disappear – small mistakes could have grave consequences. Very tightly connected systems are typically fragile and show little resilience (e.g., Drayton and Primack, 1996), which can be important as we increase the rate of disturbances. Systems with redundant and loosely connected subsystems, on the other hand, may change in many ways (particular species composition, exact spatial boundaries), yet persist relatively well. Had the natural areas been 'an inseparable part of the whole,' it is quite likely we would have lost much more biological diversity than we have.

From the perspective of human use, management, and exploitation of ecosystems, there is clearly some range of redundancy and connectedness that is optimal. But two things mean we cannot expect any particular level of redundancy – especially the level we desire – to eventuate. First, redundancy arises from interests of the redundant units, not the system as a whole. Second, there is no, or weak, feedback from ecosystem 'function' or 'health' on redundancy, except through failure of some ecosystems while others persist – and this process is not only slow, but has large random components. This suggests that ecosystem managers should not be sanguine in relying on ecosystem 'rules' inferred from whole ecosystem 'function.'[3] The optimal level of redundancy in any ecosystem is a managerial concept, not an evolved characteristic at the ecosystem level.

Two kinds of redundancy are especially important to managers. First, there is the redundancy of many similar (but not identical) subsystems (e.g., Ives *et al.*, 2000). Second, there is redundancy arising from the functional overlap of closely related species (functional redundancy of nonidentical species; cf. Naeem, 1998; McGrady-Steed and Morin, 1998; Griffiths *et al.*, 2000).[4] Both constitute risk reduction (cf. Walker, 1992; Walker, Kinzig, and Langridge, 1999) and multiple nonidentical in-use copies in our terms (see above). Both forms of redundancy may (or may not) contribute to an ecosystem's persistence in the face of external perturbations. Spatial redundancy of subsystems provides source populations for recolonization after local extinctions in nearby areas. The redundancy of species within functional groups means that, in the same area, the decline of one species may be 'compensated' (from a manager's point of view) by an increase in a different but functionally similar species. (Again, note that this does not speak to issues of biodiversity and genetic resources in any way.) Here is a further caution. Metapopulations may also exemplify a form of spatial redundancy. Relatively closely spaced populations (e.g., of fish species) may be reproductively independent.[5] Yet, as Wilson *et al.* (1999) note, this may not contribute to resilience; it can, in fact, present a danger if managers do not recognize the fact that populations have this spatially redundant structure. If we assume a single, large population, when in fact many small populations exist, we may inadvertently exterminate one after another of the small populations – we assume that recolonization will be swift (not true for a metapopulation). Under modest pressure, the redundancy is protective – but great pressure can collapse even very large systems.

Sometimes the redundancy of ecosystems, as in engineered and genetic systems, can buffer the systems – and us – from failures arising from our ignorance (e.g., of threshold effects). Perhaps because of this protection, some environmental policies tend to ignore the smaller-scale, subsystem aspects of ecosystems. This may not matter under many circumstances and for long periods of time (Low *et al.*, 1999, Wilson *et al.*, 1999); our errors may have relatively little effect in a redundant system. However, as human actions continue to erode any system (i.e., remove or degrade the subsystems that provide redundancy), there comes a point where the buffering capacity of the system is lost (Ames, Watson, and Wilson, 2000). At this point we are confronted with sudden, surprising – and usually undesirable – changes in the system.

As we noted above, sometimes we fail to recognize any impact until there are significant losses (e.g., species loss or dramatic declines in abundance). It is, in fact, hard to discern the difference between normal variability and changes that might be precursors of system collapse. For example, as some local or relatively independent fish populations are fished down, other nearby populations (often of a similar but not identical species) may grow and/or shift distribution to take advantage of newly available food sources. Compensation of this sort is common and expected, especially because large swings in species abundance are themselves common. As long as 'enough' redundant subsystems remain, 'normal' system patterns may still appear. Beyond a certain point, however, the ability of remaining subsystems to compensate (from a manager's point of view) for the loss or functional impairment of others reaches its limit. Then the view from the top is a view of a sudden and surprising decline, or even a catastrophic shift in system state (Carpenter, Ludwig, and Brock, 1999); whereas the view from below is of progressive loss of redundancy, leading to a threshold and sudden decline.

We are beginning to recognize that these subsystem losses (not only whole-system collapses) are important for issues of biodiversity. We suggest that policies and institutions that recognize, and respond to, the inherent redundancy of ecosystems are much less likely to be surprised by cumulative erosive actions. Further, there is a good possibility that institutions organized in ways that parallel the structure of the ecosystem are more likely to receive accurate and timely information about the state of the system, and to be able to respond in constructive ways (Costanza *et al.*, 2001). In other words, in multiscale systems a multiscale management hierarchy should be best suited to detect the onset of system-wide decline if change is buffered by local redundancies of some sort (Wilson, 2002).

4.3 Can redundancy reduce error and increase fit between preferences and outcomes in human decisions?

Contemporary public policy analysis assumes an individual knows all relevant options, has full information about the probability of particular outcomes of alternatives (given the actions of others), and has completely ordered preferences for outcomes. Yet such conditions are rare. The important work of Simon (1947) and Cyert and March (1963) assumed that humans have *limited* rationality that is *constrained* – constrained by the level of information present in a situation, by the limited attention that any individual can give to a myriad of potentially relevant facts, and by limits on the way that information is processed. These early arguments have been supported by considerable empirical research, especially by psychologists. We repeatedly find that decision makers overestimate their understanding of a problem, and underestimate the risk and uncertainty surrounding a problem. Kinder and Weiss (1978: 723), for example, note that 'decision-makers [are] more confident that they understand the problem and more satisfied that their policies will achieve the predicted ends than the evidence really justifies' (cited in Bendor, 1985: 292). If individuals behave with limited rationality, are organizational systems as unreliable as the individuals working within them?

Martin Landau (1969: 349), drawing on Von Neumann's (1956) work on reliability theory, argued that 'it makes a good deal of sense to regard a large-scale organization as a vast and complicated information system. It is, after all, necessarily and continuously engaged in the transmission and reception of messages.' Thus, within an administrative system, minor errors by one individual can be amplified as information is relayed, leading to major errors in final decisions. Error magnification is particularly problematic in systems or organizations that are strictly 'serial:' all subordinates report to a single supervisor who, in turn, reports to another supervisor. Yet exactly this type of system has been the favorite design of many scholars – particularly those teaching public administration. The logic of the preferred system of bureaucratic organization culminates in a central control point. 'The model which represents this dream is a linear organization in which everything is arrayed in tandem' (Landau, 1969: 354). Landau warned, however, that 'Organization systems of this sort are a form of administrative *brinksmanship*. They are extraordinary gambles. When one bulb goes, everything goes. Ordering parts in series makes them so dependent upon each other that any single failure can break the system' (*Ibid.*). Bendor (1985: 293) further analyzed the flaws in the conventional public administration advice to create streamlined decision-making systems:

> Thus, the proverb 'a chain is only as strong as its weakest link' is overly optimistic; a chain or series system is *weaker* than its weakest link. If, for example, the probability of completing acts A, B, and C is 0.9, 0.8, and 0.9, respectively, then the probability of completing the whole chain is, assuming statistical independence, $0.9 \times 0.8 \times 0.9 = 0.648$. This is less than its weakest link unless the probability of completing all the other links is one.

Landau proposed that adding 'sufficient' redundancy in administrative organization would make possible organizations that were more reliable than their individual human parts. Later, he noted that if the probability of failure in a particular system is 1 in 100, the probability of error if there were two duplicate systems would be 1 in 10 000, and if there were three such systems, it would be 1 in 1 million. Thus, 'the probability of failure decreases exponentially with arithmetic increases in duplication' (Landau, 1973: 187). Drawing on reliability theory, he cautioned that for the redundant parts of an administrative system to decrease the risk of serious errors, they need to operate independently and in such a manner '*that they cannot and do not impair other parts*' (Landau, 1969: 350). If the redundant parts were not independent, then redundancy would be not only a waste, but a dangerous addition. Note the homologies here with redundancy in natural and engineered systems.

Independence, however, does not imply a lack of overlap (cf. genetic systems, above). Drawing on the concept of equipotentiality derived from the early cybernetic analysis of biological systems (Ashby, 1960), Landau also encouraged thinking about the kinds of overlap that enable some systems to 'take over' the functions of other parts that may have been damaged. 'It is this overlap that permits the organism to exhibit a high degree of adaptability, i.e. to change its behavior in accordance with changes in stimuli' (Landau, 1969: 351). Complementary earlier work (Tiebout, 1956; Ostrom, Tiebout, and Warren, 1961) looked afresh at the multiple units of government found in many metropolitan areas. Considering a *system* of governance units in a metropolitan area, they asked whether multiple units influenced potential competition and consequent performance among these governance units.

Several mechanisms potentially increase performance because of the presence of competitive units. On the citizen-consumption side, Tiebout (1956) argued that residents could 'vote with their feet' and move to the jurisdiction that most fitted their own preferences in terms of a service/tax package. Ostrom *et al.* (1961) made a key analytical distinction between decisions to *provide* public services and decisions to *produce* these services. Once a community had decided that a service was to be provided, having multiple producers allowed public officials an opportunity to search out the most efficient set of producers for the mix of services desired by the citizens of a community. Some services would be produced by a local unit; other services would be produced by larger or other small units. Thus, competition among multiple units would generate considerably more *information about alternatives*; it would also increase the pressure to seek out the most efficient combination for a particular locality.

Substantial research on public service economies supports these analyses of redundancy (summarized in Oakerson, 1999; McGinnis, 1999a, 1999b, 2000). In 80 metropolitan areas, for example, the most efficient urban policing is found in metropolitan areas with 21 or more police departments, and the least efficient in metropolitan areas with seven or fewer departments (Parks and Ostrom, 1999). Further, efficiency is enhanced by differentiation in the services provided: by small, immediate response services, and by overlapping larger agencies that provide services such as radio communications and major homicide investigations (Parks and Ostrom, 1999). A very recent survey of over 70 empirical studies of fragmentation of urban governance found little support for the presumption that suburbs represented a costly form of redundancy: 'The extant empirical literature is scarcely a strong endorsement of the view that suburbs damage cities' (Hawkins and Ihrke, 1999: 119). More than two-thirds of the studies challenged the dominant view that suburbs were harmful; several other studies were supported only with anecdotal evidence. 'It appears that the suburban exploitation thesis has been sustained principally by studies that do not investigate benefits; that overlook evidence of benefits; or that assent to reformist claims about suburban fragmentation, commuters, and growth without systematically weighting the evidence available to test those claims' (Hawkins and Ihrke, 1999: 188, 120).

Bendor (1985) analyzed duplication (or its absence) in the planning and operation of large transportation systems, providing systematic evidence from in-depth studies of three metropolitan areas. The conclusions, while focused on urban transportation, are quite instructive for our interest in resource regimes. In general, Bendor concluded that redundancy in public services can provide higher service levels (in those public services in which it is relatively easy to measure performance and behavior can be observed) due both to the increased level of competition and to the increased reliability of such redundant systems. At the same time, there is a tendency to try to remove redundancy. Bendor argued these conditions would increase redundancy's feasibility and advantages:

1. The probability of a premature quashing of redundancy is diminished if overlapping agencies use different technologies . . . [D]ifferent technologies promote a (possibly false) expectation of functional specialization, that is, the different technologies will be deployed for different ends, whereas identical technologies make redundancy highly visible and vulnerable (p. 279).
2. If bureaus overlap rather than exactly duplicate each other's functions, redundancy is more tolerable politically (p. 280).
3. A well-established agency can mobilize its political resources to bar newcomers to its policy field. It is not accidental that both redundant cases in this study involved agencies that started almost simultaneously (p. 280).
4. Redundancy is more stable, and therefore more practical, if overlapping bureaus do not have a powerful superior close at hand.[6] For this reason, redundancy is probably more feasible among special districts than among

regular line departments because districts are less commonly embedded in hierarchies (p. 281).

5. [R]edundant agencies must retain some diversity in order to produce the full fruits of duplication. The probability of parallel agencies remaining independent is the knottiest problem in the pragmatics of redundancy theory (p. 282).

An important lesson from Bendor's research is that having multiple *non-identical* jurisdictions yields greater diversity and more flexible responses and less risk of being destroyed.

4.4 Complex adaptive systems: reducing risk through redundancy

Contemporary scholars of complex adaptive systems have integrated much earlier work. Complex adaptive systems are composed of a large number of active elements whose rich patterns of interactions produce emergent properties – which are not easy to predict by analyzing the separate system components. Holland (1995: 10) viewed complex adaptive systems as 'systems composed of interacting agents described in terms of rules. These agents adapt by changing their rules as experience accumulates.' Complex adaptive systems 'exhibit coherence under change, via conditional action and anticipation, and they do so without central direction' (Holland, 1995: 38–9). Levin (1995, 1999) successfully used complex adaptive systems to understand fragile ecosystems.

Holland pointed out that complex adaptive systems differ from physical systems that are not adaptive and that have been the foci of most scientific effort – yet, inappropriately, the physical sciences have been the model for many aspects of contemporary social science. We find it odd that social scientists have traditionally drawn more on physical analogies in developing an approach to scientific explanation than on biology and ecology. The concepts needed to understand the behavior of complex systems are not yet well developed by social scientists.

All systems face challenges that may lead them to falter or fail. Complex adaptive systems are not immune to risks, but they may have unique ways of coping with risks. In complex information systems, redundancy is seen as a major source of stability and strength as such systems are buffeted by uncertain and new events (see Axelrod and Cohen, 2000). In most information systems, such as the Internet or local area networks, current technology has only been invented within the last few decades. Thus, few precise assessments can be made of the risks they face. Innovation keeps the systems undergoing enough change for it to be hard to predict the specific risks they will face. When the sources of risk to a system are relatively independent, redundancy is a major structural attribute that reduces the overall risks to the survival of a system. As Axelrod and Cohen (2000: 107) suggest:

> The primary method of risk management for independent failures is to build redundancy into the system . . . [R]edundancy makes possible reliable traffic flows through information networks by channeling traffic around nodes that fail. In addition to redundancy, a useful design feature to deal with local failures is to avoid having any one element of the system be essential to its overall performance. This is typically achieved by making the system highly decentralized like the Internet.

In his own recommendations for devising adaptive forms of environmental management for fragile and at risk ecosystems, Simon Levin (1999: 198–206) presents 'eight commandments of environmental management.' In direct contrast to earlier views of scientific environmental management, one of Levin's 'commandments' is to 'preserve redundancy.' Not surprisingly, a second commandment is to 'maintain heterogeneity,' given that natural selection acts only on existing variability, and if we reduce variability, we reduce options for responding to future environmental changes. He stresses the close connection between the importance of redundancy and that of heterogeneity, but points out, as we have above, that the value of spare parts may only be understood when other parts are lost (Levin, 1999: 202–3):

> Redundancy is the immediate source of replacement of lost functions; heterogeneity provides the materials for adaptive responses over longer time scales . . . The essential element to understanding the importance of redundancy is to elucidate the functional substitutability of one species for another, the ecological complement to economic substitutability.

Levin is primarily concerned with redundancy of populations, but his commandment is also important for ecological subsystems (above) and for social systems. Thus, *two of the initial assumptions underlying much of modern policy are profitably contradicted in resilient complex adaptive systems*. Further, in addition to simple redundancy, having diverse structures within a complex adaptive system also helps to insure against known and unknown risks. If subunits are diversely structured, they are less likely all to be swamped by the same external risk (Holling, 1978; Gunderson, Holling, and Light, 1995).

4.4.1 Redundant resource regimes: an example

In the USA, many examples exist of dynamic resource governance systems characterized by redundancy, in which there is strong evidence of high performance.

One example is the Maine lobster fishery, which is noteworthy because of the long-term, complementary roles adopted by both local and state governance systems. Maine is organized into riparian territories along most of its coast (Acheson, 1988). Boundary rules and many of the day-to-day fishing regulations are organized by harbor gangs:

> In order to go fishing at all, one must become a member of a 'harbor gang,' the group of fishermen who go lobstering from a single harbor. Once one has gained admittance into such a group, one can only set traps in the traditional territory of that particular harbor gang. Members of harbor gangs are expected to obey the rules of their gang concerning fishing practices, which vary somewhat from one part of the coast to another. In all areas a person who gains a reputation for molesting others' gear or for violating conservation laws will be severely sanctioned. Incursions into the territory of one gang by fishers from another are ordinarily punished by surreptitious destruction of lobster gear. There is strong statistical evidence that the territorial system, which operates to limit the number of fishers exploiting lobsters in each territory, helps to conserve the lobster resource.
>
> *(Acheson, Wilson, and Steneck, 1998: 400)*

At the same time, the state of Maine has long-established formal laws that protect the breeding stock and increase the likelihood that regeneration rates will be high. 'At present, the most important conservation laws are minimum and maximum size measures, a prohibition against catching lobsters with eggs, and a law to prohibit the taking of lobsters which once had eggs and were marked – i.e. the "V-notch" law' (Acheson *et al.*, 1998: 400). Neither the state nor any of the harbor gangs has tried to limit the quantity of lobster captured. The state does not make any effort to limit the number of fishers, because this is already done at a local level. However, the state has been willing to intercede when issues exceed the scope of control of local gangs. In the late 1920s, for example, when lobster stocks were at very low levels and many local areas appear to have had substantial compliance problems, the state took a number of steps – including threats to close the fishery – that supported informal local enforcement efforts. By the late 1930s, compliance problems were largely resolved and stocks had rebounded.

In response to changes that were breaking down the harbor gang system, this was recently formalized by dividing the state into zones with democratically elected councils. Each council has been given authority over rules that have principally local impacts (e.g., trap limits, days and times fished). This formalization of local zones was followed almost immediately by the creation of an informal council of councils to address problems at higher levels. It is expected that the council of councils will be formalized soon (Wilson, 1997). Today, the state needs only about six patrol officers on the water to police the activities of 7100 lobstermen, all other fisheries, and boating, shipping, and coastal environmental laws. Clearly this is a relatively efficient redundancy. Further, the ecological impacts appear positive. During the 1990s, the fishery has grown substantially, with increased yields (Maine Department of Marine Resources, 2000). Further, the increase in yields appears to be due to an increase, not a 'mining down,' of the population.

4.5 Conclusion

We have tried to move beyond prescriptions and normative statements to some analytic considerations of how, in different systems, redundancy arises and is maintained or disappears, as a result of its benefits and costs at different levels. This is a complicated question, and so far we have no easy, singular 'answer.' The systems we described all have redundant elements, but the sources and impacts of redundancy may differ. In genetic systems, redundancy arises because alleles, in their own self-interest, manage to get duplicated. In fact, this is probably a major source of redundancy in many systems. In genetic systems, whether such duplications persist, are suppressed, or are multiplied, depends on the impact of redundancy on the functioning of the entire genome – there is some optimum level of redundancy in any particular case. In engineering systems, redundancy is designed in, typically as a risk-reduction strategy, *for the sake* of the whole system. In self-organizing systems, redundancy arises because of the self-interest of local actors. In systems of governance, additional layers of complication exist.

When will redundancy in governance enhance the efficiency of the 'whole' system? We suggest that the following conditions make redundancy advantageous. When transfer of information or actors across subsystems is inefficient or slow, redundant local systems are likely to be efficient. Similarly, when a large system or geographic region is spatially (ecologically) heterogeneous, redundant local systems may work well. In general, local systems may be best able to verify local information, address locally specific conditions, and respond rapidly. At the same time the checks and balances on local interests may work best at greater-than-local levels. The fact that 'redundant' local variations exist may mean that system-level responses can be more potent and rapid than otherwise, and/or that local variations may be able to meet unforeseen contingencies.

Further, individuals who interact with others frequently on a face-to-face basis, and know that future interactions are likely, are more apt to build trust and adopt forms of reciprocity than when interactions are more anonymous and infrequent. We note that this may work either to the advantage or to the disadvantage of the larger system (e.g., such as in Spotted Owl (*Strix occidentalis*) conservation). More experimentation can occur when local units have some autonomy to create their own rules and policies. Some experiments (each in only a small segment of a larger system) will fail, but others can learn from both the good and bad experiences.

Conversely, when a governance system is large and faces conditions that are relatively homogeneous and stable (and/or predictable), and when information and actors can be transferred rapidly, redundant local systems will be relatively inefficient. If competition among parallel units turns to destructive strategies, or if the interests of local decision makers are at sharp odds with the interests of those at other levels (e.g., Spotted Owl conservation), redundancy may escalate conflict rather than increase performance. In such cases, redundancy may be destructive unless it is embedded within larger jurisdictions with effective conflict resolution arenas.

Some tensions and trade-offs will always remain. Proponents of central or dispersed systems frequently fail to recognize these trade-offs in a relevant way. Consider the recurring debates about whether Bureau of Land Management (BLM) stocking rates should be set by Washington BLM personnel or by local BLM representatives. Both sides have some validity in their arguments; both have hidden agendas. Local, on-the-ground managers know local conditions better, can respond to them efficiently, and have more locally relevant information at hand for making decisions. On the other hand, the ability of local managers to focus only on the large-scale, long-term interests of the BLM, when these might conflict with the interests of local landowners, may be limited.

One issue that needs further clarification is the trade-off between different types of errors that can be made by governance systems. Bendor (1985: 50), for example, argued:

> Modern reliability theory distinguishes between a type one error, failing to stop an undesired event, and a type two error, failing to effect a desired one. Organizational redundancy theory has not yet incorporated this point. Landau did not discuss the question in his 1969 essay and though he subsequently (1973) discussed redundancy in the context of constitutional design, that a different kind of error is involved was not made explicit. Yet many policy sectors exhibit both types of errors. Recall, for example, that a welfare program may overlook an eligible person (error of omission) or aid an ineligible one (error of commission). A perfect welfare system would be completely reliable in both respects, but there may be trade-offs between these two kinds of reliability. Does guarding against unwanted actions nullify or vitiate efforts to ensure that desired actions occur?

Decisions about the relative impact of central, versus dispersed, decisions in any system may be difficult. Here we hope to highlight (1) the fact that not all redundancies are equivalent; (2) the relative costs and benefits of redundancy depend on political and ecological conditions; and (3) conflicts of interest exists in most systems. We need, urgently, to develop a grounded theoretical approach to the study of redundancy, for efficient and responsive management depends on matching optimal levels of redundancy to the appropriate conditions.

The presence of larger, overlapping jurisdictions is an important complement to the work of parallel, smaller-scale units. Larger units can back up smaller units in several ways: (1) providing support at times of natural disasters; (2) addressing corruption or gross inefficiency; (3) providing scientific and technical skills to complement local knowledge; (4) providing conflict resolution arenas for conflicts among parallel units; and (5) taking on functions that are generally more efficiently undertaken by larger units.

It is time to leave behind the prescriptive approach to redundancy. Instead, it is crucial to analyze the level of diversity, types of risk, and location of important information in diverse locations before making any judgment about the impact of specific kinds of redundancy in a governing system.

Notes

1. See, for example, Ostrom, Tiebout, and Warren (1961); Hirsch (1970); Ostrom, Parks, and Whitaker (1973); Niskanen (1975); Kaufman (1977); Meier (1980); Bendor (1985); Ostrom *et al.* (1988); Miranda and Lerner (1995); Bish (2001).
2. Stephens and Wikstrom, (1999: 5–6), for example, state: 'In the United States, urban regions are layered onto one of the world's most complex federal systems, with a national government, fifty states . . . 87,453 local governments as of 1997, with the number increasing over time . . . In addition, local public institutions are divided into both discrete and layered segments when it comes to how the system affects individuals and groups of citizens. Public policy is similarly fragmented. Confusion abounds. It's fair to say that this situation all too often leads to distrust and disgust with the performance of government(s).'
3. The cellular automaton 'Life' (sometimes seen as a computer screen saver) may help clarify this concept. The rules are extremely simple: depending on how neighboring pixels are occupied, a pixel will 'behave' in a certain way, generating a set of 'organisms' (agents). The original pixels are randomly scattered; over time, a stable array of agents exists. If you know the rules and watch the process, it appears delightfully clear. But if, as ecologists must, you could only consider the end arrays in all their diversity, inferring the rules would be a nightmare! Thus, it is easy for us to derive reasonable rules that we later discover are either wrong or limited in their effects. Consider the maxim that 'diversity causes stability' in ecosystems, popular a few years ago. Our favorite counter-example is the *Spinnifex* systems of central Australia. Much to managers' chagrin, this is an economically useless, very simple, non-diverse, and extremely stable ecosystem. The vast majority of the plant biomass comprises two species of *Spinnifex*, and the vast majority of animal biomass, three genera of termites. The system is extreme, with high temperatures, low soil nutrients, and very low moisture. Unless water and nutrient subsidies are applied, nothing else can persist under these conditions. Indeed, a moment's reflection suggests that while diversity (redundancy) almost certainly enhances stability, as in the examples noted, systems do not spring into existence full-blown and diverse. In fact, stability of climate, combined with moderate climatic

conditions, allows diversity to grow. If one wants to posit causality, this is the direction.

4. Note that we are talking about two separate kinds of redundancy in ecological systems: (1) spatial redundancy of similar subsystems; and (2) redundancy of species within a functional group. Both allow the system to respond to an external perturbation, but in different ways. Spatial redundancy of subsystems may provide source populations for recolonization after local extinctions nearby, depending on the connectedness of subsystems. The redundancy of species within functional groups means that, in the same area, the decline of one species may be 'compensated' (from a manager's point of view) by an increase in a different but functionally similar species. Levin (1999) notes, as we do in Table 4.1, that some redundancies are simple repetitions, while others (e.g., species redundancy in a functional group) are very close to heterogeneity/diversity. We suspect there are trade-offs between connectivity and independence. Probability of permanent local extinction increases with isolation (less connectivity), but for patchy subsystems to exist in human-dominated environments, independence is an essential characteristic. Isolated independent systems may be depauperate (unable to benefit from recolonization) compared with original (no human impact) systems, but they may still be able to function. Further, we see that redundancy contributes to function – but Walker (1992), for example, argues convincingly that maintaining ecosystem function is an excellent way to maintain species diversity (redundancy). Thus we have a positive feedback system.
5. This also raises an issue more difficult in analyzing ecosystems than, for example, engineered or genetic systems: the issues of scale in measurement and inference of redundancy (see Peterson, Allen, and Holling, 1998).
6. Bendor points out in a footnote that the proposition does not necessarily hold if there is more than one powerful superior.

References

Acheson, J.M. 1988. *The Lobster Gangs of Maine*. Hanover, NH: New England University Press.

Acheson, J.M. 1993. Capturing the commons: legal and illegal strategies. In *The Political Economy of Customs and Culture: Informal Solutions to the Commons Problem*, pp. 69–83, ed. T.L. Anderson and R.T. Simmons. Lanham, MD: Rowman & Littlefield.

Acheson, J.M., Wilson, J.A., and Steneck R.S. 1998. Managing chaotic fisheries. In *Linking Social and Ecological Systems. Management Practices and Social Mechanisms for Building Resilience*, pp. 390–413, ed. F. Berkes and C. Folke. Cambridge: Cambridge University Press.

Ames, T., Watson, S., and Wilson, J. 2000. Rethinking overfishing: insights from oral histories of retired groundfishermen. In *Finding Our Sealegs: Linking fishery People and their Knowledge with Science and Management*, pp. 153–64, ed. B. Neis and L. Felt. St Johns, Canada: ISER Press.

Andren, O. and Balandreau, J. 1999. Biodiversity and soil functioning – from black box to can of worms? *Applied Soil Ecology* 13: 105–8.

Ashby, W.R. 1960. *Design for a Brain: the Origin of Adaptive Behavior*, 2nd edn. New York: Wiley.

Axelrod, R. and Cohen, M.D. 2000. *Harnessing Complexity: Organizational Implications of a Scientific Frontier*. New York: The Free Press.

Baland, J.M. and Platteau, J.P. 1996. *Halting Degradation of Natural Resources. Is There a Role for Rural Communities?* Oxford: Clarendon Press.

Bendor, J. 1985. *Parallel Systems: Redundancy in Government*. Berkeley: University of California Press.
Bengtsson, J. 1998. Which species? What kind of diversity? Which ecosystem function? Some problems in studies of relations between biodiversity and ecosystem function. *Applied Soil Ecology* 10: 191–9.
Bish, R.L. 2001. Local government amalgamations: discredited nineteenth-century ideals alive in the twenty-first. *C.D. Howe Institute Commentary* 150 (March): 1–35.
Brookfield, J.F.Y. 1997. Genetic redundancy. *Advances in Genetics* 36: 137–55.
Budiansky, S. 1995. *Nature's Keepers: the New Science of Nature Management*. New York: The Free Press.
Callaghan, A., Guillemaud, T., Matake, N., and Raymond, M. 1998. Polymorphisms and fluctuations in copy number of amplified esterase genes in *Culex pipiens* mosquitoes. *Insect Moleular Biology* 7: 295–300.
Carpenter, S.R., Ludwig, D., and Brock, W.A. 1999. Management of eutrophication for lakes subject to potentially irreversible change. *Ecological Applications* 9(3): 751–71.
Carson, J.M. and Doyle, J. 1999. Highly optimized tolerance: a mechanism for power laws in designed systems. *Physical Review E* 60(2): 1412–27.
Cavalier-Smith, T. 1985. Selfish DNA and the origin of introns. *Nature* 315: 283–4.
Clark, A.G. 1994. Invasion and maintenance of a gene duplication. *Proceedings of the National Academy of Sciences* 91: 2950–4.
Costanza, R., Low, B.S., Ostrom, E., and Wilson, J., eds. 2001. *Institutions, Ecosystems, and Sustainability*. Boca Raton, FL.: CRC Press.
Cyert, R. and March, J. 1963. *A Behavioral Theory of the Firm*. Englewood Cliffs, NJ: Prentice-Hall.
Daily, G.C. and Ehrlich, P.R. 1995. Population extinction and the biodiversity crisis. In *Biodiversity Conservation*, pp. 45–55, ed. C.A. Perrings, K.-G. Mäler, C. Folke, C.S. Holling, and B.-O. Jansson. Amsterdam: Kluwer Academic Publishers.
Doyle, J. 1999. http://www.cds.caltech.edu/~doyle/notes/1.html
Drayton, B. and Primack, R.B. 1996. Plant species lost in an isolated conservation area in Metropolitan Boston from 1894 to 1993. *Conservation Biology* 10: 30–9.
Farmer, J.D., Packard, N.H., and Perelson, A.S. 1986. The immune system, adaptation and machine learning. *Physica D*22: 187–204.
Field, L.M. and Devonshire, A.L. 1998. Evidence that E4 and FE4 esterase genes responsible for insecticide resistance in the aphid *Myzus persicae* (Sulzer) are part of a gene family. *Biochemical Journal* 330: 169–73.
Finlay, B.J., Maberly, S.C., and Cooper, J.I. 1997. Microbial diversity and ecosystem function. *OIKOS* 80: 209–13.
Frank, D.A. and McNaughton, S.J. 1991. Stability increases with diversity in plant communities: empirical evidence from the 1988 Yellowstone drought. *OIKOS* 62: 360–2.
Freeland, S.J. and Hurst, L.D. 1998. The genetic code: one in a million. *Journal of Molecular Evolution* 47: 238–48.
Fryxell, K.J. 1996. The coevolution of gene family trees. *Trends in Genetics* 12: 364–9.
Goldstein, D.B. and Holsinger, K.E. 1992. Maintenance of polygenic variation in spatially structured poulations. *Evolution* 46: 412–29.
González-Gaitán, M., Rothe, M., Wimmer, E.A., Taubert, H., and Jäckle, H. 1994. Redundant functions of the genes knirps and knirps-related for the establishment of anterior *Drosophila* head structures. *Proceedings of the National Academy of Sciences* USA 91: 8567–71.

Graur, D. and Li, W-H. 2000. *Fundamentals of Molecular Evolution*, 2nd edn. Sunderland, MA: Sinauer, Inc.
Griffiths, B.S., Ritz, K., Bardgett, R.D., *et al.* 2000. Ecosystem response of pasture soil communities to fumigation-induced microbial diversity reductions: an examination of the biodiversity–ecosystem function relationship. *OIKOS* 90: 279–94.
Grime, J.P. 1997. Biodiversity and ecosystem function: the debate deepens. *Science* 277: 1260–1.
Grime, J.P. 1998. Benefits of plant diversity to ecosystems: immediate, filter and founder effects. *Journal of Ecology* 86: 902–10.
Groffman, P.M., and Bohlen, P.J. 1999. Soil and sediment biodiversity: cross-system comparisons and large-scale effects. *Bioscience* 49: 139–48.
Gunderson, L.H., Holling, C.S., and Light, S.S. 1995. *Barriers and Bridges to the Renewal of Ecosystems and Institutions.* New York: Columbia University Press.
Hampson, C.W. 1997. Redundancy and high volume manufacturing methods. *Intel Technology Journal Q4'97.* (Available at http://developer.intel.com)
Hawkins, B.W. and Ihrke, D.M. 1999. Reexamining the Suburban Exploitation Thesis in American Metropolitan Areas. *Publius* 29(3): 109–21.
Hawley, A. and Zimmer, B.G. 1970. *The Metropolitan Community: its People and Government.* Beverly Hills, CA: Sage Publications.
Hirsch, W.Z. 1970. *The Economics of State and Local Government.* New York: McGraw-Hill.
Hodgson, J.G., Thompson, K., and Wilson, P.J. 1998. Does biodiversity determine ecosystem function? The Ecotron experiment reconsidered. *Functional Ecology* 12: 843–56.
Holland, J.H. 1995. *Hidden Order. How Adaptation Builds Complexity.* Reading, MA: Addison-Wesley.
Holling, C.S. 1978. *Adaptive Environmental Assessment and Management.* London: John Wiley.
Holling, C.S. and Meffe, G.K. 1996. Command and control and the pathology of natural resource management. *Conservation Biology* 10(2): 328–37.
Hoxby, C. 1994. Does competition among public schools benefit students and taxpayers? Working Paper No. 4979. Cambridge, MA: National Bureau of Economic Research.
Hurst, L.D. 1996. Further evidence consistent with Stellate's involvement in meiotic drive. *Genetics* 142: 641.
Ives, A.R., Klug, J.L., and Gross, K. 2000. Stability and species richness in complex communities. *Ecology Letters* (2000)3: 399–411.
Kaufman, H. 1977. Reflections on administrative reorganization. In *Setting National Priorities: the 1978 Budget*, ed. J. Pechman. Washington DC: Brookings Institution.
Kinder, D. and Weiss, J. 1978. In lieu of rationality: psychological perspectives on foreign policy decision-making. *Journal of Conflict Resolution* 22(4): 707–35.
Landau, M. 1969. Redundancy, rationality, and the problem of duplication and overlap. *Public Administration Review* 29(4): 346–58.
Landau, M. 1973. Federalism, redundancy, and system reliability. *Publius*, 3(2): 533–42.
Lehman, C.L. and Tilman, D. 2000. Biodiversity, stability, and productivity in competitive communities. *American Naturalist* 156: 534–52.
Levin, S. 1995. Ecosystems and the biosphere as complex adaptive systems. *Ecosystems* 1: 431–6.

Levin, S. 1999. *Fragile Dominion. Complexity and the Commons.* Reading, MA: Perseus Books.
Levine, J.M. 2000. Species diversity and biological invasions: relating local process to community pattern. *Science* 288: 852–4.
Levins, R. 1992. Evolutionary ecology looks at environmentalism. Paper delivered at the Symposium on Science, Reason and Modern Democracy, Michigan State University, East Lansing, Michigan, May 1, 1992, 3–6. Quoted by Marc Landy in Civic environmentalism in the American political tradition, Presented at the Conference on Civic Environmentalism Working Group, Portland, ME, August 18, 2000.
Low, B., Costanza, R., Ostrom, E., Wilson, J., and Simon, C.P. 1999. Human–ecosystem interactions: a dynamic integrated model. *Ecological Economics* 31(2): 227–42.
MacArthur, R.H. 1955. Fluctuations of animal populations and a measure of community stability. *Ecology* 36: 533–6.
Maine Department of Marine Resources, 2000. http://janus.state.me.us/dmr/Comfish/am.lobster.htm
Maniloff, J. 1996. The minimal cell genome: 'on being the right size.' *Proceedings of the National Academy of Sciences* USA 93: 10004–6.
Margalef, R. 1969. Diversity and stability: a practical proposal and a model of interdependence. In *Diversity and Stability in Ecological Systems*, pp. 25–37. *Brookhaven Symposium in Biology* Vol. 22. Upton, NY: Brookhaven National Laboratory.
Maroni, G., Wise, J., Young, J.E., and Otto, E. 1987. Metallothionein gene duplication and metal tolerance in natural populations of *Drosophila melanogaster. Genetics* 117: 739–44.
Maynard Smith, J. 1978. *The Evolution of Sex.* Cambridge: Cambridge University Press.
McGinnis, M., ed. 1999a. *Polycentric Governance and Development: Readings from the Workshop in Political Theory and Policy Analysis.* Ann Arbor: University of Michigan Press.
McGinnis, M., ed. 1999b. *Polycentricity and Local Public Economies: Readings from the Workshop in Political Theory and Policy Analysis.* Ann Arbor: University of Michigan Press.
McGinnis, M., ed. 2000. *Polycentric Games and Institutions: Readings from the Workshop in Political Theory and Policy Analysis.* Ann Arbor: University of Michigan Press.
McGrady-Steed, J., Harris, P.M., and Morin P.J. 1997. Biodiversity regulates ecosystem predictability. *Nature* 390: 162–5.
McGrady-Steed, J. and Morin, P.J. 1998. Biodiversity, density compensation, and the dynamics of populations and functional groups. *Ecology* 81: 361–73.
Meier, K. 1980. Executive reorganization of government: impact on employment and expenditures. *American Journal of Political Science* 23(3): 396–412.
Miranda, R. and Lerner, A. 1995. Bureaucracy, organizational redundancy, and the privatization of public services. *Public Administration Review* 55(2): 193–200.
Mouchès, C., Pasteur, N., Bergé, J.B., *et al.* 1986. Amplification of an esterase gene is responsible for insecticide resistance in a California *Culex* mosquito. *Science* 233: 778–80.
Nadeau, J.H. and Sankoff, D. 1997. Comparable rates of gene loss and functional divergence after genome duplications early in vertebrate evolution. *Genetics* 147: 1259–66.

Naeem S. 1998. Species redundancy and ecosystem reliability. *Conservation Biology* 12: 39–45.
Naeem, S., Hahn, D.R., and Schuurman, G. 2000. Producer–composer co-dependency influences biodiversity effects. *Nature* 403: 762–4.
Naeem, S. and Li, S. 1997. Biodiversity enhances ecosystem reliability. *Nature* 390: 507–9.
Nagl, W. 1990. Polyploidy in differentiation and evolution. *International Journal of Cell Cloning* 8: 216–23.
Niskanen, W. 1975. Bureaucrats and politicians. *Journal of Law and Economics* 18(3): 617–44.
Norman, B. and Bean, J. 1999. A genetic algorithm methodology for complex scheduling problems. *Naval Research Logistics* 46: 199–211.
Nowak, M.A., Boerlljst, M.C., Cooke, J., and Smith, J. M. 1997. Evolution of genetic redundancy. *Nature* 388: 167–71.
Oakerson, R.J. 1999. *Governing Local Public Economies: Creating the Civic Metropolis*. Oakland, CA: ICS Press.
Ohno, S. 1970. *Evolution by Gene Duplication*. Berlin: Springer.
Ostrom, E. 1990. *Governing the Commons: the Evolution of Institutions for Collective Action*. New York: Cambridge University Press.
Ostrom, E., Parks, R.B., and Whitaker, G.P. 1973. Do we really want to consolidate urban police forces? *Public Administration Review* 33: 423–33.
Ostrom, V. 1987. *The Political Theory of a Compound Republic: Designing the American Experiment*, 2nd revised edn. San Francisco: ICS Press.
Ostrom, V. 1991. *The Meaning of American Federalism: Constituting a Self-Governing Society*. San Francisco: ICS Press.
Ostrom, V. 1997. *The Meaning of Democracy and the Vulnerability of Democracies: a Response to Tocqueville's Challenge*. Ann Arbor: University of Michigan Press.
Ostrom, V., Bish, R., and Ostrom, E. 1988. *Local Government in the United States*. San Francisco: ICS Press.
Ostrom, V., Tiebout, C.M., and Warren, R. 1961. The organization of government in metropolitan areas: a theoretical inquiry. *American Political Science Review* 55(4): 831–42.
Pankhurst, C.E., Ophel-Keller, K., Doube, B.M., and Gupta, V.V.S.R. 1996. Biodiversity of soil microbial communities in agricultural systems. *Biodiversity and Conservation* 5: 197–209.
Parks, R.B. and Ostrom, E. 1999. Complex models of urban service systems. In *Polycentricity and Local Public Economies: Readings from the Workshop in Political Theory and Policy Analysis*, pp. 355–80, ed. M. McGinnis, Ann Arbor: University of Michigan Press. Originally published in Clark, T.N. 1981. *Policy Analysis: Directions for Future Research*. Beverly Hills, CA: Sage Publications.
Peterson, G., Allen, C.R., and Holling, C.S. 1998. Ecological resilience, biodiversity and scale. *Ecosystems* 1: 6–18.
Polski, M.M. 2000. Institutional evolution and change: interstate banking reform in the United States. PhD Dissertation, Indiana University, Bloomington, IN.
Pritchett, L. and Filmer, D. 1999. What education production functions *really* show: a positive theory of education expenditures. *Economics of Education Review* 18(2): 223–39.
Risser, P.G. 1995. Biodiversity and ecosystem function. *Conservation Biology* 9: 742–6.

Rusk, D. 1993. *Cities Without Suburbs.* Washington DC: Woodrow Wilson Center Press.
Simon, H. 1947. *Administrative Behavior: a Study of Decision-Making Processes in Administrative Organization.* New York: Macmillan.
Smith, V., Chou, K.N., Lashkari, D., Botstein, D., and Brown, P.O. 1996. Functional analysis of the genes of yeast chromosome V by genetic footprinting. *Science* 274: 2069–74.
Stebbins, G.L. 1974. *Flowering Plants: Evolution above the Species Level.* Cambridge, MA: Harvard University Press.
Stephens, G.R. and Wikstrom, N. 1999. *Metropolitan Government and Governance: Theoretical Perspectives, Empirical Analysis, and the Future.* Oxford: Oxford University Press.
Symstad, A.J., Tilman, D., Willson, J., and Knops, J.M.H. 1998. Species loss and ecosystem functioning: effects of species identity and community composition. *OIKOS* 81: 389–97.
Tautz, D. 1992. Redundancies, development, and the flow of information. *BioEssays* 14(4): 263–6.
Thomas, J.H. 1993. Thinking about genetic redundancy. *Trends in Genetics* 9: 395–9.
Tiebout, C. 1956. A pure theory of local expenditures. *Journal of Political Economy* 64(5): 416–24.
Tilman, D. 1996. Biodiversity: population versus ecosystem stability. Ecology 77(2): 350–63.
Von Neumann, J. 1956. Probabilistic logics and the synthesis of reliable organisms from unreliable components. In *Automata Studies*, pp. 43–98, ed. C.E. Shannon and J. McCarthy. Princeton, NJ: Princeton University Press.
Wagner, A. 1998. The fate of duplicated genes: loss or new function? *BioEssays* 20: 785–8.
Wagner, A. 1999. Redundant gene functions and natural selection. *Journal of Evolutionary Biology* 12: 1–16.
Wagner, A. 2000. Robustness against mutations in genetic networks of yeast. *Nature Genetics* 24: 355–61.
Walker, B. 1992. Biodiversity and ecological redundancy. *Conservation Biology* 6: 18–23.
Walker, B., Kinzig, A., and Langridge, J. 1999. Plant attribute diversity, resilience, and ecosystem function. *Ecosystems* 2(2): 139–50.
Walsh, J.B. 1995. How often do duplicated genes evolve new functions? *Genetics* 139: 421–8.
Wardle, D.A., Bonner, K.I., and Nicholson, K.S. 1997. Biodiversity and plant litter: experimental evidence which does not support the view that enhanced species richness improves ecosystem function. *OIKOS* 79: 247–58.
Williams, G.C. 1975. *Sex and Evolution.* Princeton, NJ: Princeton University Press.
Wilson, J.A. 1997. Maine fisheries management initiative. In *The Social Impacts of Individual Transferable Quotas*, pp. 335–53, ed. G. Palsson. Copenhagen: TemaNord.
Wilson, J.A. 2002. Scientific uncertainty, complex systems and the design of common pool institutions. In *The Drama of the Commons*, pp. 327–59, ed. E. Ostrom, T. Dietz, N. Dolsak, P. Stern, S. Stonich, and E. Weber. Washington DC: National Academy Press.
Wilson, J.A., Low, B., Costanza, R., and Ostrom, E. 1999. Scale misperceptions and the spatial dynamics of a social–ecological system. *Ecological Economics* 31(2): 243–57.

Part II

Building resilience in local management systems

In dealing with multiple-scale systems, a useful place to start is local management systems. In the development of common property theory in the 1980s and the 1990s, the local or the community level received by far the greatest part of research attention. This was not because the local level was necessarily perceived as the most important scale of organization, but because social–ecological systems at this level provided a 'laboratory' in which principles can be generated, before they can be tested in the real world of external drivers and cross-scale interactions.

When analyzing resilience, again it makes sense to address the local level and build linkages to other scales. This approach helps simplify the analysis of change and the response to change. For example, it is easier to deal with the response and adaptation to one kind of perturbation (e.g., major hurricane), than to a perturbation complicated by an external driver (e.g., the collapse of commodity markets that previously supported an agricultural society). Also, it is easier to deal with the comparison, for example, of two local–regional forest management systems subject to the same forces of social and economic change over a period of time, than a larger system that may have come under other stresses as well. Resilience thinking helps the researcher to look beyond the static analysis of social systems and ecological systems, and to ask instead questions regarding the adaptive capacity of societies and their institutions. One way to approach these questions is to look for co-existing property rights systems, and to analyze their performance and adaptation (Chapter 5). Another way may be to investigate a given social–ecological system holistically, and to tease out the details of different kinds of adaptations that confer resilience to the system as a whole (Chapter 6). A third way is to search out cases in which there is periodic perturbation in the system (e.g., annual flood), and look specifically at how societies build resilience to enable them to live with disturbance (Chapter 7).

5

The strategy of the commons: history and property rights in central Sweden

LARS CARLSSON

5.1 Introduction

An understanding of the interdependencies between social and ecological systems in relation to the concept of resilience includes two basic dimensions, time and space (Holling, 1986; Holling, Gunderson, and Peterson, 1993). On a methodological level, the first has to do with the problem of what time frame should be used when analyzing the interplay between ecological and socio-economic systems. The contemporary debate about global warming is one example of this problem: should possible changes be assessed over years, centuries, or perhaps millennia? Because humans are social creatures, we have rather good historical records about the development of society. However, when it comes to the history of the interplay between social and ecological systems, the information is much scarcer. Uncertainties concerning the magnitude of deforestation in Africa exemplify this problem. No one really knows how much forest there was before colonization. There is a significant lack of forest data from that period (Gibson, McKean, and Ostrom, 1996). Moreover, even if such data were available, to what specific kind of behavior or decisions could observable changes be attributed?[1]

The other problem, the space problem, is exemplified by findings from an ongoing research project dealing with the Russian forest sector.[2] Along many important transportation lines in West Siberia and several other Russian forest regions, a systematic over-cut has been conducted (cf. Carlsson and Olsson, 1998; Carlsson *et al.* 1999). However, this local over-cutting has sometimes been 'compensated for,' and thereby also hidden in the statistics, by undercutting in other areas. The reality is that in regions such as the Arkhangelsk and Tomsk, most of the forest resource, approximately 60 percent, contains mature or over-mature trees. The ecological system has reached its climax, the so-called *K*-phase, and is vulnerable to sudden releases of resources (Ω-phase) through disturbances such as pests and forest fires.[3] What is the proper scope of analysis?

Should we only look at the over-cut areas or should we take the entire Oblasts, which are of the size of France, into consideration? In fact, it can be argued that both the over-cut areas and the pristine, over-mature forests can be understood in terms of resilience loss. Both systems have the potential of undergoing an ecological 'flip,' changing to completely new ecological systems. For example, in the most easily accessible areas where forests have been clear-cut and there have not been any regeneration efforts, land has turned into bogs, and pine and spruce have been replaced with aspen. In the over-mature areas, we find huge stands of dead standing trees ridden by pests and diseases and damaged by fire (Nilsson and Shvidenko, 1998). What scale of analysis should be utilized in order to provide a foundation for reliable conclusions, which link causes and effects of social activities and ecological changes?[4]

One might argue that it is desirable to consider historical records, regarding both social and ecological systems, that match the temporal scale of the renewal cycle of the ecosystem in focus. This is said in full awareness of the fact that all ecosystems contain several time frames and lack a single equilibrium. However, 'critical processes [of an ecosystem] function at radically different rates covering several orders of magnitude, and these rates cluster around a few dominant frequencies' (Holling *et al.*, 1993: 2). It can be argued that boreal forest records, which cover time periods of 100 to 200 years, might fulfill this criterion. (Following this rule, one can imagine the problem of applying a proper time frame to the analysis of a Sequoia forest.) Such records, however, are rare. The big hurdle seems to be finding forest data which provide quantitative and qualitative information on forests over long time frames, such as 100 years for a boreal forest. What is needed is not only historical data regarding changes in the forest resource (volume, density, species composition, etc.) but also information about how these might be connected to changes in contemporary forest policy and management practices.

This chapter demonstrates what insights can be gained by linking behavior to subsequent changes in an ecosystem. The study presented in this chapter is based on data covering an area of about 80 000 ha of boreal forest lands in Sweden for a period extending well over 100 years. The purpose of this chapter is to demonstrate how two different management systems affect the forest resource over a significant period of time – one the result of state ownership, the other of communal property rights.

5.1.1 Community-managed forests, living history

Sweden is one of the most heavily forested countries in the world and for centuries the utilization of this resource has been more or less regulated, first by medieval county laws and later by laws and regulations enacted by the king and the state (Mattsson and Stridsberg, 1981). The first national forestry law, a frame law, was enacted in 1903 (Stjernquist, 1973).

However, up until the mid-nineteenth century, forests had limited commercial value to the vast majority of farmers. With rapid industrialization, the situation suddenly changed. The first steam-driven sawmill began to produce marketable wood in 1850 and the first pulp plant opened in 1857 (Stjernquist, 1973). Industrialization speeded up the *Delimitation of Crown Land* process that had started in the seventeenth century. The purpose of this process was to designate, once and for all, the ownership of land, especially in the northern 'unregulated' areas. After 300 years the process is now regarded as complete.[5] The process delimitating crown land proceeded alongside another change in property rights, *The Great Redistribution of Land Holdings,* which was introduced in the mid-eighteenth century. During this period, much common land was privatized (Sporrong, 1998). However, these processes also triggered a policy resulting in the creation of vast areas under communal management and ownership. The first of these community-managed forests, named *forest commons*, was established in 1862, and all have endured up to the present day. Since this time, all subsequent units have used a similar organizational form. Collectively, the commons is the sixth biggest forest owner in the country. It should be noted that the organizational structure of these commons dates back to medieval times.

The primary reason for creating these units was to prevent forest companies from exploiting the forests in an unsustainable manner and thereby ruining the farmers. A secondary objective was to strengthen the local economy and thus establish a solid basis for taxation (Liljenäs, 1977). It can also be added that vast unpopulated areas were regarded as a potential weakness in times of war.

In this chapter we concentrate on changes in the use of forest resources on two adjacent pieces of land, Orsa Forest Common and Hamra State Park, which were both established in 1884 as a result of the previously described delimitation process (Fig. 5.1). Orsa Forest Common was composed of 53 301 ha of productive forest land and Hamra State Forest contains 25 669 ha. Both units were detached from an area defined as 'untouched pristine forest' in 1913. Consequently, the forest density in terms of number of equal-sized trees per hectare was similar in the two forest units (Fredenberg, 1913, 1924). Thus, we have two excellent cases for comparison applying a 'most similar systems design' (Przeworski and Teune, 1970). If we divide a forest area into two qualitatively equal parts, both will have similar attributes in terms of forest type, species composition, age distribution, soil condition, climate, etc. Therefore, if after 100 years we find differences in forest resources, they can be attributed

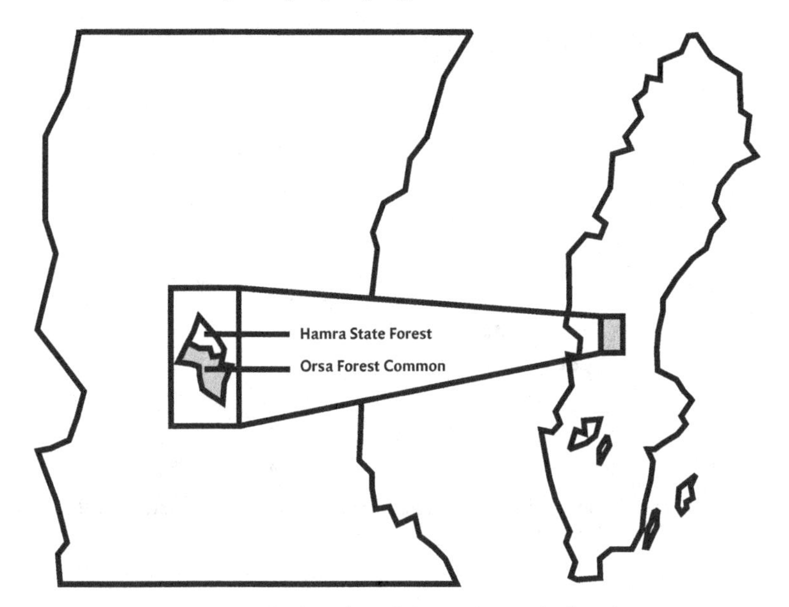

Figure 5.1 The location of the study area in Sweden.

to institutional features. In other words, if we find differences in the dependent variable, they cannot be explained by features that make the two similar. Do we find any such differences?

5.1.2 Devastation, management, and renewal

Figure 5.2 illustrates the change in forest biomass over an extended period of time. There was a dramatic reduction in the resources of the Orsa Forest Common, a 50 percent volume reduction in the 10 years prior to the turn of the nineteenth century. This period is called the 'dimension felling,' meaning that all over the country much of the old and over-mature trees were cut.[6]

However, due to deliberate regeneration, there has been a steady restoration of the resource. In the adjacent state forest, however, biomass volume continued to decrease through the next 70 years! However, neither the common nor the state forest has regained the volume it had at the turn of the nineteenth century. How can these different results be explained?

Hamra State Forest was managed by the State Forest Service in line with a centrally decided management policy based on scientific doctrines of the time (Östlund, 1993). However, it should be emphasized that both units operated under the supervision of the same type of professional foresters, educated at the same Swedish school of forestry. The manual workforce was recruited

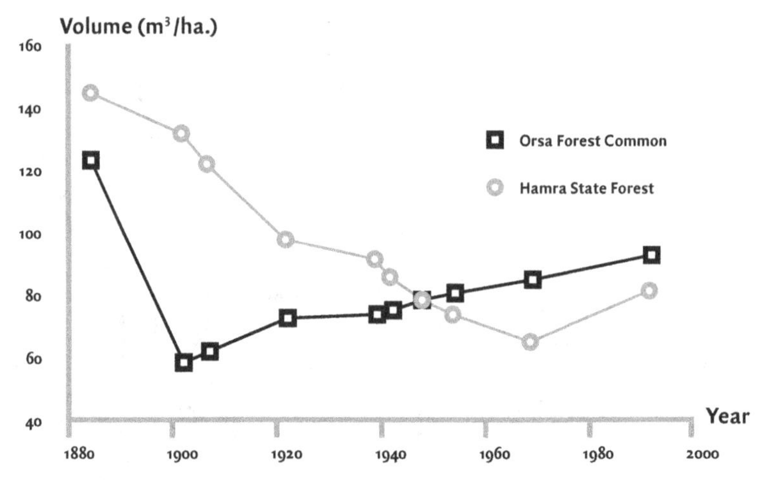

Figure 5.2 Forest resources in Orsa Forest Common and Hamra State Forest. Source: Linder and Östlund (1992: 203).

among local farmers, who provided horses and men during the cutting season. Therefore, there were no differences in the quality of labor or technology that operated in the Swedish forests at that time (Lundgren, 1984). The significant difference between the two forests is in their ownership and organization, and presumably this had decisive effects.

As with other forest commons established as a result of the delimitation of crown land, Orsa Forest Common is based on a community of single farmers comprising about 1000 individuals. Today, the ownership structure is more heterogeneous, but still farmers constitute the major group. Being a farmer was also associated with ownership of forest lands. Thus, in addition to their private lands, these farmers hold shares in a larger, jointly owned area, the forest common (Carlsson, 1995, 1999). Each farmer is the legal owner of a private piece of property, his farm, as well as a certain amount of shares in a common property. Even though many farmers jointly own the commons, they are not cooperatives. Most farms possess about the same number of shares, but larger farms often have more. It is worth noting that the shares are connected to the farm and cannot be traded. This organizational form is a mix of collective and private property rights, which may give the system some advantages, which are discussed later.

Each common is regulated by its own by-law (approved by an assembly of shareholders as well as the state authorities) and a special law governing all 33 common units. Among other things, this special law stipulates that the forest area must not be diminished and that each unit must appoint a person with professional knowledge in forestry to manage the land, typically a forester.

In each common, the assembly of shareholders elects a board from among their own members, which, together with the appointed forester, governs the unit. Although larger shareholders have more votes in the assembly, there are rules that forbid dictatorship. To a great extent, most units function as forest companies: they harvest their timber, cultivate their resources, and distribute their profits among the owners. Two principles exist for distributing the revenue, often working in parallel. The first is to distribute the profit as annual cash dividends to the farmers in accordance with their possession of shares. The second is to 'subsidize' farmers for the investments they make on their own private farms. For example, farmers are reimbursed per hectare of regenerated land for draining land, or for rebuilding barns and cowsheds. The subsidy principles are determined by the assembly of shareholders and vary among the units. In many commons a significant part of the profit is spent on the maintenance of public and private roads.

These ways of strengthening the local economy have historically been of great significance. For example, the commons have been very important contributors to the electrification of the northern countryside, the establishment of dairies, and insemination stations. The forest commons still run local sawmills, power plants, and other subsidiary enterprises for the benefit of the areas in which they operate. Over time, the policy of the commons has gradually shifted from supporting farming to supporting forestry. New policies emphasize subsidies for reforestation on a farmer's own land, the building of forest roads, and other forestry-related projects.

5.2 The behavior of the commons

The Orsa/Hamra Forest in 1884 could be characterized as a brittle, over-connected system, susceptible to sudden changes. Indeed, just such a sudden change took place in Orsa, while the transformation in the State Park was less dramatic. In order to understand why, one should consider that forestry methods in common use were those developed and governed by state forest authorities. The debate between those who advocated *patching* (cutting of small patches/glades in a stand that was naturally regenerated) and those who argued for larger cuts and deliberate regeneration programs is revealed in contemporary journals.[7] Patching had proved effective in the southern parts of Germany and Switzerland, where beech and spruce easily populated the shady openings created by this method of felling groups of trees. Since 1865, the Swedish State authorities had used this method widely, especially in the northern, forest-rich parts of the country.[8] As a consequence, artificial regeneration was uncommon. Thus, between the beginning of the 1920s until the mid-1940s there was a significant decrease in the number of seedlings and plants provided by the county forestry boards. As one observer has noted, 'The forestry boards – like the forestry experts as a whole – made propaganda for natural re-growth, to be achieved by making suitable large glades in the stands. The natural re-growth was considered preferable to forest planting on larger cut-over areas. The consequence of this of course was that the interest of forest owners in forest cultivation declined until opinion swung around' (Stjernquist, 1973: 66). By the end of the nineteenth century, large-scale management, characterized by scientific planning and the idea of 'order' (from the German *ordnung*) borrowed from eighteenth-century German forestry theories,[9] had become popular throughout Europe. At the close of the century, however, patching once again became dominant among those practicing scientific forestry (Mattsson and Stridsberg, 1981).

In the communal forest, the farmers gave their forester approval for a type of forestry that required an intensive regeneration program. First, the whole area was divided into 76 blocks and all trees thicker than 33 cm at breast height were counted. The timber at the blocks was offered to the highest bidder. As a result, vast areas were more or less clear-cut and these became subject to extensive regeneration programs. The income from the timber sales was consolidated and these funds were used for public investments such as schools, roads, and telecommunications (Kolmodin, 1953). In 1904, a contemporary observer stated that, due to its forest management, Orsa had developed from a state of poverty to 'one of the wealthiest rural communities in Europe' (*Orsa Besparingsskog 100 år*, 1980: 12).

In contrast, the state foresters at Hamra State Park more or less relied on the conventional wisdom and state-of-the-art forestry of the time, practicing patching based on a selection system following a sequence of three 20-year periods (Arpi, 1959). This type of management resulted in a more moderate harvesting, the result of which is reflected in Figure 5.2. Although no general forest law existed until 1903, this harvesting policy was, in fact, formally regulated for the state-owned forests.

Why did the farmers make such an effort to replace what had been taken away from their lands? Unlike the state managers who managed Hamra State Forest, the farmers who had established Orsa Forest Common were directly dependent on the outcome of their forest management. The inhabitants in Orsa paid no municipal taxes until the Second World War. The forest contributed all the resources that were needed for public works and services. It was regarded as their savings account, and the increment their interest rate. What would they and their successors eventually live off when all the trees were gone? The farmers lived in the forest area and presumably they also had a knowledge base to draw upon when responding to ecological feedback. Thus, it can be assumed that their successful replanting is an illustration of this knowledge and evidence that these insights are combined with a desire to generate long-term profit. The state forest can be assumed to operate according to another logic and with a significant information delay from the local level to central decision makers. This supports the colloquial insight that 'state governance does not necessarily ensure sustainable use. Given that the officials who make decisions do not have the same time horizon or interests as private owners, the general public, or the government itself, this is not surprising' (Feeny *et al.*, 1990: 11).

It should be emphasized that we do not claim that the practices employed in Orsa are 'good' and those used to manage the Hamra State Forests are 'bad.' If sustainability is defined as 'stability,' the preservation of the biomass that existed in 1880s, both groups of actors have failed. (See Holling, 1986, for a discussion of the difference between the concepts of 'stability' and 'resilience.') However, compared with the situation in 1890, the common has lost less of its volume, due, perhaps, to the conscious efforts by the farmers to regenerate their forest and thereby to halt the shrinking of the resource.

Figure 5.3 illustrates the somewhat peculiar experience of another community-managed forest in the same area. The figure covers 107 years of revenue distributed among the shareholders. The case is illustrative because it has been possible to collect historical data about direct payments to shareholders. As the Orsa Forest Common used its profit mainly for public investments, similar

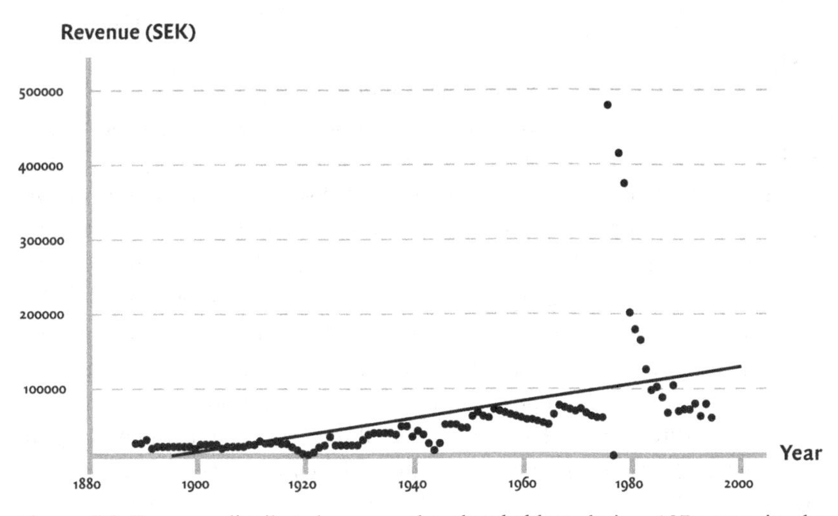

Figure 5.3 Revenue distributed among the shareholders during 107 years in the community-managed forest of Enviken (SEK, adjusted for inflation). Source: Carlsson (1995: 31).

figures from Orsa are more difficult to present. What is striking about this figure is that the economic revenue is so evenly distributed over the years. Recall that the time period covered by the figure includes a 'pre-democratic' period, the birth of liberal democracy, severe economic recession, two world wars, industrialization, as well as the birth of the information society. 'Although formal rules may change overnight as the result of political or judicial decisions, informal constraints embodied in customs, traditions, and codes of conduct are much more impervious to deliberate policies' (North, 1991: 6). The only dramatic change observed in this case is, in fact, caused by a sudden change in the taxation rules. It simply became more profitable for the commons to distribute accumulated resources among the shareholders than to have them taxed within the organization.

Although it is rare to find such detailed data from more than a single forest, recent research indicates that over a significant period of time the Swedish forest commons have generally adopted a policy based on the idea of a 'target income' (Carlsson, 1999). Such a policy has the effect that the community-managed forests have a tendency to harvest less when timber prices rise. Whether this behavior reflects a concern for the forest resource or whether it is simply an effect of the fact that farmers might prefer to sell timber from their own private lands in times with high prices is difficult to determine. Obviously, farmers have the ability to switch between two different supplies of wood in the face of changing market prices. This is coping with fluctuating prices, and enhances adaptive capacity by spreading risk through diverse management of the resource base. This mechanism both serves short-term financial need and spreads risks over a longer term.

The result is the same, however; the commons provide a sustainable yield at levels less than the annual increment (Carlsson, 1998: 85). The same type of analysis can be applied to the Orsa Forest Common and its strategy of using its forest capital to build up a foundation, the yield of which is distributed over time and for different purposes. It is in this perspective that the management practices of the Orsa farmers should be viewed. Even though the management system deliberately caused a sudden flip, a conscious regeneration or reorganization of the forest had taken place by the turn of the nineteenth century. No overuse is reported; rather, a deliberate policy to generate 'even' incomes has been practiced. It should be emphasized that the community-managed forests in Sweden are still regarded by experts as well managed and that they show an environmental concern that is in line with, or better than, those of other forest owners.[10]

Such a system, supposedly based on the judgments of farmers who live literally in the middle of their forest, is more sensitive to changes than more rigid decision systems such as that operating in the Hamra State Park. A comparison of harvesting behavior among six state parks and six community-managed forests comparable in size and located in the same county over a period of 14 years further confirms this. The comparison reveals that state management units have a tendency to act uniformly, whereas the behaviors of the community forests are much more dispersed. This probably has to do with variations in local factors that cause the community-managed units to act differently because they are in different 'phases' of their forest management. If, as research indicates, they have a target income, it would in fact be unlikely that they would have the same targets and thus the same needs for more income in the same year. Thus, their responses to changes in timber prices are different because they have different needs to reach different targets. It should also be emphasized that private forest owners act in a more 'capitalistic' way: when timber prices rise, they harvest more (Carlsson, 1995).

This hypothesis is tested in the following way. Annual harvest, measured in cubic meters of wood, is compared among the commons and state forests over a period of 14 years. Correlation coefficients are calculated by comparing each forest with all the others in the group. If a forest common (or a state forest) increases or decreases its harvest, do the others do the same?

This calculation reveals that there is no significant correlation among the forest commons. For the state forests, the situation is different. In 50 percent of the pairwise comparisons there is significant correlation ranging from 0.59 to 0.77. This means that when a state-owned management unit changes its harvesting level, it is likely that many of the other state-owned units will do the same.

5.3 Discussion

What can be learned from these examples and the contrast between state management and the collective forest management practiced by communities of farmers? One might emphasize something already discussed at length, namely that cultural, geographical, and other factors have made it beneficial for different groups to develop their own *well-tailored*, often community-based management systems which have demonstrated remarkable viability (Netting, 1981; Ostrom, 1990; McKean, 1992; McKean and Ostrom, 1995; Merlo, 1995; Berkes and Folke, 1998). In this respect, the forest commons described in this chapter are consistent with other locations (see Alcorn and Toledo, 1998, for a Mexican example). What is remarkable is that they still seem to operate quite successfully.

Private property and state property are thus only two among a number of ownership and management alternatives. In fact, communal property is often successfully combined with private property rights (Ostrom, 2000; Eggertsson, 1998). The presence of these solutions is by no means only a feature of developing countries. Community management of property also has an important role to play in advanced market economies (Buck, 1985; McKean, 1992; Arnold, 1993; McKean and Ostrom, 1995; Merlo, 1995; Eggertsson, 1998; Carlsson, 1999).[11]

The Swedish forest commons use a system with a bundle of property rights – communal property in combination with private property – that farmers can utilize. As noted, the farmers are fairly well off using this system for their livelihood. The system works because they live in the area and because governance and management systems have been adaptive to changes over time. Thus, there is a tight coupling in the context of complex systems (Levin, 1999). In fact, it has been argued that management practices in the forest commons can be analyzed in terms of transaction costs and that it is the ability to reduce these costs that explains both why the commons still exist and their relative success as competitive timber producers (Carlsson, 1999).

All successful institutional arrangements for the handling of natural resources must be adaptive. When the ecological system changes, so too should the management system. This is also true when ecological changes are caused by the existing management practice. Basic prerequisites for successful adaptation might be that rights to access and appropriate resources are well defined, that the rules are determined by the same people who are affected by them, and that monitoring and sanctioning are executed on behalf of the owners at very low costs (for a detailed discussion, see Ostrom, 1990, 1992; Bromely, 1992; Carlsson, 1999; Folke, Berkes, and Colding, 1998). Given this, we might ask what makes the farmers of Orsa and the other forest commons inclined to manage their forests in a sustainable way. The answer is found in the close linkage between the farmers and the forest common. They apply a management structure that is sensitive to changes they have caused to the ecosystem. The management of the state forest, as compared to the forest common in our example, may be understood as a means of coping with the forest resource, as shown in Figure 5.4.

Immediately after the breakdown of the brittle system, i.e., when an overmature forest had been cut, farmers initiated a deliberate regeneration program. This, combined with a harvesting policy based on a target income goal rather than profit maximizing, helped the resource recover and renew itself (arrow A in Fig. 5.4). In comparison, the policy practiced by the state caused the resource to decline for a much longer period of time (arrow B in Fig. 5.4). If forest density is a goal in itself, one can say that, in the latter case, *destruction* occurred over a longer period and density was reduced accordingly under a more rigid management system. Whether one management system is good and

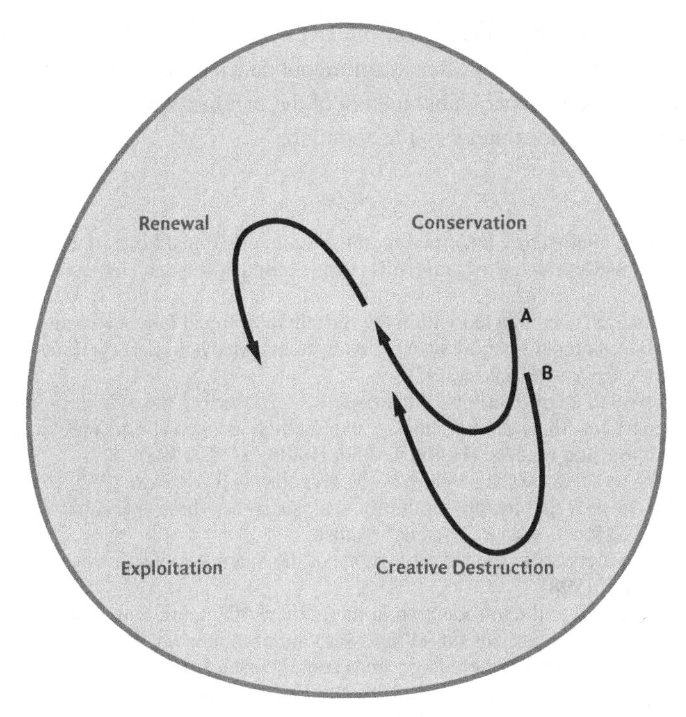

Figure 5.4 Management practice in relation to ecosystem functions.

the other is bad remains to be seen. It is a matter of two different modes of resilience; in both cases, 'the ability of [the] system to maintain its structure and patterns of behavior in the face of disturbance' is maintained (Holling, 1986: 296). Both forests have retained the characteristics of boreal forests. However, whereas state management is governed by central decisions, the behavior of the forest common is the result of local decision-making. The forest might withstand both types of management, but the commons have more to teach us in terms of successful adaptive management and thus about linking social and ecological systems. The case study discussed in this chapter demonstrates that:

- Well-organized groups of local resource users are able to successfully manage a forest resource over an extended period of time.
- Resource users that are closely connected to a resource system are in a better position to adapt to signals from the ecosystem.
- Resource systems composed of different types of property rights provide opportunities for local users to gain the benefits of each property rights system.
- In resilient social–ecological systems such as the Swedish forest commons, participants can utilize mechanisms to spread risk over time. However,

understanding how management practices affect a forest resource requires access to forest data as well as institutional data over extended periods.

- Tragedy is not an unavoidable feature of the commons; rather, its prevention is a feature of 'the strategy of the commons.'

Notes

1. This is the challenging undertaking of the the International Forestry Resources and Institutions Research Program (IFRI) that is based at Indiana University (Ostrom, 1995).
2. This research is conducted within the Sustainable Boreal Forest Resources Project at the International Institute of Applied Systems Analysis (IIASA), Laxenburg, Austria (http://www.iiasa.ac.at/).
3. This refers to a frequently used heuristic which describes the dynamics of ecosystem functions through phases: exploitation, conservation, creative destruction, and renewal (Holling, 1986; Holling *et al.*, 1993).
4. Those who advocate the so-called Gaia Hypothesis (Lovelock, 1995) would perhaps answer that the proper unit of analysis is the whole globe, because it is believed to function as a coherent organism.
5. Not when it comes to the property rights of the Saami people, however (Bengtsson, 1998).
6. With reference to the introduction, it might be worth mentioning that the situation that triggered the 'dimension felling' was similar to much of the forests in contemporary Russia where huge areas contain unexploited and over-mature stands (Nilsson and Shvidenko, 1998: 10). For a discussion about the possible benefits of introducing community management in the Russian context, see Carlsson (2000).
7. One such journal is *Skogsvårdsföreningens Tidskrift*.
8. However, due to the harsh climate here, the method was associated with negative results: see Mattsson and Stridsberg (1981).
9. For an overview of the history of German forestry, see Klose (1985).
10. This statement is based on interviews with experts from the State Forest Service in all the districts where community forests are located (Carlsson, 1995). The Swedish Commission on Collectively-Owned Forest Lands came to the same conclusion (Swedish Ministry of Agriculture, 1984: 15).
11. Compare, for instance, condominiums, neighborhood organizations, and car pools.

References

Alcorn, J.B. and Toledo, V.M. 1998. Resilient resource management in Mexico's forest ecosystems: the contribution of property rights. In *Linking Social and Ecological Systems, Management Practices and Social Mechanisms for Building Resilience*, pp. 216–49, ed. F. Berkes and C. Folke. Cambridge: Cambridge University Press.

Arnold, J.E.M. 1993. Management of forests resources as common property. *Commonwealth Forestry Review* 72(3): 157–61.

Arpi, G., ed. 1959. *Sveriges skogar under 100 år (Sweden's forests during 100 years)*. Stockholm: Domänstyrelsen.

Bengtsson, B. 1998. The legal status to resources in Swedish Lapland. In *Law and the Management of Renewable Resources,* pp. 221–43, ed. E. Berge and N.C. Stenseth. Oakland, CA: ICS Press,

Berkes, F. and Folke C., eds. 1998. *Linking Social and Ecological Systems. Management Practices and Social Mechanisms for Building Resilience.* Cambridge: Cambridge University Press.

Bromley, D.W., ed. 1992. *Making the Commons Work.* San Francisco: ICS Press.

Buck, S.J. 1985. No tragedy on the commons. *Environmental Ethics* 7: 49–61.

Carlsson, L. 1995. *Skogsallmänningarna i Sverige. (Forest Commons in Sweden.)* Research Report, TLULEA 1995: 22, Luleå University, Sweden.

Carlsson, L. 1998. The Swedish forest commons: challenges for sustainable forestry. In *Sustainability: the Challenge,* pp. 80–9, ed. A. Sandberg and S. Sörlin. Montreal: Black Rose Books.

Carlsson, L. 1999. Still going strong, community forests in Sweden. *Forestry* 72(1): 11–26.

Carlsson, L. 2000. Towards a sustainable Russian forest sector. *Natural Resources Forum* 24: 31–7.

Carlsson, L., Lundgren, N-G., Olsson, M-O., and Varakin, M. 1999. *Institutions and the Emergence of Markets, Transition in the Arkhangelsk Forest Sec*tor. IIASA, Interim Report, IR-99-021/June. Laxenburg, Austria: International Institute for Applied Systems Analysis.

Carlsson, L. and Olsson, M-O. 1998. *Institutions and the Emergence of Markets, Transition in the Tomsk Forest Sec*tor. IIASA, Interim Report, IR-98-084/October. Laxenburg, Austria: International Institute for Applied Systems Analysis.

Eggertsson, T. 1998. The economic rationale of communal resources. In *Law and the Management of Renewable Resources,* pp. 55–74, ed. E. Berge and N.C. Stenseth. Oakland, CA: ICS Press.

Feeny, D., Berkes, F., McCay, B., and Acheson, J.M. 1990. The tragedy of the commons: twenty-two years later. *Human Ecology* 18(1): 1–17.

Folke, C., Berkes, F. and Colding, J. 1998. Ecological practices and social mechanisms for building resilience and sustainability. In *Linking Social and Ecological Systems. Management Practices and Social Mechanisms for Building Resilience*, pp. 414–36, ed. F. Berkes and C. Folke. Cambridge: Cambridge University Press.

Fredenberg, K. 1913. Det ekonomiska resultatet af Hamra kronoparks förvaltning, jämfördt med samma resultat för Orsa besparingsskog åren 1884–1911. (The economic outcome of a comparison between the management of Hamra State Park and Orsa Forest Common, 1884–1911.) *Skogsvårdsföreningens Tidskrift* 6: 335–40.

Fredenberg, K. 1924. Det ekonomiska resultatet af Hamra kronoparks förvaltning, jämfördt med samma resultat för Orsa besparingsskog åren 1884–1922. (The economic outcome of a comparison between the management of Hamra State Park and Orsa Forest Common, 1884–1922.) *Skogsvårdsföreningens Tidskrift* Series B: 225–32.

Gibson, C., McKean, M., and Ostrom, E. 1996. Explaining Deforestation: the role of Local Institutions. Paper presented at the 6th annual conference of the IASCP, Berkeley, CA, June 5–9, 1996. Bloomington: Workshop in Political Theory and Policy Analysis, Indiana University.

Holling, C.S. 1986. The resilience of terrestrial ecosystems: local surprise and global change. In *Sustainable Development of the Biosphere*, pp. 292–317, ed. W. Clark and R.E. Munn. Cambridge: Cambridge University Press.

Holling, C.S., Gunderson, L., and Peterson, G. 1993. *Comparing Ecological and Social Systems.* Beijer Discussion Paper, No 36. Stockholm: Beijer International Institute of Ecological Economy.

Klose, F. 1985. *A Brief History of the German Forest, Achievements and Mistakes Down the Ages.* Eschborn: Deutche Gesellschaft für Technische Zusammenarbeit.

Kolmodin, G. 1953. Besparingsskogen (The forest common). In *Orsa*, pp. 274–371, ed. J. Boetius and O. Veirulf. Stockholm: Nordisk Rotogravyr.

Levin, S. 1999. *Fragile Dominion: Complexity and the Commons.* Reading, MA: Perseus Books.

Liljenäs, I. 1977. Allmänningskogarna i Norrbottens Län. (Forest commons in the county of Norrbotten.) Dissertation, Kungliga Skytteanska Samfundets Handlingar, No. 16. Umeå: Umeå University.

Linder, P. and Östlund, L. 1992. Förändringar i norra Sveriges skogar. (Changes in the forests of northern Sweden.) *Svensk Botanisk Tidskrift* 86: 199–215.

Lovelock, J. 1995. *The Ages of Gaia: a Biography of Our Living Earth.* Oxford: Oxford University Press.

Lundgren, N-G. 1984. Skog för export. (Forests for export.) Dissertation, Umeå Studies in Economic History 6. Umeå: Umeå University.

Mattsson, L. and Stridsberg, E. 1981. *Skogens roll i svensk markanvändning, en utvecklingsstudie. (The role of forests in the utilization of Swedish land.)* Report 32a, Umeå: Swedish University of Agricultural Sciences, Department of Forest Economics.

McKean, M.A. 1992. Management of traditional common lands (*Iriaichi*) in Japan. In *Making the Commons Work*, pp. 63–98, ed. D.W. Bromley. San Francisco: ISC Press.

McKean, M.A. and Ostrom, E. 1995. Common property regimes in the forest: just a relic from the past? *Unasylva* 46: 3–15.

Merlo, M. 1995. Common property forest management in northern Italy: a historical and socio-economic profile. *Unasylva* 46: 58–63.

Netting, R. 1981. *Balancing on an Alp: Ecological Change and Continuity in a Swiss Mountain Community.* New York: Cambridge University Press.

Nilsson, S. and Shvidenko, A. 1998. *Is Sustainable Development of the Russian Forest Sector Possible?* IUFRO Occasional Paper No. 11, ISSN 1024-414X. Vienna: International Union of Forestry Research Organizations.

North, D. 1991. *Institutions, Institutional Change and Economic Performance.* Cambridge: Cambridge University Press.

Orsa Besparingsskog 100 år. (Orsa Forest Common 100 Years.) 1980. Orsa: Orsa Sockens Allmänningsstyrelse.

Östlund, L. 1993. *Exploitation and Structural Changes in the North Swedish Boreal Forest 1800–1992.* Dissertations in Forest Vegetation Ecology 4. Umeå: Swedish University of Agriculture.

Ostrom, E. 1990. *Governing the Commons.* New York: Cambridge University Press.

Ostrom, E. 1992. *Crafting Institutions for Self-Governing Irrigation Systems.* San Francisco: ICS Press.

Ostrom, E. 1995. *The International Forestry Resources and Institutions Research Program: a Methodology for Relating Human Incentives and Actions on Forest Cover and Biodiversity.* Paper (W95I-12) presented at the Smithsonian/Man and the Biosphere Biodiversity Program (SI/MAB), International Symposium Measuring and Monitoring Forest Biological Diversity: The International Network of Biodiversity Plots, Washington DC, May 23–25, 1995. Bloomington IN: Workshop in Political Theory and Policy Analysis, Indiana University.

Ostrom, E. 2000. Private and common property rights. In *Encyclopedia of Law and Economics*, Vol. II, pp. 332–79, ed. B. Bouckaert and G. De Geest. Cheltenham, England: Edward Elgar.
Przeworski, A. and Teune, H. 1970. *The Logic of Comparative Social Inquiry.* New York: John Wiley and Sons.
Sporrong, U. 1998. Dalecarlia in central Sweden before 1800, a society of social stability and ecological resilience. In *Linking Social and Ecological Systems. Management Practices and Social Mechanisms for Building Resilience*, pp. 67–94, ed. F. Berkes and C. Folke. Cambridge: Cambridge University Press.
Stjernquist, P. 1973. *Law in the Forest, a Study of Public Direction of Swedish Private Forestry*. Lund: CWK Gleerup.
Swedish Ministry of Agriculture (Jordbruksdepartementet) 1984. *Skogsallmänningar. (Forest Commons.)* The Swedish Commission on Collectively-Owned Forest Lands, Ds Jo 1984: 15. Stockholm: Swedish Ministry of Agriculture.

6

Management practices for building adaptive capacity: a case from northern Tanzania

MARIA TENGÖ AND MONICA HAMMER

6.1 Introduction

This chapter focuses on management practices in the agroecosystem of Iraqw'ar Da/aw [irakuar da-au], the historical heartland of the agro-pastoralistic Iraqw located in the northern highlands of Tanzania. Iraqw'ar Da/aw has been inhabited since at least the late eighteenth century and is mentioned in early colonial reports for its locally developed soil and water conservation practices (Snyder, 1996). For at least 200 years, soils of limited fertility have sustained relatively high populations compared to surrounding areas, through a highly integrated agro-pastoralistic farming system (Ruthenberg, 1980).

The complexity of an agroecosystem arises primarily from interactions between socio-economic and ecological processes, and the case of Iraqw'ar Da/aw raises questions regarding the sustainability of a linked social–ecological system: why has this 'island of intensification' persisted (Widgren and Sutton, 1999)? How does it cope with the dynamics of ecosystem behavior, and how has it handled external social–ecological changes such as the post-independence rural re-settlement program in Tanzania? These social–ecological resilience issues are analyzed in this chapter, with a particular focus on management practices for coping with the dynamics of complex adaptive systems.

Sustainable management of natural resources needs to include the maintenance of biodiversity and of vital ecosystem functions and processes such as cycling of nutrients and water. Management must also include social mechanisms that receive, interpret, and process feedback signals from the ecosystems in an adaptive way (Walters, 1986; Hammer, Jansson, and Jansson, 1993; Folke, Berkes, and Colding, 1998). This sense of connectedness between humans and nature has largely been lost in conventional management approaches (Holling and Meffe, 1996). For example, the 'Green Revolution' aimed to solve the protracted food security crisis by using improved seeds, fertilizers, and pesticides, and by mechanizing agriculture. The Ujamaa villagization program in Tanzania, a comprehensive rural resettlement program, was launched in the mid-1970s. The program visioned a rural system of communities in which people were resettled in close proximity to one another to enable faster social and economic development, the latter through increased agricultural productivity and state control over production means (Kikula, 1996; Scott, 1998). However, the implementation of these large-scale programs did not take local social and ecological conditions and knowledge into account and drove people out of their evolved social–ecological systems (Pretty, 1995; Kikula, 1996). This tendency to disrupt rather than build upon existing local knowledge and experience has often worsened conditions and has led to difficulties or complete failures in introducing new knowledge and useful techniques for local resource management and utilization into local communities (Pretty, 1995; Leach and Mearns, 1996; Scott, 1998).

In this chapter, we argue that key to the persistence of the social–ecological system in Iraqw'ar Da/aw is the multitude of management practices based on local ecological knowledge and promoting and sustaining ecosystem processes and services. Combined with a nested set of institutions in which these practices are embedded, adaptive response to ecosystem dynamics is enhanced, thereby supporting resilience (*sensu* Holling, 1973) in the linked system.

Background material for this chapter was gathered during fieldwork, principally in the village of Kwermusl in the eastern part of Iraqw'ar Da/aw, using semi-structured interviews and Participatory Rural Appraisal (PRA) techniques (see Mikkelsen, 1995; Chambers, 1996; Kvale, 1996). In-depth interviews were conducted with farmers in an area where Börjeson (1999) has also performed detailed studies. These were combined with farm visits in other parts of Iraqw'ar Da/aw and key informant interviews. Existing literature on Iraqw'ar Da/aw was consulted, including dissertations, district and national reports, and other published material. The environmental history written by Lawi (1999), including a comprehensive documentation of Iraqw local knowledge, was specially useful in the production of this chapter.

The chapter begins with a description of the case-study area, some historical background, and an outline of the agroecological setting. Further, we identify and analyze local ecological knowledge as reflected in various management practices currently in use in Iraqw'ar Da/aw, and analyze how these practices are embedded in local, regional, and national institutions. Finally, we discuss key factors underlying adaptive capacity in the linked social–ecological system.

6.2 The case of Iraqw'ar Da/aw

Iraqw'ar Da/aw is located at the rim of the Rift Valley escarpment in northern Tanzania, in the Mbulu District in Arusha Region (Fig. 6.1). Iraqw'ar Da/aw

Figure 6.1 Map of Iraqw'ar Da/aw (adapted from Snyder, 1993). The Iraqw'ar Da/aw area is located in the northern part of Tanzania, near Lake Manyara. Iraqw'ar Da/aw is demarcated by the thick line and the borders of the Nou and Hassama forest reserves. Dotted lines indicate the major roads leading to the nearest town and market, Mbulu. Thin, solid lines indicate rivers and numerous rivulets of valleys. Kwermusl, the focus village of this study, is located in the central-eastern part of Iraqw'ar Da/aw.

is the core area of the Iraqw, who also inhabit most of the Mbulu District. Iraqw'ar Da/aw is the name of the area in the local Iraqw language; other names used are Mama Issara and Kainam. The area is roughly 185 km^2 and geographically isolated, surrounded by higher mountains, forests, and the Rift Valley escarpment.

Iraqw'ar Da/aw is spatially organized as seven villages grouped into two wards, Kainam and Murray. The Iraqw are agro-pastoralists, integrating agriculture with livestock. The population density of Iraqw'ar Da/aw is approximately $100/km^2$ (Börjeson, 2001), as compared with the Tanzanian mean density of $26/km^2$ (Sida, 1995). The Mbulu District has one of the fastest growing populations in Tanzania: the population increased fivefold between 1948 and 1995 (Meindertsma and Kessler, 1997). In contrast, the population increase in Iraqw'ar Da/aw was only 7 percent during the period 1957–88, due to a continuous out-migration (Snyder, 1993). Today, there is a general view in the local society that the population is growing and crowding is increasing in the area, but there are no recent data available to support this.

6.2.1 Historical development

Iraqw'ar Da/aw has been described as a small pocket of intensive agriculture based on investments in land, permanency of fields, and labor-intensive forms of cultivation and land management developed since around the late eighteenth century (Thornton, 1980; Widgren, 1999). The most common theory of the mechanisms behind the development of this agricultural system relates to geographical and political isolation, which during the nineteenth century created an overpopulation demanding intensified management (Thornton, 1980). This hypothesis is, however, questioned by Börjeson (2001), whose study reveals that the background and driving forces of agricultural intensification are difficult to disentangle.

In the early twentieth century, the Iraqw started to expand into the surrounding areas on the Mbulu plateau, into the southwest and north of Iraqw'ar Da/aw. Reasons put forward for this expansion, which continued throughout the century, are the declining power of neighboring pastoralistic tribes due to rinderpest and German pacification, and government clearings of tse-tse-infested land (Thornton, 1980). In the expansion areas, land was more available, and the land conservation practices applied in Iraqw'ar Da/aw were abandoned. Soil erosion in these parts of the Mbulu, Karatu, and Babati districts has become a considerable problem (Loiske 1995; Snyder, 1996). In the 1990s the Iraqw population was about 500 000 in Mbulu and neighboring districts (Snyder, 1996). Agriculture is mechanized and commercial to a much greater extent than in Iraqw'ar Da/aw (Snyder, 1996). Despite this, farming in Iraqw'ar Da/aw and in the expansion areas is interlinked though intricate networks of exchange and trade (Thornton, 1980; Loiske, 1999a).

6.3 The agroecosystem of Iraqw'ar Da/aw

6.3.1 Agroecological conditions

The main abiotic factors regulating the Iraqw'ar Da/aw agroecosystem are topography, soil fertility, and rainfall pattern (NSS, 1994). The landscape is characterized by steeply sloping, elongated ridges intersected by numerous valleys. The underlying bedrock is soft and highly weathered and the soils are leached, with low fertility on the slopes and slightly higher in the valleys

(NSS, 1994). The altitude ranges from 1500 m to 2300 m above sea level. Iraqw'ar Da/aw has a subhumid climate with an average annual rainfall exceeding 1000 mm, in contrast to the semi-arid parts of the Mbulu District. Iraqw'ar Da/aw has two main periods of significant rainfall and, due to the elevation, relatively low evapotranspiration and average temperatures (NSS, 1994).

Although rainfall is comparatively high, the onset and duration of rainfall may be highly variable, and the mean annual precipitation in Iraqw'ar Da/aw varies significantly from year to year (from 529 mm to 1849 mm during one 6-year period; NSS, 1994). On a larger scale, East Africa suffers from drought conditions on an irregular but recurrent basis (McGregor and Nieuwolt, 1998). In addition, El Niño–Southern Oscillation events occasionally trigger extreme amounts of precipitation in East Africa, as happened in 1997–8, causing landslides, flooding, and crop failures (Ngecu and Mathu, 1999). Other disturbances affecting agricultural success and rural livelihoods include pest outbreaks and crops or livestock pathogens (Reijntjes, Haverkort, and Waters-Bayer, 1992). In Iraqw'ar Da/aw, cutworms (*Agrotis ipsilon*) and stalk-borers (*Busseola fusca*) impose severe constraints on crop production, and diseases such as east coast fever and swarming locusts affect livestock (Snyder, 1993; Lawi, 1999).

Thus, natural resource management in Iraqw'ar Da/aw must cope with variability and unpredictability in the onset and amount of precipitation, sudden outbreaks of pest and diseases that may be coupled to drought periods (e.g., cutworms), and heavy rainstorms and fire. These kinds of disturbances are unpredictable, but may be expected as part of an environment characterized by variability. However, surprise is inevitable in any system, and the farmers of Iraqw'ar Da/aw must also be able to cope with local surprise such as the El Niño flooding and novelty such as external political intervention (Gunderson, 1999).

6.3.2 The agricultural system

The agriculture of Iraqw'ar Da/aw is non-mechanized, rainfed cultivation combined with the keeping of livestock. Production is mainly subsistence oriented and the use of chemical fertilizers and pesticides is limited, especially on subsistence crops. The farming system is labor intensive and men, women, and children participate in farm work. Women spend less time in the fields, because they also allocate considerable time to household activities (Lawi, 1999). Men are more engaged in cash-generating activities such as tree planting and the management of coffee bushes.

The agricultural landscape is a patchwork of small fields, home gardens, pasture areas, and wood lots. The average farm size in Iraqw'ar Da/aw is small, only 1.4 ha including a small pasture, according to a survey by Snyder (1996).

The pattern of rainfall allows two main planting seasons and year-round cropping (NSS, 1994). Land is traditionally classified according to inclination and by which direction it faces (ridge, eastern slope, western slope, and valley bottom – *dindirmo, intsi, genei, khatsa*). The production of maize and beans from the valley-bottom fields is crucial to the food security of the family. The west-facing slope may give two harvests per year, and the eastern slopes are more drought tolerant. To our knowledge, no survey of land allocation has been produced, but, according to Börjeson (1998), each farmer household strives to have access to all kinds of fields.

Generally, houses are located on the ridges and upper slopes, with small garden plots containing vegetables, bananas, and fruit-trees maintained near the homestead. Ridges and certain parts of the valleys, as well as areas farther away from the settlements, are also utilized for grazing on a partly communal basis. Maize is the staple crop, but sweet potatoes, beans, and wheat are also important. The farmers grow a wide variety of other crops, such as sorghum, finger millet, Irish potatoes, pumpkins, and cassava, enabling harvesting of some crops throughout the year. Important cash-crops are tobacco, coffee, bananas for local beer-brewing, and trees planted for fruits or timber (Snyder, 1996).

Livestock manure is the main source of fertilizer in Iraqw'ar Da/aw (Tengö, 1999). Cattle, goats, and sheep are commonly fed or grazed during the day and stalled at night. Manure is collected from the floor of the houses in the morning, piled and brought to the fields during planting periods. According to a household survey reported in Snyder (1996), a typical household has about five cows. In addition, households commonly keep chickens for their own consumption as well as pigs, mainly for export to the urban markets. Recycling of organic waste, through mulching, composting, and feeding cattle with residues, is another important component of the farming system.

6.3.3 Land-use changes

Although persistent as an intensive agricultural system, land use in Iraqw'ar Da/aw has not been static over time. A virtually treeless landscape in the early twentieth century became a garden-like landscape by the turn of the century, with wood lots and fruit trees (Lawi, 1999). The earlier pattern of fields on ridges and slopes and grazing areas combined with thatching grass sources as the main land use of the valleys has shifted toward an increased focus on using the valleys for farming. Although not properly investigated, this has probably decreased general access to grazing areas, and, according to Loiske (1995), elders claim that less grazing is available today and that households keep fewer cattle.

The afforestation of the area was initiated by the colonial British regime in the 1930s, and aimed to improve the fuelwood situation though planting of foreign tree species, mainly black wattle (*Acacia mearnsii*) and eucalyptus (*Eucalyptus* sp.). This project coincided with the establishment of Nou and Hassama forest reserves, to which locals had limited access.

After initial hesitation, planting of trees, especially black wattle, quickly spread in Iraqw'ar Da/aw, and private and communal woodlots became a common feature. Currently, there is no fuelwood shortage in Iraqw'ar Da/aw, and charcoal production for the market in Mbulu is a source of income.

Farmers in Iraqw'ar Da/aw claim that yields are decreasing and that it is more difficult to meet subsistence needs (Snyder, 1996; Lawi, 1999). One response to this is the expansion of maize cultivation in the more fertile valley-bottom fields. From the 1960s onwards, planting of trees for fruits and timber has been a focus of governmental campaigns in Mbulu District. The sloping fields in Iraqw'ar Da/aw are increasingly planted with trees for timber, mainly grevillea (*Grevillea robusta*), cypress (*Cypressus* sp.), eucalyptus, or fruit-trees such as pear, orange, avocado, and custard apple (Snyder, 1996). According to Snyder, farmers see tree planting as a promising land-use strategy, and grevillea is especially popular as it can be planted in the fields without competing with crops. Snyder (1996) reports that 97 percent of the households in the survey had fruit trees on their farms, and 94 percent had planted grevillea.

The range of crops used in the agroecosystem of Iraqw'ar Da/aw has varied and expanded over time. For example, maize, considered a delicacy in the 1920s, is today the most important staple crop (Lawi, 1999). Older crops, such as bulrush millet, finger millet, and sorghum, are still cultivated but on a limited scale. Irish potatoes and wheat are examples of crops introduced by the British government that were quickly incorporated into the Iraqw agricultural system. Wheat is planted on western-sloping fields, alternating with maize and beans, enabling the production of two crops per year on these fields. Farmer experimentation on a small scale with, for example, new potato varieties and soybeans is building up a wider range of crops to select from and adjust to environmental and market variability.

6.4 Local management practices in Iraqw'ar Da/aw

We have identified and listed a number of practices in use in the agroecosystem of Iraqw'ar Da/aw. To analyze these practices, we establish a framework for the identification of ecological services and functions. We distinguish between ecosystem services and processes *directly* connected to production (plant production, water and nutrient supplies), ecosystem services and processes *indirectly* connected to production (erosion control, retention of soil), and, finally, services supporting the *agroecosystem* (Table 6.1). The management

Table 6.1 *Identified management practices from Iraqw'ar Da/aw analyzed according to a framework of critical ecosystem services and processes*

	Service	Process/function	Management practices affecting the performance of processes	Performer/s	Spillover effects?
Services *directly* connected to production	Plant production	Photosynthesis	Intercropping plants with different leaf area and cover	H	No
	Water supply	Precipitation	Forest protection	C	Yes
		Groundwater recharge	Protection of water sources	C	Yes
	Nutrient supply	Nutrient recirculation	*Field*		
		Nutrient acquisition	Composting organic matter	H	No
			Applying compost as manure	H	No
			Applying cattle dung as manure	H	No
			Mulching with crop residues and weeds	H	No
			Incorporating crop residues and weeds into the soil	H/G	No
			Inter-cropping maize with beans	H	No
			Rotating beans, maize, and wheat	H	No
			Leaving N-fixating weeds in the field	H	No
			Short-term fallow	H	Yes
			Long-term fallow	H	Yes
			Trees and deeply rooted plants along and in fields	H	Yes
			Pastures		
			Baloquasi, system of communal grazing	H/G	Yes
			Rotational grazing	C	Yes
			Qasara, distribution of cattle	C	Yes

cont.

Table 6.1 (*cont.*)

Service	Process/function	Management practices affecting the performance of processes	Performer/s	Spillover effects?
		Buffer areas for dry season grazing	H/G	Yes
		Meadow as extra fodder source	G	Yes
		Weeding	C	Yes
		Burning	H/G	Yes
Pollination	Supporting populations of pollinators	Enhancing habitats by creating a patchy landscape	C	Yes
		Keeping beehives near cultivations	H	Yes
		Taboo on damaging beehives	C	Yes
Biological control	Regulating populations and species composition	Up-rooting and burning 'bad' weeds on fields and pastures	H/G	Yes
		Using certain crops to deter 'bad' weeds	H	Yes
	Supporting populations of natural enemies of pests	Promoting 'good' weeds on fields and pastures	H	No
		Inter-cropping	H	No
		Crop rotation	H	Yes
		Mulching	H	No
		Over-planting	H	No
		Picking cutworms by hand during dry periods	H	Yes
		Short-term fallow	H	Yes
		Leaving strips of vegetation along terrace edges	H	Yes
		Enhancing habitats by creating patchy landscape	C	Yes

			Taboo on birds feeding on cattle pests	C	Yes
			Burning tick-infested grass areas	G	Yes
			Work parties to remove ticks from livestock	G	Yes
Services *indirectly* connected to production	Soil formation	Weathering of rock	Planting perennial crops on contours and along terraces	H	No
		Accumulation of organic material	Mulching	H	No
			Applying compost and cattle dung as manure	H	No
			Leaving crop residues and weeds standing after harvest	H	No
			Short-term fallow	H	No
			Long-term fallow	H	No
	Erosion control	Retention of soil	Contour planting	H	Yes
			Constructing tied enhanced ridges for planting	H/G	Yes
			Mulching	H	Yes
			Keeping continuous land cover	H	Yes
			Covering bare field with grass	H	Yes
			Leaving crop residues and weeds standing after harvest	H	Yes
			Planting perennial crops on contours and along terraces	H	Yes
			Mixing crops with different rooting depths	H	Yes
			Leaving strips of vegetation along terrace edges	H	Yes
			Slaqwe, communal labor system for heavy tasks	G	Yes

cont.

Table 6.1 (*cont.*)

Service	Process/function	Management practices affecting the performance of processes	Performer/s	Spillover effects?
Water regulation	Control of run-off	Constructing cut-off drains and sluices	H/G	Yes
	Infiltration	Planting perennial crops on contours and along terraces	H	Yes
	Evapotranspiration	Terracing	H/G	Yes
		Mulching	H	Yes
		Leaving crop residues and weeds standing after harvest	H	Yes
		Constructing tied enhanced ridges for planting	H/G	Yes
		Sowing and planting in rows perpendicular to the slope	H/G	Yes
		Mixing crops with different rooting depths	H	Yes
Microclimate stabilization	Retention of water and nutrients	Mixing perennial and annual crops	H	No
	Regulation of temperature	Mixing crops with different rooting depth	H	No
	Regulation of solar radiation	Mulching	H	No
		Drainage ditches in the valley fields	H/G	Yes
		Planting on tied enhanced ridges	H/G	No
		Leaving crop residues and weeds standing after harvest	H	No
		Leaving strips of vegetation along terrace edges	H	No
		Planting shading trees	H	No

			Mulching around trees and coffee bushes	H	No
			Manual irrigation	H	No
			Leaving weeds for shade during early crop development	H	No
Services supporting the agroecosystem	Resilience	Self-organization	Enhancing species diversity of cultivated plants	H	No
		Biological diversity			
		Landscape diversity	Promoting 'good' weeds on fields and pastures	H	Yes
		Discharge of variation			
		Fluxes of nutrients, water, energy, and information	Protection of species	C	Yes
			Protection of habitats	C	Yes
			Short-term fallow	H	No
			Long-term fallow	H	No
			Landscape patchiness	C	Yes
			Pulse grazing	G	Yes
			Burning of pastures and forest areas	C	Yes
			Cattle as a mobile link	H/G	Yes

Costanza *et al.* (1997); Daily (1997); de Groot (1992); see further description in text. H = household, G = group, C = community.

practices found in Iraqw'ar Da/aw are classified according to which processes and functions of the ecosystem they interact with. The practices are also analyzed according to who performs the task and whether or not management also affects neighbors or other individuals or groups (termed spillover effects). In particular, the practices we identified cluster around the management of nutrient recycling and regulating water flow to protect the soil. Table 6.1 presents a summary of the services provided by the agroecosystem and how they are interlinked with management practices.

6.4.1 Practices directly connected to production

6.4.1.1 Plant production and water supply

By intercropping plants with different leaf area and cover, and with different sunlight needs, the efficiency of the *photosynthesis* in a field is increased (Ezumah and Ezumah, 1996). In Iraqw'ar Da/aw, maize, beans, pumpkins, and Irish potatoes are commonly found in the same fields. The water supply in wells and aquifers is good throughout the year and, even during the driest periods, the many wells and springs in the valleys do not dry out (NSS, 1994). Cropping in the valley bottoms enables utilization of moisture from the morning fog and the high water table. Long-term access to safe water is also strengthened by the traditional protection of water sources, which is connected to the belief that an evil spirit lives in aquifers and wells (Lawi, 1999). Village by-laws in Tanzania also allow the local village government to punish misuse of water sources (Box 6.1).

6.4.1.2 Nutrient supply

Nutrient recirculation in the system is enhanced by the use of livestock as nutrient processors (Smaling and Braun, 1996). In addition to cattle dung, other organic matter such as weeds, leaves, crop residues, household waste, pigs' manure, and harvested vegetation from wet valleys is composted as fertilization for the fields. Manure is brought to the field mainly during cultivation, planting or weeding, depending on manure quality and access to labor power. Multi-cropping, i.e., inter-cropping, crop rotation, and mixing crops of different duration, leads to more efficient use of available nutrients and soil moisture (Ezumah and Ezumah, 1996). It also improves *pest control* and soil conditions for growth (Reijntjes *et al.*, 1992).

Pastures are the key source for the *acquisition of nutrients* for the arable land (Tengö, 1999). Thus, careful management of cattle and the grazing areas

Box 6.1 Complementary enforcement of communal resource use

In the traditional Iraqw society, the council of elders constitutes the law enforcement in the community, while the severity of a prospective punishment is reached by consensus in *kwasleema*, the public meeting (Loiske, 1999a). Violators of the Iraqw rules, such as setting fire on hillside grazing, polluting a water source, or cultivating in a valley protected for pasture or thatching grass, can be fined beer, a goat, or even a bull. Violators can also be sanctioned by social isolation and exclusion from cooperative institutions. The most severe punishment after repeated misbehavior is expulsion from the community (Thornton, 1980; Lawi, 1999). According to Loiske (1999a), expulsion of a family was committed at least as late as 1995. The elders also act to solve conflicts in the society, whether regarding land or social issues.

In the modern system, by-laws regulate and enforce the use of some local natural resources. By-laws are recognized and approved at the district and national level, but primarily enforced at the local level, by the village council. By-laws can be the same for all villages in the country, in the district, or the ward, but they may also be developed for specific cases, for example regarding local access to resources in a forest reserve. The village council may in such cases formulate a proposal for a by-law, which will be sent to the district council for approval. If the proposal passes this instance, it is forwarded to the national level to gain legal force. In the village of Kwermusl, there are environmental by-laws concerning, for example, the protection of water sources and trees on village land and restrictions on grazing and burning. Violation of by-laws can lead to monetary fines and/or a short period in jail.

Several features of the modern by-laws and the traditional Iraqw laws overlap. For example, the thatching-grass areas in the valleys that are under the control of the local elders are also protected from cultivation and misuse by by-laws regarding the management of water sources and prohibitions on cultivating land adjacent to water sources. Informants in the realms of both the traditional and the modern systems claim that there is no conflict between the traditional and the modern enforcement systems, and that there are many similarities in which resources are protected. The main discrepancy lies in the form of the punishment. Informants also claim that a culprit who refuses punishment according to one system can be committed to the other system. Also, a person caught rule breaking may ask to be heard by one of the two systems. Thus, it seems that the two systems of enforcing the proper use of certain resources exist in parallel, with limited interaction but occasionally drawing on one another.

nurtures the *nutrient supply* for the arable land. Household access to cattle is enhanced through the cattle-distributing institution *qasara* (see explanation below), thereby increasing the capacity of farmers to restore soil fertility on their fields. Generally, five different types of pastures can be identified in Iraqw'ar Da/aw: the homestead pasture, neighborhood pastures, grazing areas within a section, distant grazing lands within the village, and the forest-based pasture in the outskirts of Iraqw'ar Da/aw. Various degrees of management and control characterize these different types, ranging from the well-tended grazing plot next to the household buildings to the far away communal lands, where practically no management is used. The farmers of Iraqw'ar Da/aw practice small-scale rotation when grazing their herds as well as seasonal rotation of pastures by reserving land for dry season grazing. Occasional burning of pastures kills parasites, removes old vegetative matter, and releases plant nutrients, allowing regeneration of vegetation more palatable and nutritious for livestock (Niamir-Fuller, 1998).

Nutrients are acquired, or re-introduced, to the system through the use of deeply rooted trees planted in or in connection to the fields. The trees use nutrients from deeper layers in the soil, which is beneficial to the crops when the trees shed their leaves (Smaling and Braun, 1996). Fallowing or the use of nitrogen-fixating plants also improves the nutrient status of a field. When a field is considered to be 'tired,' it can be left fallow for 3–4 years. Maize is commonly either inter-cropped or rotated together with beans, and several species of nitrogen-fixing weeds are left on fields and pastures (Tengö, 1999).

6.4.1.3 Biological control and pollination

Pests and pathogens impose a heavy constraint on both crop and livestock production in Iraqw'ar Da/aw. Access to veterinary services and pesticides is limited due to lack of money and extension, so farmers must rely on their own knowledge and labor working in synergy with natural processes to control the population level of unwanted species. Measures taken to prevent pest infestation on cropland are crop rotation, multiple cropping, and mulching (Reijntjes *et al.*, 1992). Attempts to control pest development include manual picking of the harmful creature such as cutworms on maize and ticks on cattle, irrigation to prevent cutworm larvae development, and burning of pastures (Lawi, 1999).

In Iraqw'ar Da/aw, the small fields and wood lots interspersed with pastures, bushlands, and tree-rich homegardens form a patchy landscape. Together with the practice of leaving strips of natural vegetation between fields as borders and erosion protection, this creates and enhances habitats that *support populations of natural enemies of pests and pollinators* (Reijntjes *et al.*, 1992; Daily, 1997). *Pollination* is an ecosystem service necessary for fruit trees, vegetables, and wild flora that is provided by domesticated bees and wild species of insects and birds (Nabhan and Buchmann, 1997). The main reason for keeping beehives is honey production, but pollination provides an additional bonus that may be important to the success of fruit production in Iraqw'ar Da/aw.

6.4.2 Practices indirectly connected to production

6.4.2.1 Soil formation

Fertile soil is a critical resource that can be considered non-renewable, due to the long time necessary for its regeneration (Baskin, 1997). The farmers of Iraqw'ar Da/aw have an intricate system of collecting cattle dung and composting residues and weeds. When applied to the field, this manure contributes to the *accumulation of organic matter* in the soil. Soil preparation of the valley-bottom fields in Iraqw'ar Da/aw includes a deep tillage mixing of soil with farm residues as well as covering the surface with uprooted vegetation, a procedure called 'making the land fat' (Lawi, 1999). In a soil survey of the Mbulu District, the topsoil in Iraqw'ar Da/aw was estimated to contain 4–6 percent organic matter content, which is considered high (NSS, 1994). High organic content of the soil promotes beneficial soil properties such as high porosity, increased cation exchange capacity, and a rich soil biota (Giller *et al.*, 1997). Thus, these practices actively create fertile soil (Netting, 1993).

Chemical fertilizers are available at nearby cities, but are expensive and thus not very accessible to the farmers of Iraqw'ar Da/aw. There is also a resistance by many farmers to using chemical fertilizers as they are considered to dry out the soil and burn the crops if the rains are not sufficient.

6.4.2.2 Erosion control, water regulation, and microclimate stabilization

The beneficial soil properties above increase *infiltration* of water and the *retention of water and nutrients* in the soil (Pimentel and Kounang, 1998). *Retention of soil* is also affected by management practices governing field preparation and planting. Planting practices in Iraqw'ar Da/aw are adjusted according to field type and crop. Sweet potatoes, for example, are planted on enhanced ridges tied at the edges that increase infiltration and control the flow of run-off, which is especially important on sloping fields. Through ridging, terracing, contour planting, and construction of sluices and cut-off drains, the farmers construct a micro-landscape of structures that dampens the effect of heavy rainfall and controls the erosive potential of surface run-off. This also contributes to *microclimate stabilization* (Reijntjes *et al.*, 1992). Even though cultivation is performed on steep areas, erosion is a limited problem in Iraqw'ar Da/aw, especially compared to other parts of the district (NSS, 1994; Meindertsma and Kessler, 1997).

The way crops are harvested by the farmers of Iraqw'ar Da/aw limits soil exposure to erosion. Crop residues of, for example, maize and sorghum are commonly left in the field after harvesting, which limits the interference of the soil microclimate. Other residues are brought to feed livestock, and the remnants are composted or brought back to the field and used as mulch. Bare soil, for example after the harvesting of sweet-potatoes, is covered with slashed weeds or stalks. Keeping the soil covered with plants or vegetative matter dampens the high kinetic energy of rainfall and binds the soil. In addition to this, the management of residues and weeds binds nutrients in organic matter, which protects it from leaching and makes the nutrient stock more manageable (Reijntjes *et al.*, 1992).

6.4.3 Services supporting the agroecosystem

Biological diversity in the system is enhanced by the wide diversity of crop species and varieties used to increase food security and enhance nutrition in the subsistence-oriented agriculture of Iraqw'ar Da/aw. The cropping pattern is adjusted in time and space, and involves selecting among a variety of crops according to season, crop characteristics, and field conditions such as inclination, fertility status, and wind exposure. In addition, there seems to be an extensive knowledge about the utilization of non-cultivated plants. For example, weeds are classified as 'good' or 'bad.' 'Bad' weeds are quickly propagating and spreading species that compete with crop needs, such as *sigin (Digitaria scalarum)*; they are uprooted and burned. 'Good' weeds that are more easily controlled are used as a tool in agriculture, for instance as shade for immature crops (Snyder, 1993), as a binder of soils after harvest, as a biomass storage medium for nutrients, and as supplementary food (Tengö, 1999). Examples of good weeds are *nii (Commelina sp.), mnafu (Solanum nigrum)*, and *tangi (Kedrostis hirtella)*. Wild vegetation is also used as an indicator of plot fertility. Species with shallow roots and broad leaves, preferably deciduous and occurring in considerable density, are considered to 'fatten the land' (Lawi, 1999); unfertile soil is indicated by poor vegetation cover and species such as *slarhama (Pteridium aquilinum)*.

In addition to the use of cultivated and non-cultivated biological diversity, Lawi (1999) describes various social taboos that conserve biodiversity in Iraqw'ar Da/aw (see Colding and Folke, 2001). One example regarding *protection of species* is a taboo on killing certain birds, such as the Ox-Pecker (*Buphagus erythrorhynchus*), which consumes ticks on cattle. Other societal prohibitions regard the destruction of large isolated trees in the landscape, mistreatment of bees and beehives, and the consumption of immature offspring of both livestock and wild species (Lawi, 1999). *Protection of habitats* occurs on both long-term and temporary bases. Certain groves and swampy valley areas with high abundance of thatching grass are examples of permanent reserves;

the seasonal restriction of grazing is an example of creating reserves that are dynamic in time and space.

Habitat protection increases *landscape diversity*, which is also enhanced by the mixed land use of fields, pasture, and wood lots on the hills of Iraqw'ar Da/aw. This pattern of patchiness intersected by *long-term and short-term fallow* areas may be an important factor in promoting the spatial resilience of the region by providing sources of seeds, plants, and other organisms necessary for regeneration after disturbances (Bengtsson *et al.*, in press). Rotational grazing and pasture burning also create landscape diversity in time and space. Intense pulsing of grazing may maintain long-term resilience and function under a wide range of climate conditions (Perrings and Walker, 1995).

An important component for building resilience of the linked social–ecological system is the local ecological knowledge generated on processes and patterns in the agroecosystem. To be able to use this accumulated knowledge in management, the practices developed must be embedded in institutions that allow for adaptive response to feedback signals. In the following sections, the institutions that embrace management practices and ecological knowledge in Iraqw'ar Da/aw are described, as well as how access to natural resources is regulated and enforced.

6.5 Institutions for managing the Iraqw'ar Da/aw agroecosystem

Traditional natural resource management in Iraqw'ar Da/aw is embedded in a political system based on consensus meetings and the authority of the elders in the society. Further, it is a part of a worldview in which spiritual beings reside in the landscape and affect the status of the natural resources. Since the independence of Tanzania, and especially through the villagization program during the 1970s, the Iraqw political system co-exists with the national political structure of governmental and administrative authority at the village and ward levels (with village councils and chairman, subvillage leaders, and a politically elected secretary). This modern political system cuts differently through the Iraqw society, and is anchored in the Christian community of Iraqw'ar Da/aw.

The significance of the elders and the belief in the worldview that they represent are declining due to the rise of modern political authorities, the spreading of Christianity in the region, and the view of young people that the traditions are old-fashioned and not compatible with development (Snyder, 1997). However, in the following section we outline the institutional framework for managing natural resources, focusing on the inherited Iraqw system rather than the modern system. Although the official authority may lie with the modern village government linked to national governmental law and enforcement bodies, it is the traditional worldview embedding the Iraqw institutions and management practices that has created the landscape and the co-evolved social–ecological system evident today. The majority of the institutions described below are still in use and represent important components for maintaining household food security and resource use.

6.5.1 Spatial and social structures

Iraqw'ar Da/aw is traditionally divided into three clearly defined spatial units, depicted in Figure 6.2. The *aya* (plural *ayemo*) is the largest political and territorial unit (Thornton, 1980). Iraqw'ar Da/aw consists of 7 *ayemo* that are each divided into sections that originally related to clans among the Iraqw, a pattern that today is diffused (Thornton, 1980; Lawi, 1999). Each section consists of a number of households, the most important social and economic units (Snyder, 1996). During the villagization program in Tanzania, the relocation of people in Iraqw'ar Da/aw was not as extensive as in other parts of Tanzania, and in many cases the earlier borders of *ayemo* and sections remained the same after the villagization (Snyder, 1993; Loiske, 1999a).

The public meeting (*kwasleema)* where decisions are made on the basis of consensus plays a central role in the Iraqw political system. The *kwasleema* is comprised of all male heads of households in an *aya*, and is the chief body undertaking collective action (Thornton, 1980). The chairman of the *kwasleema* is the *kahamusmo*, who represents the community in the outside world (Lawi, 1999). The position of the *kahamusmo* is inherited, and he also provides the link between the people and the ritual expert, *qwaslarmo*, who may be a man or a woman. The *qwaslarmo* communicates with the spirits and traditionally had a very important role in Iraqw society, particularly in periods of crisis and disruption such as drought or epidemics threatening people or livestock (Thornton, 1980).

The *aya* also has a council of wise elder men who have regular meetings regarding the ritual status of the land, drought issues, diseases, and land fertility (Lawi, 1999). Each section of an *aya* has representatives in the council of elders, and the elders of a section meet occasionally to discuss local issues (Snyder, 1993). Women and youth are also organized, with elected speakers (Snyder, 1997; Loiske, 1999a).

6.5.2 Access to natural resources

According to Lawi (1999), the Iraqw traditionally recognized three ways of gaining access to natural resources: by private development, by collective

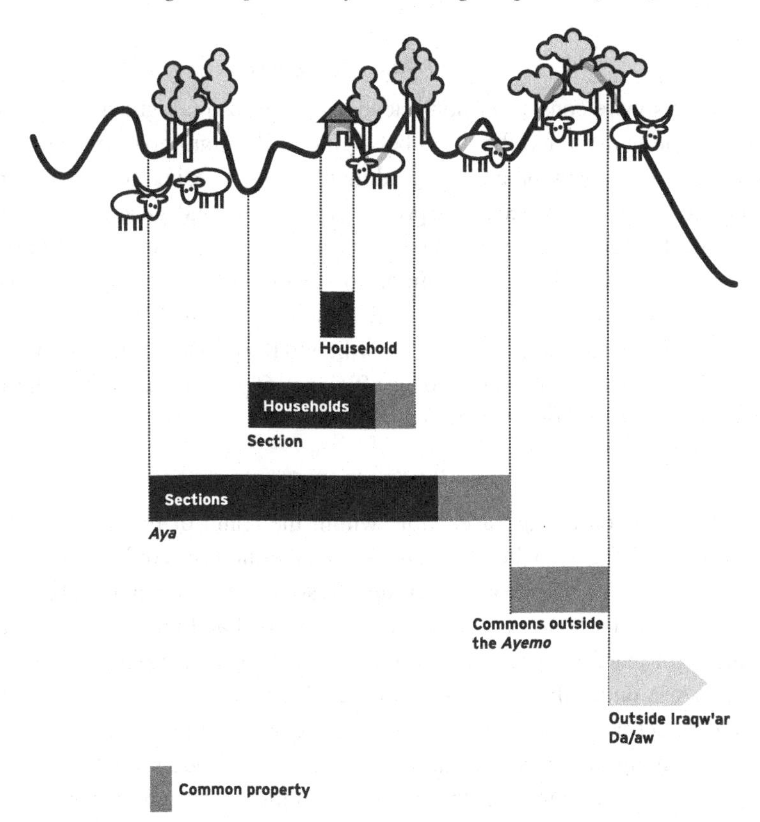

Figure 6.2 The figure outlines the landscape in Iraqw'ar Da/aw, with ridges intersected with valleys. To the east, the Rift valley Escarpment steeply slopes down some 700 m to the lowlands below. The household represents the smallest bounded social and economic unit. Every household belongs to a section. Within the section, each household has private fields and a privately maintained small pasture. In addition to this, there are areas for common use by the whole section, indicated as common property in the figure. Several sections together form an *aya* (plural *ayemo*), which is the largest territorial and the most important political unit. As in the sections, there are communal areas in the *aya* that are accessible to all inhabitants irrespective of which section they belong to. Outside the seven *ayemo* of Iraqw'ar Da/aw, further communal areas can be utilized by all people in the community. Resources outside Iraqw'ar Da/aw can be accessed through social networks (see further in the text).

control and use, and by limited access to resources in largely undomesticated or external spheres. The undomesticated spheres are the areas on the outskirts of Iraqw'ar Da/aw, accessible to all households but under limited collective control. External areas are located outside the borders of Iraqw'ar Da/aw. Natural resources in external areas can be accessed through channels of exchange and trade.

6.5.2.1 Private development

The private sphere consists of homestead buildings, a small grazing plot connected to the houses, and the fields belonging to the household. As shown in Table 6.1, management practices regarding planting, weeding, harvesting, and soil protection are controlled by the private owner of the land, generally the male head of the household, and performed by the household inhabitants. In addition to customary land rights, it is common to borrow land on a long-term basis that can be extended over generations (Börjeson, 1999); the borrower then has user rights and determines land use. According to Börjeson, this flexible system of long-term loans of land increased the efficiency of productive land use during periods of out-migration from Iraqw'ar Da/aw.

6.5.2.2 Collective control and use

Areas for common use are accessible within the realm of both the *aya* and the section (see Fig. 6.2), but there are also smaller units of common property, such as those within neighborhood groups. Resources under communal control are pastures, water sources, woodlands for fuelwood and building poles, and wet areas for thatching grass and reeds for weaving (Lawi, 1999). Communal resources are under the supervision of the elders in the sections or *ayemo*. The elders act as stewards of the communal land, monitoring its status and fertility, praying and communicating with the ritual leaders for better rains and prosperity. They announce the opening and closure of temporally restricted grazing areas and enforce the use of collective resources (see Box 6.1).

According to Lawi (1999), the local elders were also responsible for maintaining the balance between communal and private land, through permitting expansion and opening of new cropland and solving conflicts over land use. Since the 1970s, all land belongs to the village and the village council is in charge of the use and development of communal lands. Lawi (1999) claims that this transfer of power and control over local resources has deeply disrupted the local system of environmental control and management. The control that the elders have over, for example, the utilization of communal grazing areas is limited today, and an open access situation seems to be developing in the pastures on the outskirts of the *ayemo* or villages. At the section or subvillage levels, however, the elders still have some authority and, as shown in Box 6.1, traditional rules and penal codes are co-existing with the modern rules.

6.5.2.3 Undomesticated and external spheres

Outside the *ayemo* lie the undomesticated areas on the outskirts of Iraqw'ar Da/aw. These areas are utilized to a limited degree, serving as additional grazing areas and to supply construction poles and fuelwood for all inhabitants of Iraqw'ar Da/aw. The major share of the undomesticated areas is in the Hassama and Nou forests reserves. Legal access to the reserves by the local people was until recently very limited. The restrictions on utilization by the adjacent villages have been relaxed since 1998, and now limited grazing, collection of reeds and grass for handicraft, and the keeping of beehives are allowed.

The areas beyond the forest and Iraqw'ar Da/aw historically belonged to ethnic groups other than the Iraqw. Access to resources in these spheres could be gained through the exchange of goods (Lawi, 1999). As a result of outmigration during the twentieth century, the Iraqw inhabit most of the bordering areas, and each family in Iraqw'ar Da/aw belongs to an intricate network of relatives, friends, and other contacts living in the surrounding area. This is described further in the next section.

6.5.3 Institutions for cooperation and risk sharing

Within the spatial spheres of resource access outlined above, institutions for cooperation have been developed for dealing with local neighborhood resources and problems. As noted in the last two columns of Table 6.1, practices that have a spillover effect on other farmers and herders are commonly carried out in groups or by the community. This is especially true with regard to the management of pastures and soil and water conservation practices. As described above, communal lands are utilized for grazing to a large extent, and the actions of individual farmers have an obvious influence on the resource use of other farmers. A hillside used for cropping commonly includes the fields of several farmers, and management to improve the capacity of the agroecosystem to control the flow of water and retain soil and nutrients is clearly a joint interest.

Several institutions for cooperation and risk sharing exist within neighborhood groups. *Baloquasi* is a system of communal grazing with mutual herding and management of the grazing areas. Another example is *slaqwe*, a form of work party organized especially during the laborious planting periods in June–July (Loiske, 1999a). Ridging, mulching, and terracing are examples of practices from Table 6.1 that are carried out during *slaqwe*. The work party is supervised by the elders, and provides an opportunity for spreading and sharing agricultural skills and knowledge. A cooperative institution applied on a larger scale in the society is *qasara*, which concerns the distribution of cattle. According to *qasara*, cattle can be lent on a long-term basis to farmers who do not own enough cattle on their own. This optimizes access to manure and milk as well as the use of grazing areas, as a person with several cows but limited access to pastures can lend cattle to someone with better access to productive grazing lands.

One example of an institution that links Iraqw'ar Da/aw to the larger Iraqw society is the *inaoha* described by Loiske (1999a) and Lawi (1999). *Inaoha* is a system of maize exchange between farmers of Iraqw'ar Da/aw and friends and relatives in surrounding areas. Climatological factors create a displacement in time between the harvest periods of Iraqw'ar Da/aw and the drier and hotter lowland areas. When crops are harvested in the lowlands, the farmers from Iraqw'ar Da/aw visit their lowland contacts, who share some of their harvest. This favor is repaid to lowland farmers during the harvest period in Iraqw'ar Da/aw. The exchange is performed also during periods when food availability is abundant, to secure the social ties for periods of food scarcity. *Inaoha* relies on a long-term commitment: farmers cannot expect to get the same amount in immediate return, but can expect support and sharing of scarce resources in times of crisis. The *inaoha* contacts are also used for the exchange and trade of other goods such as livestock, tobacco, salt, and iron hoes. Loiske (1999b) claims that an Iraqw family can belong to a network of five to 25 other families and that their channels of exchange include different ecological zones to ensure access to goods during lean periods.

In Iraqw'ar Da/aw a farmer thus has a number of alternatives when responding to variations in resource availability. For example, a farmer who has low fertility on his field can choose to lay some land fallow or use composted manure to increase the nutrient status. There is also the possibility to borrow cattle to increase access to nutrient-rich manure, or to borrow fertile land from someone less needy (Börjeson, 1999). If the rains are late, a farmer household can rely on the diversity of private crops and field types, but also on the security network of *inaoha*.

Also, the feedback mechanisms are in a nested setting. Feedback can be interpreted using ecological knowledge accumulated both by individuals and within the various societal spheres, as the institutional memory provided by the elders' experience. For example, pastures are monitored and managed on an individual, private level and also in a larger context, nested from neighborhoods up to the village level. The different types of pastures create possibilities for micro-mobility and spatial and temporal flexibility in resource use, practices that have been found among pastoralists in sub-Saharan Africa (Niamir-Fuller, 1998). Even though collective management and action at a larger scale are limited today in Iraqw'ar Da/aw, there are still possibilities to respond to feedback using successive spheres of use intensity, and information can be spread among the different users.

In addition, trial-and-error monitoring enforces or discourages the continued use of management practices, e.g., ridges for planting. Snyder (1996) describes a situation in which some young farmers removed the terraces on their fathers' land to allow them to cultivate another row of crops. However, all the planted seedlings were washed away during the rain periods that followed, and the benefit of terracing became obvious. In the last section of this chapter we discuss how institutions, including the management practices, may contribute to the adaptive capacity of a linked social–ecological system.

6.6 Factors promoting adaptive capacity in Iraqw'ar Da/aw

Iraqw'ar Da/aw has persisted as an intensive system in a variable climate during periods of social changes such as the villagization following independence (Snyder, 1996; Lawi, 1999; Börjeson, 2001). This indicates adaptive capacity and flexible management able to achieve sustainable use of local resources. Two main factors promoting adaptive capacity found in Iraqw'ar Da/aw are: (1) the extensiveness of management practices directed toward functioning of the system rather than merely resource output, including indirect processes provided, for example, by wild plant species and distant grazing and forest areas; and (2) a decentralized but nested system of institutions that buffers local disturbance and allows for response to feedback signals on several levels.

6.6.1 Management practices for agroecosystem function

In our analysis of the agroecosystem management in Iraqw'ar Da/aw, we claim that the multitude of management practices used by the farmers sustain and enhance the critical ecosystem processes and services necessary for the agro-ecosystem, listed in Table 6.1. The behavior of dynamic ecosystems has been described as having two general phases: the 'frontloop' and the 'backloop' (Berkes and Folke, 1998). These terms relate to the adaptive cycle of eco-system dynamics developed by Holling (1986). The frontloop describes the succession of ecosystems, with phases of exploitation and conservation, whereas the backloop represents the rapid processes of release and renewal triggered by a disturbance. During the backloop, disturbance may cause the system to shift from one stability domain, with a specific set of structuring and reinforcing processes, to another. By allowing and enhancing critical processes that keep the agroecosystem within a particular stability domain, a flip to a domain less valuable to the farmers in Iraqw'ar Da/aw may be prevented (Holling, 1995).

As elaborated above, the management practices in use in Iraqw'ar Da/aw create structures that facilitate processes such as plant growth, nutrient circulation, and decomposition. The Iraqw farmers take advantage of local variability in time and space by using a multitude of crops and varieties and by mixing land use instead of trying to homogenize the landscape and control variation. We suggest that during the frontloop phases, these management practices strengthen and increase the efficiency of critical processes for biomass production, such as the provision of light, water, and nutrients. Also, the farmers build capacity to adapt to, or decrease the effect of, expected disturbances such as heavy rainstorms, pests, or drought. Such practices, that frame the impact of disturbances, are directed toward soil and water conservation, biological control, and the diversification of crops and fields.

Management of the backloop, the phase of regeneration and succession in ecosystems, may decrease capital loss after a disturbance. It can also help utilize resources such as soil nutrients, organic matter, or moisture during a new phase of exploitation. A majority of the management practices in Table 6.1 are directed toward fast processes at a small or intermediate scale, such as seasonal crop production on individual fields, protection of soil and crops from heavy downpours, and rotational grazing of cattle. However, soil and water conservation practices, although directed toward fast and intermediate processes, also affect slower variables, such as the long-term generation of fertile soils.

In managing their resources, the farmers take advantage of small-scale disturbances and even create them, for example by burning pasture and harvesting crops from a field. These disturbances trigger a release, wherein capital such as nutrients and moisture, earlier bound in biomass, is set free and thereby accessible for crops or cattle. Several management practices also concern capturing released capital for reorganization such as multi-cropping, which takes advantage of nutrients released after the harvest of one crop, and mulching practices that conserve moisture. There are also practices or social mechanisms that nurture sources of renewal during the reorganization phase (Berkes and Folke 1998), for example the various types of resource and habitat reserves, and the creation of a patchy landscape.

Agricultural diversification at the scales of both the field and the landscape may have long-term benefits through the enhancement of functional diversity and structural complexity (Swift, 1997). In systems like Iraqw'ar Da/aw, diversity functions as a means of building resilience in the social system through enhanced food security, and in the ecological system by performing and maintaining critical functions. The small-scale mosaic of different landscape units may create spatial resilience for disturbances such as fires, pest outbreaks, and periods of prolonged droughts (Bengtsson et al., in press). Also, the diversity found on the species level in the agricultural landscape, including both cultivated plants and wild biodiversity and ecosystems, is a tool for a well-functioning and productive agroecosystem (Janzen, 1999). The mental system boundaries in Iraqw'ar Da/aw not only embrace the cultivated parts of the agroecosystem, but also recognize the interdependence and usefulness of the wild diversity within the system (Lawi, 1999). Additionally, Loiske (1999b) emphasizes the diversification within the regional socio-economic exchange network, regarding access to both different kinds of goods and different ecological zones. Socio-economically, the households in Iraqw'ar Da/aw combine crop production with the ranching of livestock and commonly also timber production and small-scale handcraft businesses (Snyder, 1996). Hence, the Iraqw seem to be practicing ecosystem management across scales, from the individual plot to that of entire landscapes. This requires nested institutional and organizational arrangements (Ostrom, 1990).

6.6.2 The institutional framework – decentralized but nested

The institutions outlined in earlier sections and in Box 6.1 constitute a nested set of institutions for monitoring and response related to natural resources. The household is part of the section and *aya* (village), as well as smaller neighborhood cooperative institutions. In addition, the grid of social ties, based on kinship or other personal relations, increases livelihood security through institutions like *inaoha* (food exchange) or *qasara* (cattle loans). These features embedded in the Iraqw society link the local farmer to a wider societal and ecological scale that embraces different socio-economic and agroecological settings. By taking advantage of the heterogeneity of local climate and land use, the capacity to cope with and buffer local disturbance is increased.

Tight feedback loops, indicating the outcome of actions within a perceivable time span, are essential for any adaptive change (Levin, 1999). In Iraqw'ar Da/aw, the multitude of management practices performed by household members in the private sphere has created a finely tuned system for reading and responding to changes in the landscape. In addition, the decentralized management of the private field is linked to other components of the system with a varying degree of collective control. As shown in Table 6.1, the control of management and enforcement is lifted to a higher, collective layer of the society when managing critical forcing functions in the ecosystem, such as nutrient recycling and water control. The successive spheres of collectively controlled land use provide a buffer capacity that relies on risk sharing among households at various levels in the society.

The management of cultivated land and grazing areas in Iraqw'ar Da/aw can be described as adaptive management by local people (Berkes, Colding, and Folke, 2000). We suggest that the careful and finely tuned directing of processes and fluxes, while preserving choice and opportunity through buffer mechanisms, builds on accumulated knowledge gained through trial and error over several centuries in Iraqw'ar Da/aw. Such accumulated trial-and-error knowledge is a prerequisite for preserving opportunities for adaptive capacity. However, drawing solely on knowledge based on experience at the local scale may create a social–ecological system less prepared for novelty. The possibility to acquire and filter new ideas and knowledge from other areas is vital. Many of the so-called indigenous soil and water conservation practices gathered from sub-Saharan Africa by Reij, Scoones, and Toulmin (1996) turned out to be developed elsewhere but efficiently incorporated and adopted into the local system. One way of being adaptive is to be open to influences from outside, but only to accept changes consistent with the original system. A solid base of ecological knowledge relying on a finely tuned trial-and-error use of the supporting ecosystem may provide the necessary platform for filtering external ideas. In Iraqw'ar Da/aw, indications of such filtering of external influences include the adoption of new crops on a small scale, incorporated into the overall cropping pattern of the agricultural system, as well as changes in land-use patterns in response to the decline of soil fertility.

6.6.3 Lessons learned – future perspectives

The timing, frequency, and magnitude of disturbance pulses are important factors influencing the structure and dynamics of ecosystems (Holling, 1986). Experience of regular disturbances may lead to an adjusted management scheme that factors in disturbances (see Chapter 7). In Iraqw'ar Da/aw, we conclude that the agroecosystem has been resilient toward commonly experienced disturbances such as drought periods, pest outbreaks, and erratic rainfall. The occurrence of disturbances is irregular, but the irregularity is part of a pattern that is recognized and expected. Also, these disturbances are occurring at temporal and spatial scales perceivable by the local farmers, who thus have a possibility to adjust their management and to build buffer capacity for recurrent disturbances. The nestedness of institutions on different scales creates a memory in time and space and widens the scope of what sort of feedback can be perceived by the society.

The traditional institutions are rooted in a specific worldview that can be considered a slowly changing variable in a linked social–ecological system. The community elders have traditionally been the guardians of the community's well-being, both via ritual pathways to secure rain and avoid diseases and through monitoring the resource base and solving conflicts in the society. In doing this, the elders represent an institutional memory that maintains and archives local ecological knowledge and can provide information necessary during periods of crisis and change (Folke and Berkes, 1998).

The modern village government seems to encourage private development at the expense of common property resources, which may shrink the possibilities for risk sharing and flexible resource management. Lawi (1999) claims that the recent treatment of land issues by the village council lacks transparency to the local people, leading to insecurity and suspicion. This loss of trust and of institutions for collective action and control may lead to an erosion of social resilience. If local ecological knowledge is fading along with the authority of the elders, the capacity to respond to future changes and disruption of the social–ecological system in an adaptive way may be threatened.

However, the modernized part of the society, including the village government and the Christian community, is linked to the outside world to a higher degree. The focus is on developing the society by embracing schools, health care, and increased cash-crop production. Such linkages may supply new knowledge to the system that improves its capacity to cope with novel disturbances such as market fluctuations and government policy changes. Also, as described in Box 6.1, modern and traditional worldviews co-exist and are intertwined. There is a potential for modern institutions to acknowledge and incorporate traditional techniques and knowledge. The degree to which this is done depends largely on the openness and capacity of individual leaders in both spheres.

For the social–ecological system to persist, the integrity of locally adapted systems in which management practices and knowledge are embedded needs to be protected, but not isolated, from external driving forces (Folke and Berkes, 1998). The strength of local knowledge may be in its inspiration to a multitude of management practices that reveal the complexity of the system. This holistic picture may be lost at higher levels. However, the local scale is nevertheless linked to a wider ecological, social, and economical scale on regional, national, and global levels. It is not enough to have suitable management institutions, however excellent, at the local level if they are not nested in institutions at the regional and national scales. The future survival of the agroecosystem and embedded ecological knowledge of Iraqw'ar Da/aw lies in farmers' ability to adjust to the demands of commercialization and modernization by building upon, not abandoning, existing social and ecological capital.

Acknowledgements

We would like to thank Carl Folke, Fikret Berkes, and three anonymous reviewers for constructive comments on the manuscript. In Tanzania, many people have contributed to the outcome of the fieldwork; in particular we are indebted to Basili Awett, Ward Executive Officer of Murray Ward, and Deogratius Hilu, Agriculture and Livestock Officer at Mbulu District. Also, we would like to thank Lowe Börjeson, Anna-Carin Andersson, and Simon Alexanderson for their assistance in fieldwork and discussions. This work was funded by the Swedish International Development Cooperation Agency (Sida) and the Swedish Council for Forestry and Agricultural Research (SJFR).

References

Baskin, Y. 1997. *The Work of Nature: How the Diversity of Life Sustains Us.* Washington DC: Island Press.

Bengtsson J., Angelstam P., Elmqvist T., *et al.* press. Reserves, resilience and dynamic landscapes. *Ambio.*

Berkes, F., Colding, J., and Folke, C. 2000. Rediscovery of traditional ecological knowledge as adaptive management. *Ecological Applications* 10: 1251–62.

Berkes, F. and Folke, C., eds. 1998. *Linking Social and Ecological Systems. Management Practices and Social Mechanisms for Building Resilience.* Cambridge: Cambridge University Press.

Börjeson, L. 1998. *Landscape, Land Use and Land Tenure in Mama Issara, Tanzania. Mapping a 'Traditional' Intensive Farming System.* Minor Field Study Report. Uppsala: Swedish University for Agricultural Studies.

Börjeson, L. 1999. Listening to the land. The Iraqw intensive farming system as told by a hill and its inhabitants. In *'Islands' of Intensive Agriculture in the East African Rift and Highlands: a 500-Year Perspective*, pp. 56–73, ed. M. Widgren and J. Sutton. Working Paper Series No. 42. Stockholm: Environment and Development Studies Unit, Stockholm University.

Börjeson, L. 2001. Geography and history of an intensive farming system in Tanzania. Unpublished licentiate dissertation in Human Geography, Stockholm University.

Chambers, R. 1996. *Whose Reality Counts? Putting the First Last.* London: Intermediate Technology.

Colding J. and Folke, C. 2001. Social taboos: 'invisible' systems of local resource management and biological conservation. *Ecological Applications* 11: 584–600.

Costanza, R., d'Arge, R., de Groot, R., *et al.* 1997. The value of the world's ecosystem services and natural capital. *Nature* 387: 253–60.

Daily, G.C. 1997. *Nature's Services. Societal Dependence on Natural Ecosystems.* Washington DC: Island Press.

de Groot, R. 1992. *Functions of Nature.* Amsterdam: Wolters-Noordhoff.

Ezumah, H.C. and Ezumah, N.N. 1996. Agricultural development in the age of sustainability: crop production. In *Sustaining the Future: Economic, Social, and Environmental Change in Sub-Saharan Africa*, pp. 215–44, ed. G. Benneh, W.B. Morgan, and J.I. Uitto. Tokyo: United Nation University Press.

Folke, C. and Berkes, F. 1998. *Understanding Dynamics of Ecosystem–Institution Linkages for Building Resilience.* Beijer Discussion Paper Series, No. 112. Stockholm: The Beijer International Institute of Ecological Economics.

Folke, C., Berkes, F., and Colding J. 1998. Ecological practices and social mechanisms for building resilience and sustainability. In *Linking Social and Ecological Systems. Management Practices and Social Mechanisms for Building Resilience*, pp. 414–36, ed. F. Berkes and C. Folke. Cambridge: Cambridge University Press.

Giller, K.E., Beare, M.H., Lavelle, P., Izac, A.-M.N., and Swift, M.J. 1997. Agricultural intensification, soil biodiversity and agroecosystem function. *Applied Soil Ecology* 6: 3–16.

Gunderson, L.H. 1999. Resilience, flexibility and adaptive management – antidotes for spurious certitude? *Conservation Ecology* 3(1). Available online: URL: http://www.consecol.org/vol3/iss1/art7

Hammer, M., Jansson, A.-M., and Jansson, B.-O. 1993. Diversity change and sustainability – implications for fisheries. *Ambio* 22: 97–105.

Holling, C.S. 1973. Resilience and stability of ecological systems. *Annual Review of Ecology and Systematics* 4: 1–23.

Holling, C.S. 1986. The resilience of terrestrial ecosystems: local surprise and global change. In *Sustainable Development of the Biosphere*, pp. 292–320, ed. W.C. Clark and R.E. Munn. Cambridge: Cambridge University Press.
Holling, C.S. 1995. What barriers, what bridges? In *Barriers and Bridges to the Renewal of Ecosystems and Institutions*, pp. 3–34, ed. L. Gunderson, C.S. Holling and S.S. Light. New York: Columbia University Press.
Holling, C.S. and Meffe, G.K. 1996. Command and control and the pathology of natural resource management. *Conservation Biology* 10(2): 328–36.
Janzen, D. 1999. Gardenification of tropical conserved wildlands: multitasking, multicropping, and multiusers. *Proceedings of the National Academy of Science USA* 96(11): 5987–94.
Kikula, I.S. 1996. *Policy Implications on Environment*. Dar es Salaam: Dar es Salaam University Press.
Kvale, S. 1996. *Interviews: an Introduction to Qualitative Research Interviewing*. Thousand Oaks, CA: Sage Publications.
Lawi, Y.Q. 1999. May the spider web blind witches and wild animals: local knowledge and the political ecology of natural resource use in the Iraqwland, Tanzania, 1900–1985. Unpublished doctoral thesis, Boston University.
Leach, M. and Mearns, R. 1996. *The Lie of the Land. Challenging Received Wisdom on the African Environment*. London: Villiers Publications.
Levin, S.A. 1999. *Fragile Dominion*. Reading, MA: Perseus Books.
Loiske, V.-M. 1995. The village that vanished. The roots of erosion in a Tanzanian village. Unpublished doctoral thesis, Stockholm University.
Loiske, V.-M. 1999a. *Natural Resource Management Institutions among the Iraqw of Northern Tanzania*. Working Paper Series No. 35. Stockholm: Environment and Development Studies Unit, Stockholm University.
Loiske, V.-M. 1999b. Persistent peasants: the case of the Iraqw intensive agriculture in Tanzania. In *'Islands' of Intensive Agriculture in the East African Rift and Highlands: a 500-Year Perspective*, pp. 44–55, ed. M. Widgren and J. Sutton. Working Paper Series No. 42. Stockholm: Environment and Development Studies Unit, Stockholm University.
McGregor, G.R. and Nieuwolt, S. 1998. *Tropical Climatology*. Chichester: Wiley and Sons.
Meindertsma, J.D. and Kessler, J.J. 1997. *Towards a Better Use of Environmental Resources. A Planning Document of Mbulu and Karatu Districts, Tanzania*. Mbulu District Council, Tanzania.
Mikkelsen, B. 1995. *Methods for Development Work and Research: a Guide for Practitioners*. New Dehli: Sage Publications.
Nabhan, G.P. and Buchmann, S.L. 1997. Services provided by pollinators. In *Nature's Services. Societal Dependence on Natural Ecosystems*, pp. 133–50, ed. G.C. Daily. Washington DC: Island Press.
Netting, R. 1993. *Smallholders, Householders. Farm Families and the Ecology of Intensive, Sustainable Agriculture*. Stanford: Stanford University Press.
Ngecu, W.M. and Mathu, E.M. 1999. The El-Nino-triggered landslides and their socio-economic impact on Kenya. *Environmental Geology* 38 (4): 277–84.
Niamir-Fuller, M. 1998. The resilience of pastoral herding in Sahelian Africa. In *Linking Social and Ecological Systems. Management Practices and Social Mechanisms for Building Resilience*, pp. 250–84, ed. F. Berkes and C. Folke. Cambridge: Cambridge University Press.
NSS, 1994. *Land Resources Inventory and Land Suitability Assessment of Mbulu District, Arusha Region, Tanzania*. Reconnaissance Soil Survey, Report R5.

Tanga, Tanzania: Ministry of Agriculture, National Soil Service, Mlingano Agricultural Research Institute.

Ostrom, E. 1990. *Governing the Commons: the Evolution of Institutions for Collective Actions*. Cambridge: Cambridge University Press.

Perrings, C. and Walker, B.H. 1995. Biodiversity loss and the economics of discontinuous change in semi-arid rangelands. In *Biodiversity Loss: Ecological and Economic Issues*, pp. 190–210, ed. C.A. Perrings, K.-G. Mäler, C. Folke, C.S. Holling and B.-O Jansson. Cambridge: Cambridge University Press.

Pimentel, D. and Kounang, N. 1998. The ecology of soil erosion in ecosystem. *Ecosystems* 1(5): 416–26.

Pretty, J. 1995. *Regenerating Agriculture. Policies and Practices for Sustainability and Self-Reliance*. London: Earthscan.

Reij, C., Scoones, I., and Toulmin, C., eds. 1996. *Sustaining the Soil. Indigenous Soil and Water Conservation in Africa*. London: Earthscan.

Reijntjes, C., Haverkort, B., and Waters-Bayer, A. 1992. *Farming for the Future, an Introduction to Low-External-Input and Sustainable Agriculture*. London: MacMillan.

Ruthenberg, H. 1980. *Farming Systems in the Tropics*. Oxford: Oxford University Press.

Scott, J.C. 1998. *Seeing Like a State*. New Haven: Yale University Press.

Sida, 1995. *Change in Tanzania 1980–1994*. Stockholm: Swedish International Development Co-operation Agency.

Smaling, E.M.A. and Braun, A.R. 1996. Soil fertility research in Sub-Saharan Africa: new dimensions, new challenges. *Communications in Soil Science and Plant Analysis* 27(3–4): 365–86.

Snyder, K.A. 1993. Like water and honey. Moral ideology and the construction of community among the Iraqw of Northern Tanzania. Unpublished doctoral thesis, Yale University.

Snyder, K.A. 1996. Agrarian change and land-use strategies among Iraqw farmers in Northern Tanzania. *Human Ecology* 24(3): 315–40.

Snyder, K.A. 1997. Elder's authority and women's protest: the Masay ritual and social change among the Iraqw of Tanzania. *Journal of the Royal Anthropological Institute* 3(3): 561–76.

Swift, M. 1997. Agricultural intensification, soil biodiversity and agroecosystem function in the tropics. *Applied Soil Ecology* 6(1): 1–2.

Tengö, M. 1999. Integrated nutrient management and farming practices in the agro-ecological system of Mama Issara. Unpublished Honors thesis in Natural Resources Management, Stockholm University.

Thornton, R.J. 1980. *Space, Time and Culture among the Iraqw of Tanzania*. New York: Academic Press.

Walters, C. 1986. *Adaptive Management of Renewable Resources*. New York: MacMillan.

Widgren, M. 1999. Islands of intensive agriculture in eastern Africa: the social, ecological and historical contexts. In *'Islands' of Intensive Agriculture in the East African Rift and Highlands: a 500-Year Perspective*, pp. 3–14, ed. M. Widgren and J. Sutton. Working Paper Series No. 42. Stockholm: Environment and Development Studies Unit, Stockholm University.

Widgren, M. and Sutton, J., eds. 1999. *'Islands' of Intensive Agriculture in the East African Rift and Highlands: a 500-Year Perspective*. Working Paper Series No. 42. Stockholm: Environment and Development Studies Unit, Stockholm University.

7

Living with disturbance: building resilience in social–ecological systems

JOHAN COLDING, THOMAS ELMQVIST, AND PER OLSSON

7.1 Introduction

Disturbances such as fire, cyclones, and pest outbreaks create variation in natural systems and ecosystem renewal that may be important for the maintenance of biological diversity. Many natural disturbances are inherent in the internal dynamics of ecosystems, and often set the timing of ecosystem renewal processes fundamental for maintaining resilience in ecosystems (Holling *et al.*, 1995).

By disturbance we mean 'any relatively discrete event in time that disrupts ecosystem community or population structure and changes resources, substrate availability, or the physical environment' (White and Pickett, 1985: 7). We distinguish between abiotic and biotic disturbances. Abiotic disturbances are those where the direct cause of disturbance is generated by nonbiotic agents. Examples include fires, hurricanes, volcanic eruptions, earthquakes, flooding, and drought. Examples of biotic disturbances include insect and pest attacks, predators, invasion of exotic species, and the grazing and browsing of herbivores.

Conventional resource management, based on economic production targets, commonly seeks to reduce natural variation in target resources, because fluctuations impose problems for the industry dependent on the resource (Holling and Meffe, 1996). Control of resource stock variability and flows can be achieved in a number of ways. For instance, by increasing financial investments in technologies for harvesting, a modern fishing industry can invest in larger fleets and more effective gear in order to maintain an even flow of production. Maintenance of high and even flows of monoculture crops in large-scale agriculture may be achieved by investing in various energy inputs, such as insecticides, pesticides, and irrigation.

Such management practices reduce the effect of natural disturbance. This may be effective in the short run, but over time may reduce resilience in management systems and surrounding ecosystems by making them more vulnerable or less able to accommodate novel surprises and disturbances that cannot be anticipated in advance (Baskerville, 1995; Holling and Sanderson, 1996). For example, over time, industrial fisheries based on quota management may considerably reduce the survival capacity of fish stocks, because effective negative-feedback response mechanisms and effective institutions are lacking (Finlayson and McCay, 1998).

We characterize management by command-and-control as 'S-phase management' in the context of the adaptive renewal cycle (Holling, 1986), for which external energy inputs exclude natural disturbance and maintain the system in a configuration of 'optimality', i.e., in the climax stage of the conservation phase (see Fig. 1.2 in Chapter 1). Such management is based on the notion of a single equilibrium in natural systems, rather than multiple equilibria (Holling and Sanderson, 1996). Management systems that fail to understand the role of disturbance for maintaining ecosystem structure and function may become both ecologically and economically brittle.

Small-scale societies often have a reduced capacity for substituting their direct reliance on local products with credits in financial markets and also lack access to sophisticated technology. They therefore rely heavily on sustainable management of their local ecosystems for survival. Hence, they are strongly motivated to develop practices and structures that may reduce the effect of disturbance. We term this type of management 'backloop management,' because it indirectly considers the release–reorganization phases in the adaptive renewal cycle developed by Holling (1986) (see Fig. 1.2 in Chapter 1). In backloop management, natural disturbances become an integrated part of manipulating and modifying the natural resource base, and managers actively respond to episodic or rare events using flexible institutions and management practices that reduce the risk that large-scale ecological crises will occur. Local, decentralized institutions in many settings are key to adapting to disturbance, endogenizing it, and even utilizing it, because they tend to be flexible and are able to respond more quickly to risk and uncertainty than centralized institutions (Scoones, 1999).

In this chapter we review adaptive ecological and social strategies employed by local resource users to cope with natural disturbance and environmental variability. We first review risk management as practiced by local communities in three cases for coping with disturbance: cyclones in Samoa, floods in Bangladesh, and droughts in arid and semi-arid Africa. We go on to describe how the protection of ecosystem structures including human-induced, small-scale disturbances may be a tool for ecosystem management, and how these practices can build resilience, or buffer capacity, to cope with disturbance. The next section deals with the importance of disturbance for the learning that generates ecological knowledge. We conclude with a summary of the insights provided by these cases for improved management of natural resources and ecosystems.

7.2 Management strategies for coping with natural disturbance

7.2.1 The effects of large-scale disturbance on agricultural systems in Samoa

The role of biodiversity for building long-term ecosystem sustainability, particularly in the presence of large-scale disturbances, has been discussed in a number of recent publications (e.g., Holling, Berkes, and Folke, 1998; Folke, Berkes, and Colding, 1998). It has been hypothesized that resilient ecosystems need to contain a large number of species, many of which seem to be unimportant for the structure and function of the system in the general course of events but may play crucial roles in reorganization and restructuring processes after disturbances (Holling *et al.*, 1995). However, most studies and experimental tests of diversity–resilience relationships have dealt with noncultivated systems and species (e.g., Tilman, Wedin, and Knops, 1996); the role of species diversity for sustainability in agricultural systems has received less attention (Matson *et al.*, 1997). Based on a case study from Polynesia, we discuss how maintaining a high degree of species diversity in agricultural systems (advanced polyculture) may be part of a deliberate strategy deployed by local farmers to reduce their vulnerability to tropical cyclones.

7.2.1.1 Polynesian polyculture

As in many tropical areas, traditional agriculture in Polynesia can be characterized as an advanced polyculture, with annual crops mixed with a large number of shrub and tree species – also referred to, for example, as agroforestry, tree gardens, and multicropping (Kirch, 1991). In addition to securing a diverse food supply, this highly variable and sophisticated system provides a large number of goods and services to local farmers. The large number of trees and shrub species present provide shade, erosion control, soil improvement, wind protection, and weed/disease control, as well as being a source of timber, fuel wood, weapons, ornaments, medicines, dyes, fabric, oils, rubber, objects of religious and mythological value etc. (Clarke and Thaman, 1993). Recent evaluations of advanced polycultures in the Pacific have argued that they are highly productive systems with a strong positive net energy yield, while at the same time being independent of fertilizers and pesticides. Maintenance is usually based on renewable resource inputs, and polycultures may provide important refuges for many endangered wild species of plants and animals in severely deforested areas (Clarke and Thaman, 1993).

Historically, this sustained-yield agricultural system may have developed partly as an adaptation to the degradation caused by humans themselves during the initial colonization phase of the islands (Kirch, 1997). An initially rapidly growing human population may have reached carrying capacity relatively soon after colonization (Kirch, 1997). This may have resulted in widespread famines (e.g., Kirch, 1991) and a pressing need to develop more sustainable agricultural methods (Clarke and Thaman, 1993).

The global market economy, governmental policies, agricultural subsidies, and decreased tenurial security contribute to a decline in the practice of advanced polyculture throughout the Pacific (Clarke and Thaman, 1993; Clarke, Manner and Thaman, 1999). However, within Polynesia, there is considerable variation among areas. In Samoa, for example, polyculture is still practiced to a significant extent, whereas in other areas, such as the Cook Islands and French Polynesia, it has virtually disappeared and has been replaced by cash crop monocultures (Clarke and Thaman, 1993). The reasons for this regional variation are likely to be complex and involve colonial history, the extent of outside subsidies, and market relations, as well as the specific environmental conditions under which the farmers operate. We focus on one such environmental condition, namely infrequent but severe tropical cyclones affecting agricultural practices.

7.2.1.2 Cyclones and their effects on crops

Cyclones are unpredictable events in both a temporal and a spatial sense, and in cyclone-prone areas agricultural production may be severely reduced at variable intervals. In Samoa, located in the western part of Polynesia, cyclonic storms are relatively frequent, with more than 40 recorded since 1831. Severe cyclones may occur at intervals of 20–30 years (based on records from the last 160 years). In the early 1990s, two very severe cyclones were recorded in the Samoan archipelago within a 22-month period. On February 1–3 1990, cyclone Ofa, the most severe storms in more than 160 years, hit the islands, with the eye passing about 80 km west of the island of Savai'i, Western Samoa. Winds recorded at up to 216 km/h caused severe forest damage. A SPOT-SAT satellite false color image, taken 2 weeks after the storm, showed that only small patches comprising less than 1 percent of the primary forest retained normal foliage on the eastern third of Savai'i. This area includes the Tafua peninsula, which, just prior to the cyclone, had been established as a rain forest preserve under indigenous control (Cox and Elmqvist, 1991). Twenty-two months later, on December 6–8 1991, tropical cyclone Val struck the islands, a storm of comparable intensity with the eye passing in a north–south direction directly over Savai'i. Again, forest damage was extensive, with more than 90 percent of the primary forest defoliated (Elmqvist *et al.*, 1994).

A post-cyclone study of the storm effects on agricultural production was made in the traditional village of Tafua (Lindberg and Mossing, 1996). This study and one by Clarke (1993) revealed evidence of significant variation in the extent of crop damage (Table 7.1). In general, cash crops were damaged more than subsistence crops. Also, within each crop species, damage varied among genetically distinct varieties (coconut, *Cocos nucifera*) or among trees of different sizes (breadfruit, *Artocarpus altilis*) or derived from cuttings or seeds (cocoa, *Theobroma cacao*) (Table 7.1). Recovery periods varied greatly, both within and among crops, from minor damage and short recovery periods (taro), to major damage and short recovery periods (banana), to major damage and long recovery periods (breadfruit) (Table 7.1). Taro, a major crop that suffered only minor damage and had a short recovery period, suffered substantial damage 6 months after the cyclone due to insect outbreak. Later, in 1993, the taro blight *Phytoptera colocasiae* destroyed 90 percent of all taro. To summarize, observations revealed that from these studies it is possible to conclude that:

- the most important cash crops were among the most damaged;
- the crop species that tended to best survive the cyclones (taro) was subsequently severely reduced by insect attacks and fungal pathogens;
- a minor crop (yams) became the most important crop for an extended period of time when taro was lacking; and
- in the absence of outside subsidies, farmers took very high risks by investing exclusively in monocultures of cash crops or taro.

In areas where tropical cyclones occur with some frequency a diverse set of crop species and cultivars may reduce the risk of a total loss of food supply. This idea was supported in interviews with farmers in Samoa. In a survey shortly after the cyclones, 19 farmers in Tafua were asked 'What could you do to reduce the effects of another cyclone?' The three most frequent responses were: (1) pray, (2) diversify my crops, and (3) work harder (Lindberg and Mossing, 1996). It was also evident from these interviews that planning, at least in a European short-term sense of command and control (Holling and Meffe, 1996), does not exist. If the future is perceived as intrinsically unpredictable, short-term planning probably makes much less sense as a tool to avoid crises than polyculture as a viable bet-hedging strategy. This thinking may well be representative of subsistence farmers who must cope with severe crises without, or with only limited, subsidies from the outside (Lockwood, 1971). Paulson and Rogers

Table 7.1 *The effect of cyclone Ofa on crop production and recovery rates*

	Diet (%) Importance	Damage (%)	Recovery rate (months)	Comments
Staple food				
Taro (*Colocasia esculenta*)	39–43	Minor (<5)	2–4	Severely attacked by *Spodoptera litura* 6 months after cyclone
Banana (*Musa spp.*)	16–23	Major (90–100)	9–10	Damage differed between varieties
Breadfruit (*Artocarpus altilis*)	16–19	Major (50–90)	10–15	Large trees recovered very slowly; fermented fruits traditionally used as emergency food
Taamu (*Alocasia macrorrhiza*)	15–19	Minor (<5)	5–7	Traditionally used in times of food shortage
Yam (*Dioscorea sp.*)	Minor	Minor (<5)	5–7	Became the staple in areas with no taro
Commercial crops				
Coconut (*Cocos nucifera*)	Major	Moderate–major	>10 (partial)	Damage varied spatially and between varieties
Cocoa (*Theobroma cacao*)	Major	Moderate–major (20–30)	>24 (partial) (need shade for good production)	Damage varied spatially and between varieties; trees derived from cuttings more susceptible than those from seeds

Adapted from Clarke (1993) and Lindberg and Mossing (1996).

(1997) concluded from their study in Samoa that village agriculture has shown great resilience in the face of both market forces and the cyclones and that local knowledge reservoirs about polyculture contributed to a rather rapid recovery from the cyclones.

7.2.1.3 Local institutions for crisis management

The various ecosystems and species associated with local commons are often managed by local-level institutions that regulate access and use rights to resources in time and space. Such institutions can be defined as codes of conduct that define practices, assign roles, and guide interactions, the set of rules actually used. Institutions can also be defined as humanly devised constraints structuring human interaction. In Samoa, the maintenance of polyculture as a sustain-yield agricultural system is embedded in a sophisticated institutional structure, including land tenure and a reciprocal gift-giving system. In a traditional village, a number of chiefs of different ranks (*matai*), each representing an extended family, form a village council (*fono*). The *fono* determines the overall focus of agriculture and the use of communal land. Part of the traditional role of the chief is to organize community response to periodic environmental disasters. Each extended family cultivates a house lot and plantation lots with defined boundaries, and land control is traditionally linked to a specific chief title. Beyond the plantation lots are family reserve lands, in which only portions are cultivated at any given point in time. Further away from the village center are village lands, ultimately controlled by the village council and serving as a buffer. On this land a whole village may come together to plant taro (used in village ceremonies), to support some village project, or to provide the village with more food after a crisis. This social system, with land buffers at different levels of organization and control from the family to village levels, provides flexibility in food production and builds resilience in the social–ecological system.

Another strategy to cope with unpredictable disturbances is the use of techniques for emergency food storage. A tradition in many Polynesian islands is the use of a sophisticated technique for fermenting breadfruit in pits (Ragone, 1991). This fermentation process made long-term food storage possible in a hot and humid climate. Pit size varied from 1 m in depth to up to 9 m in large storage pits, and these would keep the fermented breadfruit in good condition for a year, or in some cases for decades (Ragone, 1991). Pit fermentation is rarely used today in Polynesia, but was practiced in several villages in Savai'i immediately following the cyclones (Lindberg and Mossing, 1996). The cyclones thus resulted in a revitalization of this food storage technique and provided an opportunity for young people to learn the different steps and procedures involved in making a good fermentation pit. Hence, large-scale disturbances may strengthen social organization and revitalize the practice of traditional techniques and methods in agriculture, food preparation, or house construction. For example, villagers in Tafua were asked whether they perceived that anything good came as a result of cyclone disturbances, and the majority gave a positive answer (Lindberg and Mossing, 1996).

Throughout the Pacific, human societies have had to adapt to pulses of natural disturbances such as cyclones, drought, tsunamis, and flooding (Kirch, 1997). In many cases, anthropogenic impacts such as clear-felling of forests on steep slopes have made human societies even more vulnerable to flooding events and may have further enforced the use of terrace construction and other means of hill slope stabilization (Kirch, 1997). Such 'landscape enhancements' (Spriggs, 1997) increase the resilience of the landscape and food production security. The interviews with Samoan farmers support the notion that the widespread tradition of simultaneously growing several crop species and cultivars may in fact be a system maintained as part of a strategy to increase resilience in the face of large, unpredictable disturbances.

In Polynesia, cyclones decrease in frequency when moving from Western into Central and Eastern Polynesia. This spatial and temporal variability provides the opportunity for an interesting comparative study. Components of resilience and sustainability in agriculture may be analyzed by comparing agricultural management in areas frequently and severely damaged (several times within a human generation, as in Samoa) with areas very seldom hit (once every second or third generation, as in Eastern Polynesia). In Eastern Polynesia, polyculture has largely been replaced by monocultures of cash crops (Clarke and Thaman, 1993). In the event of a severe cyclone, farmers would be forced to rely on outside subsidies for survival. In contrast, Samoan farmers retain the ability to recover from a cyclone without outside help, through the practice of polyculture and related practices.

7.2.2 Char-dwellers of Bangladesh

Schmuck-Widmann (1996) describes several crisis-response strategies developed among the 2 million people living on the Jamuna chars in Bangladesh, focusing on the 6000 char-dwellers living in the Gabsara union in the district of Tangail. There, the Jamuna River divides into several channels that flow around both large and small 'chars' or temporary islands.

Several flood control measures at the Jamuna have decreased the stability of the chars. For example, at the end of the 1960s, the *Brahmaputra Right Embankment* (BRE) was completed, which channels the Jamuna in order to protect banks and stabilize the river. This contributed to the widening of the river and its splitting into several channels that constantly changed course, thus in turn changing living conditions on the chars (Schmuck-Widmann, 1996).

In 1988, the flood was so strong that two-thirds of the surface area of Bangladesh was under water for several weeks, causing considerable material damage. In 1989, the G7-group decided to impose flood prevention measures, resulting in the *Flood Action Plan* (FAP), financed by the World Bank and 14 donor nations. Various river structures, such as embankments and polders, were set up to protect humans and industrial facilities from floods, control water levels, and increase agricultural production (FPCO, 1994, 1995).

Such flood control measures may be sensible, but the FAP planners failed to take into account the fact that rural populations have learned to adapt their agriculture production systems to the yearly floods (Schmuck-Widmann, 1996). Yearly floods are perceived as a normal part of agriculture in Bangladesh because they irrigate and fertilize cultivation fields. In addition, fishery depends on the floods. Because 90 percent of Bangladeshis make their living from agriculture, the FAP construction project is an example of how engineering knowledge and traditional ecological knowledge can conflict with each other (*ibid.*). In the case of the char-dwellers, the FAP construction project has had unforeseen consequences for the Jamuna River morphology, leading to increased char erosion. Based on the work of Schmuck-Widmann (1996), we have synthesized some of the major risk management strategies used by the char-dwellers to cope with increased environmental uncertainty.

7.2.2.1 Chars and floods

Chars are islands made up of sediment deposits. They may be washed away by floods in less than a year or remain stable for decades. Only on older and inhabited chars do occasional trees grow. The Jamuna River dries up almost completely during the winter months, when char farmers cultivate part of the riverbed. The floods start at the beginning of the monsoon season in June; high water levels last until the end of September. However, the water level can vary considerably during this time. Within days, the water level can rise so abruptly that chars are partially or completely washed away. This means that the char-dwellers must move within a few days and begin a new life on other chars. According to Schmuck-Widmann (1996), *flexibility* and *innovative ability* are basic requirements for living on the chars. Char-dwellers have built up a store of knowledge that enables them to adjust to floods and erosion. When water levels fall, some parts of the chars break off and land is eroded. Char-dwellers can predict changes in the river arms and the creation of new land by observing water levels throughout the year and through frequent travel and exchange of observations with others in their region.

The char-dwellers report declining living conditions since the 1970s, which have been characterized by a series of abnormal floods or lack of flooding. Char-dwellers have been forced to move almost continually. An average char-dweller has moved eight times by the age of 44; in the past, inhabitants were uprooted from their chars only periodically. The problem today is not floods but a change in river morphology. The construction of embankments and dikes by FAP has altered water flows, erosion and sedimentation patterns, and fish habitat.

The majority of the char-dwellers are farmers. Cultivating rice, harvested three times a year, provides the yearly basic food staple. Each of the three rice strains is adapted to dominant weather conditions during its growth season. Thus, harvest success is dependent on the floods and rains. Rice is largely grown for subsistence needs, because char land is not suitable for large-scale rice cultivation. Fish constitute the main protein source apart from pulses.

Floods are essential for agriculture, and char-dwellers have adapted their farming methods to them. In years with abnormally low levels of flooding, harvests are severely affected. In the summer of 1994, a severe drought caused failure of the *amon* rice crop. *Amon* grows with the floods, and the high water levels needed for the rice did not occur. Insects and rats largely destroyed the crops that the char-dwellers did manage to grow. According to the farmers, the pests multiplied because the floods did not destroy them.

7.2.2.2 Risk management mechanisms among char-dwellers

Below we review the three major risk management mechanisms among Bangladeshi char-dwellers, based on the Schmuck-Widmann study.

1. *Polyculture and animal husbandry.* Three soil types determine crop cultivation. On catkin land (the least valuable and most sandy land), farmers grow catkin grass (locally known as *kaisha* or *kash*) and have learnt to cultivate groundnut. On fertile soil, all rice types, wheat, spices, vegetables, mustard, and other types of crops are grown. On intermediate fertile and sandy soils, pulses, sesame, linseed, sweet potato, some rice types, and jute types are grown. This mixed cultivation system provides for both a varied diet and food security, as weather conditions may be unsuitable for a particular crop, but ideal for another.

   Char-dwellers also raise capital by breeding and buying animals, which they sell for money in times of crisis. Animals include cows, goats, sheep, hens, doves, and ducks. On average, char-dwellers own more animals than mainland dwellers. They only use a small portion of their livestock products themselves, selling the rest to acquire material goods, animals, land, food, and medicine during bad floods.

2. *Erosion-buffering practices.* In response to erosion, char-dwellers reclaim land and promote soil fertility by cultivating catkin grass. This practice forms the basis of char land management and also provides the most important building material and fodder during floods. Catkin grass can grow on the sandy soil and survives periods of droughts as well as abnormally severe floods. It is therefore a 'multiple-disturbance-tolerant' species.
3. *Institutional risk-spreading mechanisms.* Common law allows any char-dweller to harvest catkin grass, regardless of who owns the land. The removal of catkin for pasture land and crop cultivation is a hard job. Therefore, anyone who clears catkin land may use it for 2 years, after which the field returns to its original owner. This allows farmers without land a temporary opportunity to farm for free. As catkin only grows on chars and riverbanks, it has a market on the mainland. Common law also specifies that anyone has the right to settle on land owned by another. Only three of the 20 households questioned were living on their own land, indicating frequent resettlement due to erosion and floods (Schmuck-Widmann, 1996).

   Land owned by a family is often dispersed over a wide area. Owning land spread over several locations may reduce the risk that erosion damage will affect the whole family. Farmers unable to utilize their land because it is under water temporarily switch to other sources of income, such as paid work, handicrafts, and fishing.

Social structures and cultural traditions ensure that char-dwellers help each other during crises. So-called *resource to solidarity duties of relatives* on the mainland is facilitated by the wide and connected network of relatives with other villages and with other chars. Parents take care that such networks are formed through children's marriages. Marriages between char-dwellers and mainland dwellers represent a strategy for providing an escape route to relatives unaffected by the floods. After a failed harvest, villagers exchange food.

7.2.3 Risk management mechanisms among African pastoralists

African pastoralists have traditionally used risk management mechanisms to build buffer capacity in their management system. The warm, arid ecosystem of African pastoralists is constantly changing. Arid and semi-arid African ecosystems have a low net primary productivity and high variability in ecosystem structure and productivity. Increases in spatial and temporal variability are caused by decreases in rainfall. The most limiting factor is water, but as the ecosystem becomes more humid, soil nutrients become the limiting factor (Niamir-Fuller and Turner, 1999). It is therefore impossible to define a stable equilibrium state; hence the arid ecosystem can be characterized as a 'multiple equilibrium' system, making the degree of predictability low (Niamir-Fuller and Turner, 1999).

In the past, pastoral ecosystems were relatively resilient despite severe drought episodes. This may have been due to 'lower human-population density (but not necessarily livestock density); land-tenure security vested in customary communal institutions; mobility of animals; and traditional natural resource management and improvement techniques' (Niamir-Fuller and Turner, 1999: 22).

These systems have changed, sometimes to the point that they are barely recognizable (Niamir-Fuller and Turner, 1999). Livestock mobility has declined, and sedentarization has greatly increased, leading to increased ecological and economic vulnerability in the pastoral systems of dryland Africa, and has resulted in severe land degradation in the semi-arid zones. Sedentarization has increased continuous grazing around settlements, resulting in reduced vegetation cover and diversity and soil degradation. It has also led to an invasion of unpalatable plants due to lower grazing pressure in distant pastures (*ibid.*). Generally, pastoral systems no longer have control over their socio-political environment, and are therefore unable to collectively adapt their production systems to ecological changes. Only in ecologically marginal areas, untouched by market forces, do traditionally managed systems still exist (*ibid.*).

7.2.3.1 Diversification and mobility

Diversification is an important risk management strategy used by many pastoral groups. This includes managing a diverse mix of livestock species within the same household. For example, an appropriate mix of herding animals is used to utilize different vegetation types and patches in a dynamic fashion (Niamir-Fuller, 1998). This strategy may also reduce the risk of various perturbations (Niamir-Fuller and Turner, 1999). Herd splitting, where surplus animals are sent on transhumance (i.e., regular seasonal movement of livestock between well-defined pasture areas) also effectively reduces the risk of overgrazing in the home base area and takes advantage of better pastures elsewhere (Niamir-Fuller, 1998).

The key determinant among African pastoralists for coping with landscape variability and disturbance has been *mobility* (Niamir-Fuller and Turner, 1999). Disturbance types may include disease outbreaks including tsetse and other insect-borne diseases, droughts, and potential raids (Bassett, 1986). Mobility among pastoralists is of two types: micro-mobility, or daily movements around the village or camp, and macro-mobility, across long-distance routes and seasonal grazing areas. The progressive widening of grazing radius around wells as the wet season advances is one example of micro-mobility that contributes to building resilience in grazing areas used by pastoralists. This practice is employed by the Maasai of Kenya, who leave enough forage around wells for the dry season (Niamir-Fuller, 1998). Although 'the mobility paradigm' is currently recommended as a strategy to combat ecosystem degradation, it often entails profound changes in government policies and in some cases customary institutions (Niamir-Fuller and Turner, 1999).

Pulsed grazing, which results from the mobility practices of some pastoralist groups, contributes to the capacity of the semi-arid grasslands of Africa to function under a wider range of climatic conditions relative to permanent livestock ranching. Constraining livestock movement by way of ranching, often facilitated by the drilling of boreholes to provide water to cattle year round and over wider areas, may expose grasses to continuous grazing pressure to which they are not adapted (Hudak, 1999). The combination of heavy grazing and active fire suppression management may push a grassland savanna system beyond the threshold into a relatively stable state, i.e., from an inherently unstable grassland system into a relatively stable thorn woodland, a state not readily reversed through natural processes (Hudak, 1999). Gradual resilience loss in a grassland savanna may become severe before the need for remediative action is realized and actions to ameliorate the situation are taken (*ibid.*). If the capacity of the ecosystem to deal with pulses is reduced, for example through overgrazing, an event that previously could be absorbed can flip the grassland ecosystem into a relatively unproductive state (from the pastoralists' perspective), dominated and controlled by woody plants for several decades (Walker, 1993).

7.3 Disturbance-buffering structures and disturbance as tools for ecosystem management

Many local communities adapt to and even depend on natural disturbances for their survival, such as flooding for the irrigation and fertilization of cultivation fields among the char-dwellers. Local communities may also protect ecosystem structures to reduce the effects of unpredictable natural disturbances. Two examples of such structures are temporarily protected areas such as 'buffer zones' and permanently protected areas such as 'sacred groves.'

Sahelian rangeland pastoralists in arid and semi-arid Africa use buffer zones that are protected from grazing except in emergencies such as during prolonged drought (Niamir-Fuller, 1998). Buffer zones are shared areas where permanent claims are not recognized among different tribes (Niamir-Fuller, 1999). Rangeland pastoralists also use 'range reserves,' which are fixed, well-defined areas that a pastoral group establishes within its own annual grazing area in order to provide a 'savings bank' of forage in the event of drought.

Permanent habitat protection is commonly found among local resource users from various parts of the world (Hughes and Chandran, 1998; Colding and Folke, 2001). Strong religious beliefs and social conventions often enforce such protection. In many places these habitats are viewed as sacred, and local communities enforce strict institutions such as taboos for their protection (Colding and Folke, 2001). Whole forests, forest patches, coast stretches, rivers, or ponds may be protected in this way.

In India sacred groves are a common feature in rural areas (e.g., Ramakrishnan, Saxena, and Chandrashekara, 1998). Sacred groves also exist in Africa, for example in Ghana, Nigeria, and Kenya (Colding and Folke, 2001). A sacred grove may consist of a patch of trees or entire forests, ranging from a hectare up to a few square kilometers,often set aside as a protected area for religious purposes (Gadgil and Vartak, 1976; Ramakrishnan, 1998). In South America, so-called 'spirit sanctuaries' exist among the Kuna of Panama (Chapin, 1991), the Tukano of Brazil/Colombia (Chernella, 1987), and the Cocnucos and Yanaconas of Colombia (Redford and MacLean Stearman, 1993). For example, the Tukano set aside areas of the forested river margin where fishing is not allowed. Any deforestation of the river edge is prohibited and fishing is permitted along as little as 38 percent of the total river margin (Chernella, 1987). The Tukano also impose taboos on landing on particular islands. The result is a management system that distinguishes between human use areas and animal refuge areas.

Such 'social fencing of ecosystem types' (Ramakrishnan, 1998: 4) provides for a number of ecological services upon which humans depend (Daily, 1997). These include maintenance of landscape patchiness, preservation of biological diversity, provision of habitat for threatened species, regulation of local hydrological cycles, prevention of soil erosion, pollination of crops and plants, preservation of locally adapted crop varieties, and protection from wind and fire (Colding and Folke, 2001). In the Western Ghats of India, for example, the evergreen sacred groves can act as natural fire breaks, sheltering tracts of fire-sensitive species and safeguarding water sheds in landscapes of slash and burn and fire-prone secondary forests (Chandran, Gadgil, and Hughes, 1998). Gadgil and Vartak (1976) report that sacred groves in the Western Ghats supply timber in the event of emergency such as the destruction of an entire village settlement by fire.

Socially fenced ecosystem types may also be critical to the renewal of ecosystems after disturbance. In some parts of the world, sacred groves are the only primary forests remaining (Gadgil and Vartak, 1976; Dorm-Adzobu,

Ampadu-Agyei, and Veit, 1991; Wilson, 1993). Hence, they may play a critical role as sources for restoring degraded ecosystems, as well as providing habitats for species important in the recolonization of disturbed ecosystems, such as pollinators and seed dispersers. Keystone species may also be associated with sacred groves in India, such as those of the genus *Ficus* and *Quercus* (Ramakrishnan, 1998). Hence, the protection of such species may be highly functional for the maintenance of biological diversity and for building resilience in a landscape.

As noted by Folke *et al.* (1998), local resource users may also actively create small-scale disturbances in the landscape. Such practices create smaller cycles of ecosystem renewal that allow local users to make use of a wide range of species that exist in the various stages of ecological succession. For example, traditional agroforestry practices such as shifting cultivation create forest gaps and enable people to produce crops or enhance the supply of wild foods without disrupting natural renewal processes, when these practices are properly applied (Folke *et al.*, 1998). Shifting cultivation may create patchiness in the landscape, resulting in a mosaic of forest, fallow, and gardens (Orejuela, 1992). African herders behave like a disturbance, by following the migratory cycles of herbivores from one area to another (Folke *et al.*, 1998).

Another example of creating small-scale disturbance is the use of fire, as practiced by some traditional groups (Berkes, Folke, and Gadgil, 1995). Until the late 1940s, Amerindians of Northern Alberta, Canada, regularly used fire to open up clearings (meadows and swales), corridors (trails, traplines, ridges, grass fringes of streams and lakes), and windfall forests (Lewis and Ferguson, 1988). These clearings increased habitat area for species such as ungulates and waterfowl, while preventing invasion by shrub species. In northwestern North America, native peoples practiced landscape burning to encourage the growth of berry and root crops and seed production (Gottesfeld Johnson, 1994). Modern Amerindian burning is currently only occurring on federally designated Indian reserve lands, because these lands are not subject to Provincial Forest Service regulations that prohibit fire management (Gottesfeld Johnson, 1994).

Australian aborigines possessed detailed technical knowledge of fire, and used it to improve the feeding habitat for game and to assist in the hunt itself (Lewis, 1989). Villagers of the Aravalli hills in India commonly practice 'fire bathing,' whereby they set fire to forest lands to please a local hill god (Pandey, 1998). This ritual helps recycle soil nutrients and enhance the growth of grass.

Creating small-scale disturbances in the landscape can also be important for reducing the effects of large-scale natural disturbances. For example, controlled burning of grass and deadwood reduces the spread of accidental, large-scale fires by preventing the slow build-up of fuel (Gottesfeld Johnson, 1994).

7.4 Learning by trial and error

In this chapter, a number of ecological and social strategies related to natural disturbance have been described. These are likely to have arisen in local communities as a result of ecological monitoring and the long-term experience of local resource users with environmental variability and disturbance. In reference to learning, it may be useful to distinguish between what Rappaport (1979: 97) has termed the 'cognized' and 'operational' models of nature. The former refers to how people perceive and interpret nature, the latter to understanding how nature really operates using methods of the objective sciences, in particular the science of ecology. A similar distinction is made by Gunderson (1996: 17–18) when describing how adaptation relates to learning. Gunderson refers to the cognized model as the 'schema' and the operational model as the testing of schemas through 'real' world experiments.

Gunderson (1996) also makes the important point that people can adapt to natural systems without necessarily learning. In fact, institutions for the management of natural systems may work as long as the system is not subject to significant changes such as unforeseen, large-scale disturbances. If a significant system change occurs for which no operational understanding of phenomena exists, there is a high probability that what Hilborn (1992) calls 'reactive' learning will occur. This type of learning is slow because no operational monitoring and evaluation mechanisms exist and people act as 'fire-fighters.' Eventually, learning may take place, but at considerable cost, or learning may not take place because institutions may be unable to adjust to new environmental conditions. Such a phenomenon is often referred to as 'cultural inertia' or the inability to respond to changing environmental conditions (Boyd and Richerson, 1985).

We argue that in cases of cultural inertia there is a critical lack of fit between the cognized and operational models of nature. Institutions are unable to respond and adjust to dynamic feedback from ecosystems because they are not based on adequate knowledge about the processes and functions of ecosystems and the impacts of applied resource management practices. Thus, a functioning environmental feedback system is missing due to institutions based on the wrong content, such as social dogma and myths.

Ecological monitoring and experience with environmental variability and disturbance probably play a key role in the development of many ecological and social practices dealt with in this chapter. For example, frequently recurring natural disturbances may revitalize local-level management practices and their linked social mechanisms such that cultural inertia is reduced. Many of the ecological and social practices described here probably represent rules of thumb for managing local natural resources and services provided by local ecosystems. Knowledge about environmental phenomena in these local communities appears to be based on practical experience acquired through trial and error over several generations, and involves traditional or local ecological knowledge (Gadgil, Folke, and Berkes, 1993; Berkes, Colding, and Folke, 2000; Berkes and Folke, 2002).

7.5 Synthesis and conclusions

This chapter provides examples of ecological and social practices for coping with environmental variability and natural disturbance. It describes a number of risk-spreading strategies used by local communities in various disturbance-prone geographical settings to avoid large-scale social–ecological crises. The chapter also describes how habitat protection, shifting cultivation, and fire management can play a role in mitigating the effects of natural disturbance and how such practices may contribute to providing a flow of resources and services on which local communities depend.

These examples represent ways of responding to environmental variability and natural disturbances at the local level instead of blocking them out, as is often done in conventional resource management (Holling and Meffe, 1996). This dichotomy in the perception of disturbance is reflected in the different ways in which char-dwellers of Bangladesh and policy experts view the flooding of the Jamuna River. Rural people of Bangladesh, including char-dwellers, consider annual floods as normal and have developed ecological and social strategies to live with this disturbance. By contrast, the experts who designed the FAP perceive annual flooding as a constraint to development and have devised measures to prevent flood damage.

The perception and understanding of the role of natural disturbance for ecosystem renewal have consequences for the management strategies developed in a society. People in the local communities described in this chapter can be referred to as 'ecosystem people' (Dasmann, 1988) in the sense that they depend directly on functioning local ecosystems for their survival and often lack sophisticated technologies to block out natural disturbance. Hence, in order to avoid large-scale social–ecological crises, the members of these societies are strongly motivated to develop strategies that deal with environmental uncertainty and variability locally. This characteristic appears to be key to maintaining resilience in any system (Levin, 1999).

By contrast, in many large-scale Western societies, social hardship resulting from resource management failures is often mitigated by investment in insurance and options provided for by the capital market. While such measures may be important in helping people survive periods of crises, they do not necessarily

Table 7.2 *Social and ecological practices in relation to the release and reorganization phases of the adaptive renewal cycle*

	Release phase Evoking small-scale disturbances	Putting the brakes on release	Reorganization phase Nurturing sources of renewal
Ecological practices	Shifting cultivation[a]	Emergency crops[f,g]	Protection of habitat[b,g,i,j]
	Pulse grazing[b]	Erosion control[h]	Protection of keystone species[h,g]
	Fire[c,d,e]	Polyculture[f,h]	
		Preserving wind breaks[f]	
		Preserving saving banks[b,g]	
		Diversifying livestock[b]	
		Disturbance-tolerant species[f,h]	
		Multiple-disturbance-tolerant species[h]	
Social practices	Mobility[b]	Reciprocal gift giving[f,h]	Social fencing by taboos[k,i,l]
	Rituals[e]	Saving money[h]	
		Alternative sources of income[h]	
		Emergency jobs[h]	
		Marriage strategies[h]	
		Stocking food[f]	
		Mobility[h,b]	
		Keeping livestock to sell[h]	
		Flexible user rights[h,b]	
		Dispersed land ownership[f,h,b]	
		Village council (fono system)[f]	
		Social fencing by taboos[l,g,i]	

[a]Mexican *milpa* (Alcorn and Toledo, 1998); [b]African pastoralists (Niamir-Fuller, 1998, 1999; Niamir-Fuller and Turner, 1999); [c]Amerindians of Alberta, Canada (Lewis and Ferguson, 1988); [d]Australian aborigines (Lewis, 1989); [e]Aravalli Hill tribes of India (Pandey, 1998); [f]People of Savai'i, Western Samoa (Clarke, 1993; Clarke and Thaman, 1993; Lindberg and Mossing, 1996); [g]Sikkimese buddhists of India (Ramakrishnan, 1998); [h]Char-dwellers of Bangladesh (Schmuck-Widmann, 1996); [i]Tukano people of Brazil/Colombia (Chernella, 1987); [j]Kpaa Mende people of Sierra Leone (Lebbie and Guries, 1995); [k]Kuna of Panama (Chapin, 1991); [l]Cocnucos and Yanaconas of Colombia (Redford and MacLean Stearman, 1993).
Also indicated are examples of indigenous groups for which these practices have been described.

lead to ecological learning from environmental feedback, or to adjustments of applied management practices.

Conventional resource management has largely concerned itself with the exploitation and conservation phases of the adaptive renewal cycle (Berkes and Folke, 2002). However, because the ability of an ecosystem to reorganize is determined by the effectiveness of the release and reorganization phases (Holling, 1986; Folke and Berkes, 1998), resource management needs to focus on practices that contribute to the renewal capacity of the ecosystem in which resources are extracted and exploited. Table 7.2 summarizes most of the ecological and social practices described in this chapter. These concern the release and reorganization phases, the backloop of the adaptive renewal cycle (see Fig. 1.2 in Chapter 1).

In a world characterized by rapid social and ecological change, it is critical to generate knowledge about the functioning of ecosystems in all phases of the adaptive renewal cycle. There is a need to generate knowledge about processes and functions to understand ecosystem behavior and to approach social and ecological systems as a single coupled and dynamically complex system (Gunderson, Holling, and Light, 1995; Christensen *et al.*, 1996; Berkes and Folke, 1998; Dale *et al.*, 1998; Levin, 1998). The ecological and social practices described in this chapter appear to derive from learning by trial and error and represent responses to environmental feedback. Hence, they may be the types of practices advocated by promoters of 'adaptive management' (Holling, 1986; Walters, 1986; Gunderson *et al.*, 1995). Learning from local communities with long-term experience in environmental variability and uncertainty in many parts of the world may yield valuable rules of thumb for managing complex ecosystems such that resilience and options for human welfare are not reduced.

References

Alcorn, J.B. and Toledo, V.M. 1998. Resilient resource management in Mexico's forest ecosystems: the contribution of property rights. In *Linking Social and Ecological Systems. Management Practices and Social Mechanisms for Building Resilience*, pp. 216–49, ed. F. Berkes and C. Folke. Cambridge: Cambridge University Press.

Baskerville, G. 1995. The forestry problem: adaptive lurches of renewal. In *Barriers and Bridges to the Renewal of Ecosystems and Institutions*, pp. 37–102, ed. L. Gunderson, C.S. Holling, and S.S. Light. New York: Columbia University Press.

Bassett, T.J. 1986. Fulani herd movements. *Geographical Review* 76: 233–48.

Berkes, F., Colding, J. and Folke, C. 2000. Rediscovery of traditional ecological knowledge as adaptive management. *Ecological Applications* 10(5): 1251–62.

Berkes, F. and Folke, C., eds. 1998. *Linking Social and Ecological Systems. Management Practices and Social Mechanisms for Building Resilience.* Cambridge: Cambridge University Press.

Berkes, F. and Folke, C. 2002. Back to the future: ecosystem dynamics and local knowledge. In *Panarchy: Understanding Transformations in Systems of Humans and Nature*, pp. 121–46, ed. L.H. Gunderson and C.S. Holling. Washington DC: Island Press.

Berkes, F., Folke, C., and Gadgil, M. 1995. Traditional ecological knowledge, biodiversity, resilience and sustainability. In *Biodiversity Conservation*, pp. 281–300, ed. C.A. Perrings, K-G. Mäler, C. Folke, C.S. Holling, and B-O. Jansson. Dordrecht: Kluwer Academic Press.

Boyd, R. and Richerson, P.J. 1985. *Culture and the Evolutionary Process*. Chicago: University of Chicago Press.

Chandran, M.D.S., Gadgil, M., and Hughes, J.D. 1998. Sacred groves of the Western Ghats of India. In *Conserving the Sacred for Biodiversity Management*, pp. 211–32, ed. P.S. Ramakrishnan, K.G. Saxena, and U.M. Chandrashekara. New Delhi: Oxford and IBH Publishing Co. Pvt. Ltd.

Chapin, M. 1991. Losing the way of the great father. *New Scientist* 131: 40–4.

Chernella, J. 1987. Endangered ideologies: Tukano fishing taboos. *Cultural Survival* 11(2): 50–2.

Christensen, N.L., Bartuska, A.M., Brown, J.H., *et al.* 1996. The report of the Ecological Society of America Committee on the Scientific Basis for Ecosystem Management. *Ecological Applications* 6(3): 665–91.

Clarke, T. 1993. The effects of a cyclone on crops. *Journal of South Pacific Agriculture* 1(1): 66–77.

Clarke, W.C., Manner, H.I., and Thaman, R.R. 1999. *Agriculture and Forestry. The Pacific Islands. Environment and Society*. Honolulu, HI: Bess Press.

Clarke, W.C. and Thaman, R.R., eds. 1993. *Agroforestry in the Pacific Islands: Systems for Sustainability*. Tokyo: United Nations University Press.

Colding, J. and Folke, C. 2001. Social taboos: 'invisible' systems of local resource management and biological conservation. *Ecological Applications* 11: 584–600.

Cox, P.A. and Elmqvist, T. 1991. Indigenous control of rain forest preserves. *Ambio* 20: 317–21.

Daily, G. 1997. *Nature's Services: Societal Dependence on Natural Ecosystems*. Washington DC: Island Press.

Dale, V.H., Lugo, A.E., MacMahon, J.A., Steward, T.A., and Pickett, S.T.A. 1998. Ecosystem management in the context of large, infrequent disturbances. *Ecosystems* 1: 546–57.

Dasmann, R.F. 1988. Towards a biosphere consciousness. In *The Ends of the Earth: Perspectives on Modern Environmental History*, pp. 277–88, ed. D. Worster. Cambridge: Cambridge University Press.

Dorm-Adzobu, C., Ampadu-Agyei, O., and Veit, P.G. 1991. *Religious Beliefs and Environmental Protection: the Malshegu Sacred Grove in Northern Ghana.* The Ground Up Case Study Series 4. Washington, DC: Center for International Development and Environment, Word Resources Institute.

Elmqvist, T., Rainey, W.E., Pierson, E.D., and Cox, P.A. 1994. Effects of tropical cyclones Ofa and Val on the structure of a Samoan lowland rain forest. *Biotropica* 26: 384–91.

Finlayson, A.C. and McCay, B.J. 1998. Crossing the thresholds of ecosystem resilience: the commercial extraction of northern cod. In *Linking Social and Ecological Systems. Management Practices and Social Mechanisms for Building Resilience*, pp. 311–37, ed. F. Berkes and C. Folke. Cambridge: Cambridge University Press.

Folke, C. and Berkes, F. 1998. *Understanding Dynamics of Ecosystem–Institution Linkages for Building Resilience*. The Beijer Discussion Papers, No. 112. Stockholm: The Beijer Institute of Ecological Economics, The Royal Swedish Academy of Sciences.

Folke, C., Berkes, F., and Colding, J. 1998. Ecological practices and social mechanisms for building resilience and sustainability. In *Linking Social and Ecological Systems. Management Practices and Social Mechanisms for Building Resilience*, pp. 414–36, ed. F. Berkes and C. Folke. Cambridge: Cambridge University Press.

FPCO (Flood Plan Coordination Organization). 1994. Report on the Flood Action Plan. Draft. Dhaka: Ministry of Water Resources.

FPCO. 1995. Bangladesh Water and Flood Management Strategy. Dhaka: Flood Plan Coordination Organization.

Gadgil, M., Folke, C., and Berkes, F. 1993. Indigenous knowledge for biodiversity conservation. *Ambio* 22: 151–6.

Gadgil, M. and Vartak, V.D. 1976. The sacred groves of Western Ghats in India. *Economic Botany* 30: 152–60.

Gottesfeld Johnson, L.M. 1994. Aboriginal burning for vegetation management in northwest British Columbia. *Human Ecology* 22: 171–88.

Gunderson, L.H. 1996. *A Primer on Adaptive Environmental Assessment and Management*. St Lucie, FL: CRC Press.

Gunderson, L., Holling, C.S., and Light, S.S. 1995. *Barriers and Bridges to the Renewal of Ecosystems and Institutions*. New York: Columbia University Press.

Hilborn, R. 1992. Can fisheries learn from experience? *Fisheries* 17(4): 6–14.

Holling, C.S. 1986. Resilience of ecosystems; local surprise and global change. In *Sustainable Development of the Biosphere*, pp. 292–317, ed. W.C. Clark and R.E. Munn. Cambridge: Cambridge University Press.

Holling, C.S., Berkes, F., and Folke, C. 1998. Science, sustainability and resource management. In *Linking Social and Ecological Systems. Management Practices and Social Mechanisms for Building Resilience*, pp. 342–62, ed. F. Berkes and C. Folke. Cambridge: Cambridge University Press.

Holling, C.S. and Meffe, G.K. 1996. Command and control and the pathology of natural resource management. *Conservation Biology* 10: 328–37.

Holling, C.S. and Sanderson, S. 1996. Dynamics of (dis)harmony in ecological and social systems. In *Rights to Nature: Ecological, Economic, Cultural, and Political Principles of Institutions for the Environment*, pp. 57–86, ed. S. Hanna, C. Folke, and K-G. Mäler. Washington DC and Covelo: Island Press.

Holling, C.S., Schindler, D.W., Walker, B.W., and Roughgarden, J. 1995. Biodiversity in the functioning of ecosystems: an ecological synthesis. In *Biodiversity Loss. Economic and Ecological Issues*, pp. 44–83, ed. C. Perrings, K-G. Mäler, C. Folke, C.S. Holling, and B-O. Jansson. Cambridge: Cambridge University Press.

Hudak, A.T. 1999. Rangeland mismanagement in South Africa: failure to apply ecological knowledge. *Human Ecology* 27: 55–78.

Hughes, J.D. and Chandran, M.D.S. 1998. Sacred groves around the Earth: an overview. In *Conserving the Sacred for Biodiversity Management*, pp. 69–86, ed. P.S. Ramakrishnan, K.G. Saxena, and U.M. Chandrashekara. New Delhi: UNESCO, Oxford & IBH Publishing Co. PVT. LTD.

Kirch, P.V. 1991. Polynesia agricultural systems. In *Islands, Plants and Polynesians. An Introduction to Polynesian Ethnobotany*, pp. 113–33, ed. P.A. Cox and S.A. Banack. Portland, OR: Dioscorides Press.

Kirch, P.V. 1997. Introduction. The Environmental History of Oceanic Islands. In *Historical Ecology in the Pacific Islands*, pp. 1–21, ed. P.V. Kirch and T.L. Hunt. New Haven: Yale University Press.

Lebbie, A.R. and Guries, R.P. 1995. Ethnobotanical value and conservation of sacred groves of the Kpaa-Mende in Sierra Leone. *Economic Botany* 49: 297–308.

Levin, S. 1998. Ecosystems and the biosphere as complex adaptive systems. *Ecosystems* 1: 431–6.

Levin, S. 1999. *Fragile Dominion: Complexity and the Commons*. Reading, MA: Perseus Books.

Lewis, H.T. 1989. Ecological and technical knowledge of fire: aborigines versus park managers in Northern Australia. *American Anthropologist* 91: 940–61.

Lewis, H.T. and Ferguson, T.A. 1988. Yards, corridors and mosaics: how to burn a boreal forest. *Human Ecology* 16: 57–77.

Lindberg, P. and Mossing, A. 1996. The effects of cyclones on agriculture in Western Samoa. Department of Physical Geography, Umeå University, Sweden.

Lockwood, B. 1971. *Samoan Village Economy*. London: Oxford University Press.

Matson, P.A., Parton, W.J., Power, A.G., and Swift, M.J. 1997. Agricultural intensification and ecosystem properties. *Science* 277: 504–9.

Niamir-Fuller, M. 1998. The resilience of pastoral herding in Sahelian Africa. In *Linking Social and Ecological Systems. Management Practices and Social Mechanisms for Building Resilience*, pp. 250–84, ed. F. Berkes and C. Folke. Cambridge: Cambridge University Press.

Niamir-Fuller, M. 1999. Toward a synthesis of guidelines for legitimizing transhumance. In *Managing Mobility in African Rangelands. The Legitimization of Transhumance*, pp. 266–90, ed. M. Niamir-Fuller. London: Intermediate Technology Publications Ltd.

Niamir-Fuller, M. and Turner, M.D. 1999. A review of recent literature on pastoralism and transhumance in Africa. In *Managing Mobility in African Rangelands. The Legitimization of Transhumance*, pp. 18–46, ed. M. Niamir-Fuller. London: Intermediate Technology Publications Ltd.

Orejuela, J.E. 1992. Traditional productive systems of the Awa (Cuaiquer) Indians of Southwestern Colombia and neighboring Ecuador. In *Conservation of Neotropical Forests. Working from Traditional Resource Use*, pp. 58–82, ed. K.H. Redford and C. Padoch. New York: Columbia University Press.

Pandey, D.N. 1998. *Ethnoforestry: Local Knowledge for Sustainable Forestry and Livelihood Security.* New Delhi: Himanshu.

Paulson, D.D. and Rogers, S. 1997. Maintaining subsistence security in Western Samoa. *Geoforum* 28: 173–87.

Ragone, D. 1991. Ethnobotany of breadfruit in Polynesia. In *Islands, Plants and Polynesians. An Introduction to Polynesian Ethnobotany*, pp. 203–20, ed. P.A. Cox, and S.A. Banack. Portland, OR: Dioscorides Press.

Ramakrishnan, P.S. 1998. Conserving the sacred for biodiversity: the conceptual framework. In *Conserving the Sacred for Biodiversity Management*, pp. 3–16, ed. P.S. Ramakrishnan, K.G. Saxena, and U.M. Chandrashekara. New Delhi: Oxford and IBH Publishing Co. Pvt. Ltd.

Ramakrishnan, P.S., Saxena, K.G., and Chandrashekara, U.M., eds. 1998. *Conserving the Sacred for Biodiversity Management*. New Delhi: Oxford and IBH Publishing Co. Pvt. Ltd.

Rappaport, R.A. 1979. *Ecology, Meaning, and Religion*. Berkeley, CA: North Atlantic Books.

Redford, K.H. and MacLean Stearman, A. 1993. Forest-dwelling native Amazonians and the conservation of biodiversity: interests in common or in collision? *Conservation Biology* 7: 248–55.
Schmuck-Widmann, H. 1996. *Living with the Floods. Survival Strategies of Char-Dwellers in Bangladesh.* Berlin: FDCL.
Scoones, I. 1999. Ecological dynamics and grazing-resource tenure: a case study from Zimbabwe. In *Managing Mobility in African Rangelands. The Legitimization of Transhumance*, pp. 217–35, ed. M. Niamir-Fuller. London: Intermediate Technology Publications Ltd.
Spriggs, M. 1997. Landscape catastrophe and landscape enhancement: are either or both true in the Pacific? In *Historical Ecology in the Pacific Islands*, pp. 80–104, ed. P.V. Kirch and T.L. Hunt. New Haven: Yale University Press.
Tilman, D., Wedin, D., and Knops, J. 1996. Productivity and sustainability influenced by biodiversity in grassland ecosystems. *Nature* 379: 718–20.
Walker, B.H. 1993. Rangeland ecology: understanding and managing change. *Ambio* 22: 80–7.
Walters, C. 1986. *Adaptive Management of Renewable Resources.* New York: McGraw Hill.
White, P.S. and Pickett, S.T.A., eds. 1985. Natural disturbance and patch dynamics: an introduction. In *The Ecology of Natural Disturbance and Patch Dynamics*, pp. 3–13, ed. S.T.A. Pickett and P.S. White. Orlando, FL: Academic Press.
Wilson, A. 1993. Sacred forests and the elders. In *Indigenous Peoples and Protected Areas*, pp. 244–8, ed. E. Kemf. London: Earthscan Publications Ltd.

Part III

Social–ecological learning and adaptation

Introduction

Given that some level of uncertainty always exists in complex systems, decision makers need to continuously monitor and integrate appropriate ecological, social, and economic information into management. Such adaptive management, whereby policy making is seen as an iterative experiment, acknowledges uncertainty, rather than assuming it away. Carrying out adaptive management requires a great deal of information to provide feedback to the manager regarding the consequences of the policy experiment. In addition to some of the conventional kinds of ecological and economic data, adaptive management requires qualitative information in the form of feedback from the social–ecological system to indicate the direction in which management should proceed.

Where does the information for adaptive management come from? Some of it comes from conventional science and social science, but some of it can also come from the knowledge held by the resource users themselves. Many local and traditional knowledge systems are characterized by the use of local ecological knowledge to interpret and respond to environmental feedback to guide the direction of resource management. These local management systems have something in common with adaptive management – they emphasize feedback learning and address uncertainty that is intrinsic to all systems. How do we access and use local and traditional knowledge, and what kinds of arrangements are necessary to bring together the full spectrum of knowledge pertinent to a problem?

The three chapters in this section provide insights into these questions. Chapter 8 explores the role of local ecological knowledge in complex systems management, and concludes that a key issue is to share knowledge in the form of 'adaptive co-management.' Chapter 9 explores resource management as problem solving in which the solutions are not technical but require stakeholder participation in a collaborative effort; adaptive management does not provide a set recipe but a collaborative process for learning-based and negotiated problem solving. The context of Chapter 10 is cross-cultural: co-management is based on mutual learning and, once again, on joint problem solving in a kind of adaptive dance as in Gunderson's Chapter 2.

8

Exploring the role of local ecological knowledge in ecosystem management: three case studies

MADHAV GADGIL, PER OLSSON, FIKRET BERKES, AND CARL FOLKE

8.1 Introduction

Local resource users have come to play an increasingly significant role in the ecosystem approach to resource and environmental management. The way it is being organized, its relationship to the institutionalized, professional science, and its role in catalyzing new ways of managing environmental resources have all become important subjects (Kellert *et al.*, 2000; Gadgil *et al.*, 2000; Olsson and Folke, 2001). Local ecological knowledge is a central component of such management regimes, and in this chapter we present three case studies in an attempt to explore its role. These case studies deal with three contrasting socio-economic, cultural, and political settings: that of Sweden, a relatively equitable and homogeneous society; of Canada, a society with a gulf between the Euro-Canadians and the indigenous people; and of India, a highly stratified society but with strong traditions of learning and democracy conducive to the development of participation in resource management.

The development of local knowledge in management appears to have been motivated in two distinctive ways. On the one hand, it may attempt to complement the more general knowledge developed by professional science, with site-specific, contextualized knowledge generated by local users through local observations and experiments. On the other hand, local ecological knowledge may be an attempt to challenge those manifestations of professional science that tend to serve relatively narrow, vested interests. The first motivation dominates in the Swedish case study, where different levels of governance collaborate in a relatively smooth fashion and with similar value systems. The second is significant in the divided societies of Canada and India, where the professional scientific establishment does not share the economic interests and cultural values of the subordinate classes of the society. In particular, the knowledge system of the citizens from subordinate classes comes from a stream where empirically validated knowledge and beliefs commingle and that professional scientists therefore view with great skepticism. As a result, developing a mutually supportive relationship between citizen knowledge and establishment science poses a more difficult challenge in these countries.

Using three case studies from diverse cultures and environments, this chapter explores the quality of that local knowledge and how it is used (if at all) in management practices, and whether such knowledge is recognized by authorities and used in co-management. In particular, we explore the role of local knowledge to deal with change, and how it can improve the knowledge base to respond to change adaptively. We refer to this as adaptive co-management because it combines the adaptive management perspective of Holling (1978) with the idea of co-management or the sharing of management power and responsibility between government and local resource users (e.g., Pinkerton, 1989). In the Swedish case, the issue is lake acidification caused by long-range transport of anthropogenic emissions of sulfur and nitrogen. In the Canadian case, it is oceanographic change related to large-scale hydroelectric development; and in the Indian case, it is the loss of biological diversity in rural areas. The Swedish case focuses on a local fishing association that has the legal right to manage fish and crayfish populations. Here, local ecological knowledge was mobilized to respond to a need for improved management. Consequently, local resource users' institutions for fish and crayfish management and watershed management were formed. The Canadian case is the Hudson Bay Bioregion project, organized by the Inuit community of Sanikiluaq, involving 28 Inuit and Cree communities scattered around the vast region. This project aimed to build an integrated regional-scale knowledge base of environmental change from the point of view of aboriginal people, drawing upon the observations of hunters and fishers. In India, a group of ecologists, science teachers, and students in undergraduate colleges and workers from non-governmental organizations (NGOs) has worked with local communities in 52 clusters of villages distributed in different ecological zones of the country to document people's ecological knowledge, perceptions of ongoing ecological changes, forces driving these changes, and prescriptions for prudent management of environmental resources. The exercise has generated many valuable insights into these issues and in some cases promoted community-based initiatives at good resource management.

8.2 Crayfish management in Lake Racken

Acidification of lakes adversely affects recreational fisheries. Coping with this environmental challenge promotes a whole series of information and ecosystem management initiatives in the Lake Racken watershed in the municipality of

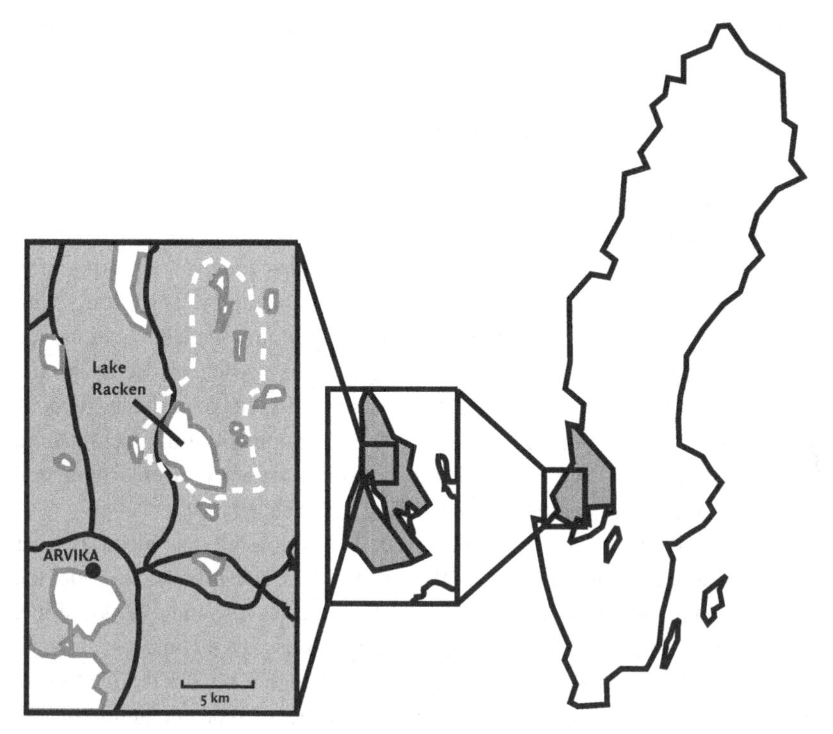

Figure 8.1 To the right, a map of Sweden with Värmland County in gray; in the middle the municipality of Arvika; and to the left Lake Racken, where the watershed is indicated by a dotted line.

Arvika, Värmland County, Western Sweden (Fig. 8.1; see Olsson and Folke, 2001).

In Sweden, governmental resource management agencies have progressively created an arena for the involvement of local people in liming programs to counteract acidification and in the management of fish and crayfish. Local fishing associations commonly play such a role in involving people in many parts of Sweden, managing vast numbers of lakes, rivers, and streams. In the early 1980s, the area for management by local fishing associations was extended from lakes and rivers to include whole watersheds. This extension was founded on a decision at the national level and carried out by the Swedish Land Survey. At this time, fishing associations were only involved in management through communication by developing proposals and requests about management to authorities. Decisions were still taken at the national level of authority. In 1994, fishing associations were provided the right to make decisions concerning fishing and fish conservation. However, some decisions are still taken at the national level, such as a ban on certain fishing methods, and permission is required for the stocking and transfer of fish and shellfish between different water bodies.

The fishing association currently managing Lake Racken watershed was initiated in the early 1980s by a few individuals with the common goal to address the problem of acidification arising in the tributaries of Lake Racken. This environmental change triggered a joint effort by these individuals of sharing, combining, and developing their individual experience and knowledge. At this time there were no locally coordinated management practices incorporating ecological knowledge about species or ecosystem dynamics. This group, with the goal to counteract the acidification process by liming the lake, later moved toward communal fish and crayfish management in a watershed context and today they manage several different species, including noble crayfish (*Astacus astacus*) and brown trout (*Salmo trutta*). The development of the local fishing association enhanced the potential to deal with a diversity of environmental variability, including situations never before experienced.

The crayfish population of the lake declined drastically within a 20-year period, between the mid-1960s and the late 1980s. During the last decade, the local fishing association has taken various measures to respond to this decline. The crayfish of Lake Racken are not currently a source of income for the local community, but crayfishing is an important and highly regarded social event and the main reason why many are fishing. It is a ritual repeated annually, and the whole event contributes to the enjoyment of consuming the locally caught shellfish, which are considered a delicacy.

8.2.1 Deploying ecological knowledge

Olsson and Folke (2001) studied the ecological knowledge among members of Lake Racken fishing association related to crayfish and crayfish management. Their study revealed that they possess knowledge and understanding about species and their biology as well as knowledge of specific ecological processes and functions and how these affect crayfish. The knowledge ranges from the level of individual crayfish to the watershed and is related to complex dynamic ecosystems and cross-scale interactions.

For example, people are aware of the problems of acidification in the area and its temporal and spatial variation. Some infer that crayfish suffer greater acid shock in the parts of the lake where tributaries join it because of the sudden influx of acidic water following spring snowmelt. There is also knowledge of how acidification affects different species and their freshwater habitat. For example, there is an awareness of how acidification affects water quality, including calcium levels, alkalinity, and metals, and local people are engaged in measurements of the water quality throughout the watershed. There is also awareness that the strings that attach the eggs to the female crayfish are broken off in more acidic waters. Locals have observed declines in other species sensitive to acidification, such as roach and trout, and increases of species more tolerant of acidic conditions, such as perch. They dissect the stomachs of fish such as perch to identify crayfish predators and also observe predation by other species, which in combination with knowledge about crayfish feeding habits results in a local awareness of interactions between trophic levels.

Knowledge varies among local users, with two people being particularly knowledgeable. These two individuals play a key role in systematic monitoring, including pH, alkalinity, metals, and several indicator species such as insects, mollusks, and fish. One of them is a biology teacher and the other is a technician at the Lake Racken waterworks where drinking water is treated and distributed to the citizens of Arvika. These individuals are key stewards in the sense of systematic monitoring and various management initiatives, but others in the community also play important roles. Some contribute with knowledge and experience concerning organizational and institutional response in relation to social, economic, or ecological change. Individuals might have contacts with key individuals in other similar local organizations or at other organizational levels (e.g., municipalities, county, national) that facilitate co-management and information sharing. Individuals involved in fishing and hunting contribute with knowledge and experience from almost daily observations of the area and provide information such as the kind of species preying on crayfish, fluctuations in the populations of these predators, algae growth, and other changes in the crayfish habitat.

The occasional observations and systematic monitoring at the local level are complemented with information from scientific studies and from surveys of the area carried out by authorities at the municipality and county levels. Specific studies also include those on mercury levels conducted by a high school in Arvika. The two key stewards represent the key link in transferring such information and knowledge to the local decision-making process of the fishing association. Information is communicated to the other members of the fishing association both informally and during formal meetings.

In addition, there is also an exchange of information and knowledge among similar groups. When acidification was discovered in Lake Racken area by one of the stewards, people of a neighboring watershed were consulted to help the Lake Racken community to form a liming group. Fishing associations often share information during formal meetings of the Arvika Fishing Circle, which includes fishing associations within the municipality of Arvika. Sometimes representatives from authorities and companies participate in these meetings.

This implies that there is not only an input to the local communities from outside sources, but also a feedback of practical implementation of ecological knowledge to stakeholders at other institutional and organizational scales. Most fishing associations are members of the Värmland County Fishing Association, which is a branch of the Swedish Association for Owners of Fishing Rights. Decisions about fishing and fish management can only be made at the local level by the fishing associations, and organizations at other levels function as information facilitators, organizers, and lobbyists.

Knowledge about resource and ecosystem dynamics in relation to the crayfish population among local resource users in the Lake Racken area is thus a mix of external knowledge from various governmental and NGOs and from scientific findings and knowledge generated through occasional observation and systematic monitoring at the local level. In the Lake Racken area, scientific and formal information is contextualized and combined with locally generated knowledge. In this sense, scientific and local knowledge intermingle.

8.2.2 Management practices and institutional dynamics

The responsibility for fish and crayfish management has been shared among local fishing associations, municipality, and national government since 1994 (Olsson and Folke, 2001). Thus, management practices for fish and crayfish management observed locally are embedded in institutions at different organizational levels, which constitute a nested set of institutions. Members of the Lake Racken fishing association monitor ecosystem change through a bundle of indicators and respond to nature's dynamics to secure and enhance the productivity of fish and crayfish populations. This is an ongoing process in which local ecological knowledge is used to re-evaluate and reshape management practices and the rules they are embedded in for improved performance.

For example, the local fishing association has tried different management practices to increase the crayfish stock. As a first effort to do something about the low crayfish catches, the association proposed a 3-year closure of fishing. The authority granted the application and there was no fishing between 1990 and 1993. This was just before the devolution of management rights to local fishing associations in 1994. Since then, the local fishing association has changed the size regulation from 9 cm to 10 cm, and has changed the harvesting time from 2 consecutive days in early August to 2 widely separated days at the beginning and end of the month. The reason for having 2 days of crayfishing with a couple of weeks in between is to provide an opportunity to estimate the population size. The process involves catch and release, with marking of crayfish with a white dot on the carapace, a practice performed by a limited number of people.

The local fishing association is still involved in liming of the watershed. However, since the grants for liming have been subject to a VAT tax, the municipality of Arvika has taken over the application and administration because they can deduct this tax. The fishing association and the municipality of Arvika also cooperate in rewarding people for catching mink, but decisions regarding hunting are not decentralized to the same extent as fishing. For instance, the regulations for hunting mink are controlled at the national level. National law also requires people to be careful with and clean fishing gear, boats, and other equipment that is moved between different waters. This is to prevent the spread of a fungal disease among crayfish that since 1907 has taken a heavy toll on the native noble crayfish population (*Astacus astacus*) in Sweden.

Olsson and Folke (2001) also identify individual practices, encouraged by the local fishing association for improved crayfish management, which are not embedded in formal institutions. These are only recommendations and a person will not be sanctioned for not following them. They include improving habitat to provide shelter, increasing food availability, increasing crayfish aggregation, moving crayfish between localities to prevent inbreeding, and enhancing the crayfish stock by selective fishing, for which people are advised to remove large males and throw back females.

8.2.3 Possible pathways

A crisis or major change like the acidification and decrease of the crayfish population triggers an opportunity to reorganize, i.e., to create, re-evaluate, and reshape management practices, rules, and organizational structure. Decisions in times of reorganization will direct the linked social–ecological systems into a certain trajectory or pathway. Some decisions can reduce flexibility and limit future options, whereas others may do the reverse.

The institutional and organizational changes and the efforts of key individuals in the Lake Racken area have led to an increased capacity to cope with the acidification threat. They have also led into a trajectory of developing an ecosystem management approach, generating knowledge about cross-scale interactions ranging from the watershed level to the individual crayfish. However, some people have the opinion that the recovery rate of the crayfish population is too slow, which has resulted in members of the fishing association discussing alternative management strategies. These involve (a) stocking with reared noble crayfish from a commercial hatchery, and (b) members of the local fishing association running a hatchery with the purpose to stock with noble crayfish reared from the egg-bearing females of Lake Racken. These alternative options are believed by some to fulfill the goal of attaining a large crayfish population more quickly, but they may also counteract the process of being alert in responding to environmental feedback and further alienate people from their ecosystems (Olsson and Folke, 2001).

The Swedish case shows how a local organization and its members that are given a chance, organize themselves, monitor and observe changes in their social–ecological environment, and use their knowledge to create, re-evaluate, and reshape local institutions. The devolution of management rights in 1994 was a step in the direction of co-management, using existing organizational structures and the potential of local fishing associations to manage fish resources. It provides an arena where local and scientific knowledge can complement each other.

8.3 Hudson Bay bioregion

The second case study, from the Hudson Bay area of Canada, was triggered by concern over the environmental impact of large-scale power generation, in this case through strings of hydroelectric dams in northern parts of three Canadian provinces, Quebec, Ontario, and Manitoba. The officially sanctioned environmental impact studies focused on one project at a time. For example, the Government of Quebec's environmental assessment, completed in 1993, addressed only the specific impacts of the Great Whale or James Bay II project. The indigenous people of the region, with extensive dependence on hunting, were concerned that there were cumulative impacts of the network of hydro projects on the environment of the region as a whole that were being ignored. Some sporadic initiatives by government departments and university researchers to assess cumulative impacts did not progress very far. Moreover, governments were reluctant to authorize studies of cumulative impacts of existing and proposed development. It is in this context that local people have assumed the role of challenging the biases of the establishment resource management by generating ecological information based on detailed local-level observations, combined across the region to provide a regional level of understanding.

8.3.1 Initiating the community-based project

The process was initiated by the hunters in the area affected by the huge (15 000 MW) James Bay I project in the belief that they were already seeing impacts that had been ignored. The tiny Inuit (Eskimo) community of Sanikiluaq on the Belcher Islands in eastern Hudson Bay, which is downstream from the coastal currents generated by the plume of the La Grande River on which four dams are located (Martini, 1986), complained about changes in sea-ice pattern and currents. These, in turn, were affecting marine mammal and seabird populations.

Receiving little satisfaction from the government but getting good support from its neighbors, Sanikiluaq took the lead to organize a project involving 28 Inuit and Cree communities around Hudson Bay. Carried out between 1992 and 1995, the project aimed to build an integrated regional-scale picture of environmental change from the point of view of aboriginal people, drawing upon the day-to-day and year-to-year observations of hunters and fishers. The project was supported by a northern-oriented national NGO, the Canadian Arctic Resources Committee, which did the fund raising and helped with the logistic support for the project. Aboriginal leaders provided the intellectual direction of the work, and back-up was provided by a number of leading northern scientists with both government and university affiliations.

The study was carried out through six regional meetings that brought together hunters and other knowledgeable people from the six regions of Hudson/James bays. Much of the information was collected on maps and digitized for Geographic Information Systems (GIS) analysis. A second series of meetings helped to verify the information and fill the gaps. Two workshops with scientists helped consolidate the information and formulate ways of presenting the material. Progress reports were issued in 1995 and the final report in 1997 (McDonald, Arragutainaq, and Novalinga, 1997).

8.3.2 Compiling local ecological knowledge

Some of the findings were related to the effects of hydroelectric development (e.g., the strings of reservoirs attracting migratory geese inland), but other findings may have been related to climate change (Fast and Berkes, 1998). On Southampton Island, for example, local people reported that snow was arriving before the freshwater freeze up, creating a different kind of lake ice. Whale Cove reported that snow had increased but that it melted earlier than it did in the past. Chesterfield Inlet, Southampton Island, and Arviat all reported increasingly more erratic weather, such as snow melting in May but blizzards occurring as late as June (McDonald *et al.*, 1997). Sanikiluaq hunters reported recent changes in currents, sea ice and winterkill of common eiders (*Somateria mollissima*). The work of the biologists Robertson and Gilchrist (1998) provided cross-verification by corroborating Inuit observations of changes in regional sea-ice conditions in eastern Hudson Bay as related to eider winterkill.

The overall picture that emerged from the Hudson Bay bioregion study was an accelerated pace of environmental change in Hudson Bay, with large-scale changes in goose migration patterns and in the sea ice and currents of the bay.

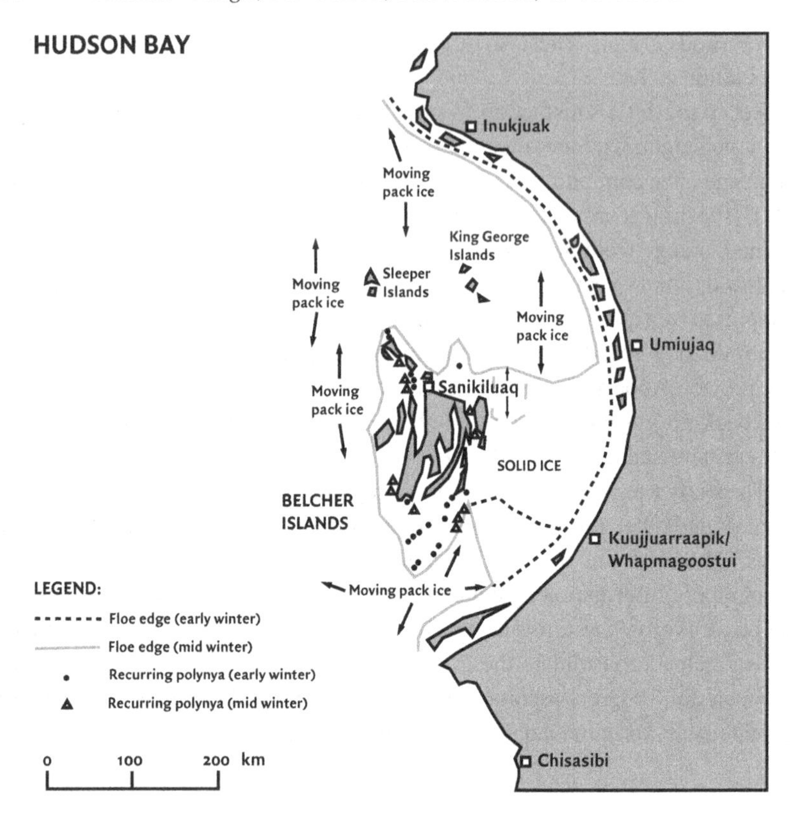

Figure 8.2 Eastern Hudson Bay sea-ice conditions in the period about 1920–70. Adapted from McDonald *et al.*, (1997).

The changes were narrated in quotations and through maps, which provided details of changes based on indicators monitored by indigenous experts but rarely measured by Western scientists, thus showing how traditional knowledge can complement scientific data (Fenge, 1997).

It is generally thought that traditional knowledge complements scientific data by providing local information. However, one of the significant aspects of the Hudson Bay bioregion case was the use of traditional knowledge for the assessment of impacts and environmental change over a large area. Figures 8.2 and 8.3 show the details of changes in the ice pattern before and after the 1970s when the plume of water from the hydro development project started to change the oceanography of the region of Sanikiluaq.

The figures show that the few hundred Inuit residents of the area have knowledge of the distribution of ice and current features of an area of some $600 \times 600\ km^2$. They know where the ice floe edge is in winter (because that is where people hunt seals) and how it changes from early winter to mid-winter.

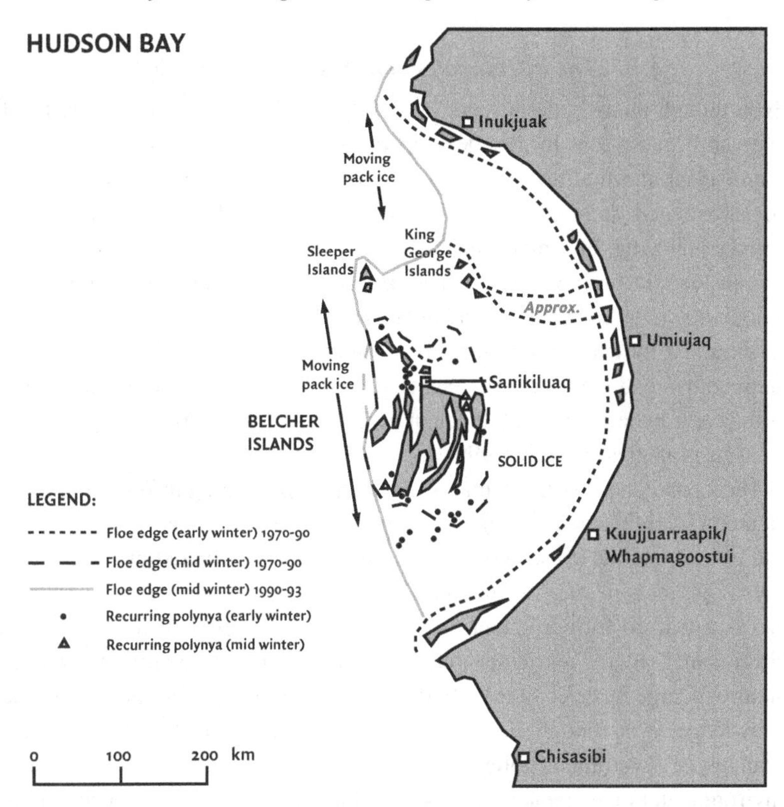

Figure 8.3 Eastern Hudson Bay sea-ice conditions in the period 1970–93. Adapted from McDonald *et al.*, (1997).

They know where the polynyas are. These are the permanent open-water areas (surrounded by solid ice) where various species, such as the eider ducks, congregate. The figure of the earlier period shows that people, as interviewed in 1993, had no trouble reconstructing the environment, as known in the period, 1920–70. Also, by constructing pre-impact and post-impact maps of ice and currents, they were able to show the major changes. Note, for example, the reduction in the number of mid-winter polynyas, constricting the critical habitat of species such as the eider and resulting in winterkills.

Figures 8.2 and 8.3 show the details of only one community out of the 28 involved in the study. The region-wide maps in the study report were compiled by putting together six area maps, each of which aggregated the overlapping knowledge of adjacent communities (McDonald, *et al.*, 1997). The resulting maps are more detailed than the ice maps produced by oceanographers, not only for the local areas, such as around the Belcher Islands as shown in the figures, but also for the Hudson and James bays as a whole.

8.3.3 The integration of local ecological knowledge

Even though many authors regard traditional knowledge to be merely locally relevant because it is locally developed, the case shows the feasibility of the use of local and traditional knowledge synoptically over a large area. This is not an isolated finding; as the book *Sacred Ecology* documents, many traditional knowledge and management practices are in fact common enough to be considered as principles. For example, many groups of arctic and subarctic indigenous peoples, from Labrador to Alaska, monitor the fat content of the caribou that they hunt (a major food species across the north). This provides them with a readily observable qualitative index of the health of the animal and of the caribou herd, an index that can be monitored over time to help decide on hunting strategies (Berkes, 1999).

The Canadian case shows that, in addition to generating information to supplement scientific data, local environmental knowledge can provide several additional benefits: (a) selecting critical variables not covered in scientific surveys (e.g., inshore currents and ice characteristics); (b) generating hypotheses that can lead to further scientific studies (e.g., the relationship between environmental change and eider duck winterkill); and (c) compiling a synoptic picture of large-scale change by building and combining regional maps of local knowledge. Even though the report of the project itself did not result in tangible changes in government policy, the findings have been used for resource and environmental management purposes, and to provide insights into how to deal with uncertainty and surprise, as in climate change (Fast and Berkes, 1998). Perhaps more significant for policy making, the knowledge of the indigenous groups has been used by the co-management boards set up by previous land claims agreements, the *James Bay and Northern Quebec Agreement* of 1975 and the *Nunavut Agreement* of 1993. These agreements recognize the right of Cree and Inuit peoples to participate in the management of their living resources, conservation planning, land use planning, and impact assessment. Indigenous groups won these rights, not because they were able to show the legitimacy of their knowledge, but because of legal challenges to the authority of Canadian federal, provincial, and territorial governments as sole authority on resource and environmental matters.

The use of peoples' knowledge in the co-management bodies created by these native land claims agreements is significant. As Kendrick discusses in Chapter 10 of this volume, 'co-managing knowledge' within these bodies may be characterized as a mutual learning system. How the two kinds of knowledge can be used together is a matter of mutual education, respect, and trust building. The particular caribou co-management board described by Kendrick is the oldest such board in Canada, and it has a longer track record than the co-management institutions under the *Nunavut Agreement*, which covers most of Kendrick's case study area. What generalizations can be offered regarding the co-management of different kinds of knowledge in the northern Canadian experience?

One conclusion from the Canadian experience is that indigenous knowledge holders and scientists can grow to appreciate one another's knowledge. However, there are limits to the acceptability of each kind of knowledge by the other. Scientists require empirical validation of traditional knowledge and have well-articulated rules of evidence. They tend to be skeptical of knowledge generated without scientific measurements and quantification. Also, scientists are uncomfortable with the belief component of traditional knowledge, which can be characterized as a knowledge–practice–belief complex (e.g., Gadgil, Berkes, and Folke, 1993). For their part, holders of traditional knowledge do not consider that science is capable of verifying their knowledge. In the case of conflicts between the two kinds of knowledge (e.g., whether hunting should be allowed for a particular species), they consider their knowledge to be superior because it is grounded in local observations. They have a profound skepticism of the 'book knowledge' of scientists and regard claims of (scientific) expertise with disdain, unless the individual scientist is personally known to have studied that particular area for years.

8.4 People's science movements in India

India – a relatively poor, highly stratified society, yet with a strong democratic system and rich traditions of learning – has a vibrant participation in resource management movement, commonly called PSM or People's Science Movement, organized as an All India People's Science Network. The movement first took root in the early 1960s in the state of Kerala, where the statewide PSM, known as KSSP, has over 40 000 individual members. Its initial motivation was to wean people away from the traditional cultures that were viewed as mired in superstitions and responsible for the continued depressed status of the subordinate classes. Its mission was thus to communicate professional science to all citizens, especially the poor and uneducated, and to eradicate superstitions. It did not consider local ecological knowledge to be of any value. It saw traditional practices of nature conservation, such as the protection of sacred groves or sacred fig trees (*Ficus* spp.), as superstitions that needed to be eradicated (Zachariah and Sooryamoorthy, 1994).

These PSMs began to re-examine their wholehearted championship of professional science in the early 1970s when confronted with environmental issues such as water pollution emanating from the Mavur Rayon factory or the submersion of Silent Valley, a biodiversity-rich, hilly region by a hydroelectric project. In these cases the assessments of environmental impacts by the scientific establishment were clearly inadequate and biased. PSMs then added independent, professionally unbiased evaluations of the environmental, social, and economic impacts of such activities as an additional significant concern (Gadgil and Chandran, 2000). Initially, these assessments were undertaken primarily by scientists from academic institutions and communicated to the broader public, but did not involve any element of local ecological knowledge.

8.4.1 Co-management of forest resources

Independently of these activities of the PSMs, but prompted by the broader awareness of environmental challenges, resource management agencies began to set up in the 1970s systems of co-management of forest and water resources. These systems assigned to local people some additional share in the resource in return for assistance in guarding the resources. The local people, however, had little role in planning and decision making; nor was their local ecological knowledge put to any use. However, the setting up of these systems of co-management triggered the spontaneous establishment of thousands of village forest committees (VFCs) in many parts of the country, especially in the state of Orissa with a large concentration of tribal population (Poffenberger and McGean, 1996; Saxena, 1997).

One such VFC was set up in the Dhani Panchayat (the equivalent of a village council) in 1987 in response to large-scale commercial felling. The VFC organized complete protection of a forest area of 840 ha in extent. However, this protection was not exercised as a rigid prescription. It was relaxed to permit some cattle grazing after 3 years of good regeneration; at the same time some of the poorest households were permitted limited levels of harvest of fuelwood. Thus local ecological observations were being employed to organize a regime of adaptive management. There was, however, no involvement of professional scientists and their scientific information and that of professional resource managers in this system (Panigrahi and Rao Giri, 1996; Nayak, Rao Giri, and Singh Neera, 1996).

8.4.2 Intellectual property

In the 1990s there were important developments, which began to change the attitude of professional scientists toward knowledge held by people with no contact with formal science. Thus scientists at the Tropical Botanical Garden and Research Institute in Thiruvantapuram in Kerala developed a new drug called Jeevani from a rain-forest herb, *Trichopus zeylanicus*, on the basis of information supplied by two members of the Kani tribe. In light of the provisions for benefit sharing embodied in the Convention on Biological Diversity, they shared half the royalty received from a drug company with a trust set up by the tribe to which the informants belonged (Pushpangadan, Rajasekharan, and George, 1998).

With such examples of effective management of natural resources by ordinary, mostly illiterate citizens, as well as proven commercial value of their knowledge, PSMs began to re-evaluate their perceptions of local knowledge. While they remain committed to the eradication of superstitions, they now acknowledge that some practices such as the conservation of sacred groves may serve a very useful social purpose, even though they had originally been implemented through the belief that a forest deity would punish any violators of taboos to remove wood from the grove. In fact, PSMs as well as professional scientists began to survey and document such refugia (Ramakrishnan, Saxena, and Chandrashekara, 1998). Their rationale was also examined further and shown to include secular motivations as well (Gadgil, Hemam, and Reddy, 1998).

8.4.3 Local-level resource mapping

With the PSMs' interest in reaching large masses of people, they became involved in a major literacy drive that was launched in the late 1980s against the backdrop of about 50 percent of the Indian population remaining illiterate 40 years after independence. Amongst their objectives was to reach 100 percent literacy in parts of the already highly literate state of Kerala, and in this KSSP achieved a considerable measure of success. In general, the literacy campaign was notable for inducting substantial local voluntary effort to supplement the usual highly centralized bureaucratic development projects commonly prevalent in India.

A significant fallout of the literacy campaign and the interest in generating literature for neoliterates was the Panchayat Level Resonance Mapping (PLRM) program, again initiated by KSSP. Panchayat denotes the lowest tier of self-government in India, and PLRM was an exercise involving citizens, primarily villagers, many of them newly literate, in mapping the land and water use in their localities. The program was guided by scientists from the Centre for Earth Science Studies, Thiruvananthapuram, and the citizens' own knowledge was not a matter of interest to the program. However, its relevance was gradually realized during the course of the PLRM exercises. A well-known example of this was the experience in Kaliassery Panchayat in Kannur district. Here there were recurring flood problems that often claimed human lives. The participatory mapping exercise revealed an old drainage channel that had been blocked. The mapping exercise prompted people to take voluntary action to restore the structure, which has subsequently mitigated the flood problem (Gadgil and Chandran, 2000).

8.4.4 Peoples' Biodiversity Registers

All these developments led in 1995 to a serious attempt to examine whether local ecological knowledge can be combined effectively with professional scientific knowledge and then deployed to support systems of adaptive co-management throughout the country. This is being attempted through a program called 'People's Biodiversity Registers' (PBRs). It is a program of documenting the understanding of lay people, primarily rural and forest-dwelling communities, of living organisms and their ecological setting. The information recorded relates to present status as well as changes over recent years in distribution and abundance, factors affecting distribution and abundance including habitat transformations and harvests, known uses, and economic transactions involving these organisms. The document also records the perceptions of local people concerning ongoing ecological changes, their own development aspirations, and finally their preferences as to how they would like the living resources and habitats to be managed. The experience of preparation of these 50-odd PBRs has been most positive, with considerable enthusiasm generated amongst teachers and students in educational institutions, amongst NGO activists, as well as among members of local communities (Gadgil *et al.*, 2000).

Some very interesting developments have been triggered in the course of preparation of the PBRs. One such happy experience comes from Himachal Pradesh. Nanj, a village on the bank of the River Sutlej, witnessed a novel community initiative during the course of study. The village was an active participant in the literacy movement during 1992–3 and the people were exposed to a variety of issues relating to natural resource management. As a consequence, a heavily degraded patch of forest was enclosed by consensus to prevent harvesting. The regeneration has been extremely good and promising. During the literacy campaign, a blackboard had been painted on a wall at a public place in the village for open classes and dissemination of information. Between 1994 and 1996, the blackboard had fallen into disuse. It was revived again during the PBR documentation to display the gist of information collected. This resulted in public debates on the issues raised by the information and in turn on conservation actions. One such debate centered on the species *Kambal*. This is a multipurpose tree found up to the mid-Himalayas. It is considered to be a good fuelwood and its leaves are used as green manure in ginger cultivation. It was pointed out on the blackboard that due to excessive pressure of both fuelwood and manure collection, the *Kambal* had been reduced to a bush in the forest, leading to declining availability of both fuelwood and manure. After many days of discussion in front of the blackboard, it was decided that leaf manure for ginger was a higher priority. As other fuelwood species were available in the forest, the extraction of *Kambal* would be restricted to leaves for green leaf manure and the bushes would be pruned in such a way that one or two shoots would be permitted to grow. At the same time, a few progressive farmers decided to experiment with agricultural crop residues as a substitute for *Kambal* leaves for manure. Over 1 year, they demonstrated that there was no difference in the yields from the two kinds of manure and subsequently more farmers turned to crop residues as this meant lower labor inputs. As a consequence, *Kambal* is now flourishing in the forest and, due to careful pruning and good rootstock, will grow back to trees in a few years time (Gadgil *et al.*, 1998).

An account of the experience appeared in the Annual Survey of Environment for 1998 published by Hindu, one of the leading English language newspapers of south India. A large number of people from all over India have expressed an interest in undertaking PBR exercises in their own area as a result of this exposure. Similar interest has been expressed from Brazil and South Africa as well. More concretely, the government of India, in the Biological Diversity Bill tabled in the parliament in June 2000, has specifically entrusted to the village councils the responsibility of documenting biodiversity resources, knowledge, and conservation efforts. Further, the bill provides for direct sharing of the royalties from the commercial application of these efforts with individuals or groups of people. Although the bill does not specifically mention the village documents as the basis for benefit sharing, it would become eventually imperative for the government to do so.

8.5 Co-managing knowledge

These three experiences from Sweden, Canada, and India point to a great challenge for the scientific community. The ecological knowledge of tribal, peasant, herder, or any other resource-user group is of relevance in the context of systems of adaptive co-management. In particular, it incorporates knowledge derived from historical observations of 'natural experiments' and their dynamics (e.g., succession following a fire event) of these systems. Because it is difficult to systematically conduct properly planned and replicated experiments in complex systems, local observations of such experiments can be of significant value. This is particularly true for situations of change and dealing with change in an adaptive fashion. Hence, the incorporation of local and traditional knowledge into adaptive co-management becomes particularly important.

However, co-managing different kinds of knowledge is fraught with pitfalls. It is true that both local knowledge and scientific knowledge are based on empirical observations and the need to interpret and understand the world. But there are major differences in these two kinds of knowledge. Western science has very specific rules about the admissibility of evidence and turning observations into hypotheses. By contrast, local knowledge can be broadly characterized as knowledge–practice–belief systems (Gadgil *et al.*, 1993; Berkes, 1999). Local knowledge often blends knowledge and belief without clear distinction. Given the strong tradition of skepticism of Western science, it is difficult for those trained in this stream to deal with local knowledge. Often the tendency is to try to tease the knowledge and belief components apart, and then to try to assimilate into science that which is empirically valid (e.g., Mackinson and Nøttestad, 1998; Colding and Folke, 2001).

From one point of view, it is important to identify and use the empirically valid component of local knowledge. From another point of view, however, to do so is to miss the point. Is local knowledge merely marginalized knowledge? Can it overcome its marginalization only by 'fitting' into the framework of 'establishment science?' Teasing apart knowledge and belief undermines the very process by which local knowledge is practiced. Local knowledge, without its belief component, is out of context. Some people who work closely with local and traditional knowledge systems have commented that Western science has a tendency to try to reduce traditional knowledge to either 'myth' or 'data.' The alternative view is to respect the integrity of each knowledge system within its own framework and worldview, and to bring together knowledge systems by treating them as equal but different (Berkes, 1999).

Further problems arise because of the acceptance of so-called post-modernist critiques of science. This critique rejects the claims of modern science to be a special knowledge system with far stronger links to objective reality than any other knowledge system. In fact, some post-modernists view science as just another belief system. We believe this critique not to be valid; modern science is indeed a far better organized system of elaborating knowledge of the world, and the exercise of bringing local knowledge together with scientific knowledge must acknowledge the significance of this aspect. At the same time, it needs to be acknowledged that modern science has little of the wealth of detailed context-specific observations of the dynamics of complex ecological systems. Knowledge of how to respond to disturbance and how to build resilience for enhancing adaptive capacity is still in its infancy. Such knowledge, site specific and often embedded in management practices of local resource users, exists as a part of the knowledge systems of tribal peoples, peasants, herders, and fishers in many parts of the world (Berkes and Folke, 2002).

This chapter explores the potential of combining scientific knowledge with local knowledge in a process of adaptive co-management. In the Swedish case study, adaptive co-management manifests itself in the establishment of local fishing organizations that blend scientific and local ecological knowledge, which is supported by governmental institutions. In the Canadian case study, such adaptive co-management is achieved by the use of indigenous knowledge in the formal land claims agreements between Cree and Inuit peoples of the North and governmental agencies. In the Indian case study, adaptive co-management is expressed through joint forest management programs, the recognition of the intellectual property rights of local people, and local-level resource mapping with the intent of protecting biodiversity hot spots such as sacred groves.

These case studies and others indicate that involving local people and designing an institutional and organizational structure for incorporating local ecological knowledge in ecosystem management are desirable. Local institutions need to be integrated or nested within institutions at other organizational levels to be able to match social and ecological processes at various scales. For example, the institutional and organizational structure observed in the Swedish case makes possible the process of constant testing of rules in relation to ecological and socio-economic factors at different scales. This in turn creates feedback loops at different scales, and a cross-scale institutional dynamic that we argue is necessary to consider in adaptive co-management.

An important part of adaptive co-management is to stimulate further the possibility for local organizations to interact with each other and with organizations at other levels. That is, we argue that adaptive co-management would be enhanced by linking institutions both horizontally (across space) and vertically (across levels of organization), using the terminology of Ostrom *et al.* (2002). Information sharing and conflict resolution across scales are important for ecosystem management. The flow of information, interactions, and other linkages between and among organizations for adaptive co-management are areas that should be investigated further. For example, the Swedish case pinpoints the functional role of different individuals to facilitate the flow of information. Different individuals at different organizational levels play various key roles in the processes of creating, re-evaluating, and reshaping management practices, rules, and organizational structure in relation to ecological and social dynamics. The quality of the knowledge and understanding that these stewards possess is of crucial importance for which trajectory is chosen for the linked social–ecological system.

As explored at the workshop on the role of local and regional assessments in an international ecosystem assessment (Winnipeg, September 1999), ecosystem assessments by local people can fulfill several important objectives. These objectives include promoting participatory processes; creating new information to share across scales; making optimal use of existing knowledge; developing indicators of change and resilience to monitor ecosystem dynamics; and transforming existing institutions toward ecosystem management. Such ecosystem assessments by local people can create alliances between owners of formal and informal knowledge, as in the India case. They can establish links among governments, local users, and scientists, as in the Swedish case. They can create new information about local ecosystem conditions, to be shared vertically (from local to national levels) and horizontally (among regional groups of indigenous peoples), as in the Canadian case. Such exercises can build on pluralistic approaches, help monitor change, and create a vision of desirable environmental futures, through a process of adaptive co-management.

References

Berkes, F. 1999. *Sacred Ecology: Traditional Ecological Knowledge and Management Systems*. Philadelphia and London: Taylor & Francis.

Berkes, F. and Folke, C. 2002. Back to the future: ecosystem dynamics and local knowledge. In *Panarchy: Understanding Transformations in Systems of Humans and Nature*, pp. 121–46, ed. L.H. Gunderson and C.S. Holling. Washington DC: Island Press.

Colding J. and Folke, C. 2001. Social taboos: 'invisible' systems of local resource management and biological conservation. *Ecological Applications* 11: 584–600.

Fast, H. and Berkes, F. 1998. Climate change, northern subsistence and land based economies. In *Canada Country Study: National Cross-Cutting Issues*, Vol. VIII, pp. 205–26, ed. N. Mayer and W. Avis. Downsview, Ontario: Environment Canada.

Fenge, T. 1997. Ecological change in the Hudson Bay bioregion: a traditional ecological knowledge perspective. *Northern Perspectives* 25(1): 2–3.

Gadgil, M., Acharya, S., Barman, R., *et al.* 1998. Where are the people? *The Hindu Survey of the Environment '98*. pp. 107–37. Chennai: The Hindu Press.

Gadgil, M., Berkes, F., and Folke, C. 1993. Indigenous knowledge for biodiversity conservation. *Ambio* 22: 151–6.

Gadgil, M. and Chandran, M.D.S. 2000. M.K. Prasad: scientist as a social activist. *Frontline* 17: 84–7.

Gadgil, M., Hemam, N.S., and Reddy, B.M. 1998. People, refugia and resilience. In *Linking Social and Ecological Systems. Management Practices and Social Mechanisms for Building Resilience*, pp. 30–47, ed. F. Berkes and C. Folke. Cambridge: Cambridge University Press.

Gadgil, M., Seshagiri Rao, P.R., Utkarsh, G., Pramod, P., Chatre, A., and Members of the People's Biodiversity Initiative. 2000. New meanings for old knowledge: the People's Biodiversity Registers Programme. *Ecological Applications* 10: 1307–17.

Holling, C.S., ed. 1978. *Adaptive Environmental Assessment and Management.* London: Wiley.
Kellert, S.R., Mehta, J.N., Ebbin, S.A., and Lichtenfeld, L.L. 2000. Community natural resource management: promise, rhetoric, and reality. *Society and Natural Resources* 13: 705–15.
Mackinson, S. and Nøttestad, L. 1998. Combining local and scientific knowledge. *Reviews in Fish Biology and Fisheries* 8: 481–90.
Martini, I.P., ed. 1986. *Canadian Inland Seas.* Amsterdam: Elsevier.
McDonald, M., Arragutainaq, L., and Novalinga, Z., compilers. 1997. *Voices from the Bay: Traditional Ecological Knowledge of Inuit and Cree in the Hudson Bay Bioregion.* Ottawa: Canadian Arctic Resources Committee and Municipality of Sanikiluaq.
Nayak, P.K., Rao Giri, Y., and Singh Neera, M. 1996. India's emerging experiences with joint forest management (mimeographed report). Bhuvaneshwar, Vasundhara, Orissa: Vasundhara.
Olsson, P. and Folke, C. 2001. Local ecological knowledge and institutional dynamics for ecosystem management: a study of crayfish management in the Lake Racken watershed, Sweden. *Ecosystems* 4: 85–104.
Ostrom, E., Dietz, T., Dolsak, N., Stern, P.C., Stonich, S., and Weber, E.U., eds. 2002. *The Drama of the Commons.* Washington DC: National Academy Press.
Panigrahi, R. and Rao Giri, Y. 1996. Conserving biodiversity: a decade's experience of Dhani Panch Mouza people. Bhuvaneshwar, Orissa: Vasundhara.
Pinkerton, E., ed. 1989. *Co-operative Management of Local Fisheries: New Directions for Improved Management and Community Development.* Vancouver: University of British Columbia Press.
Poffenberger, M. and McGean, B., eds. 1996. *Village Voices, Forest Choices: Joint Forest Management in India.* New Delhi: Oxford University Press.
Pushpangadan, P., Rajasekharan, S., and George, V. 1998. Benefit sharing with Kani tribe: a model experimented by Tropical Botanic Garden and Research Institute (TBGRI). Presented at Medicinal Plants for Survival: International Conference on Medicinal Plants, February 16–19. Bangalore, India: National Institute of Advanced Studies.
Ramakrishnan, P.S., Saxena, K.G., and Chandrashekara, U.M., eds. 1998. *Conserving the Sacred for Biodiversity Management.* New Delhi: Oxford and IBH Publishing Co. Pvt. Ltd.
Robertson, G.J. and Gilchrist, H.G. 1998. Evidence of population declines among common eiders breeding in the Belcher Islands, Northwest Territories. *Arctic* 51: 378–85.
Saxena, N.C. 1997. *The Saga of Participatory Forest Management in India.* Bogor, Indonesia: CIFOR Publications.
Zachariah, M. and Sooryamoorthy, R. 1994. *Science for Social Revolution? Achievements and Dilemmas of a Development Movement.* New Delhi: Vistaar Publications.

9

Facing the adaptive challenge: practitioners' insights from negotiating resource crises in Minnesota[1]

KRISTEN BLANN, STEVE LIGHT, AND JO ANN MUSUMECI

9.1 Introduction

The chapter draws lessons and insights from interviews with practicing resource managers involved in leading diverse groups of primary interest groups through resource management crises and change. Each of these management efforts was perceived by the interviewed practitioners and others as experimenting with new ways to recouple and renew social–ecological systems. They represent a nested set of local and regional experiments within one organizational context, a state resource management agency that was intentionally trying to reorganize through novel approaches to management and citizen involvement (Fig. 9.1). All of the cases profiled were characterized by involvement of multiple stakeholders with competing interpretations, values, and goals for the resource system, and reflected a conscious design to engage citizens in creating alternative platforms for resource negotiation (Woodhill and Röling, 1998). In each case, practitioners were experimenting with learning to function differently, outside traditional norms of leadership.

The goal of this study was to identify management practices and frameworks that are founded on knowledge and understanding of dynamics in both human and ecological systems, and to identify the key elements contributing to adaptive response. In this chapter, we develop a matrix based on the release and reorganization phases of the Holling adaptive cycle in an attempt to classify the 'tacit understanding,' or intuitive guiding principles, which emerged in interviews. Practitioners articulated principles loosely, drawing metaphors from systems theory and chaos theory, organization and change management, and ecosystem management. The chapter explores whether and how practices based on these guiding principles contributed to creating adaptive capacity and resilience in social–ecological systems.

Figure 9.1 Department of Natural Resources ecosystem management case studies at multiple scales.

9.1.1 Rationale

Conventional resource management has been characterized as a crisis-response model, because in constraining a managed system to optimize for a few narrow targets, it often invites larger and larger external feedbacks that ultimately compromise the resilience of the system. Over time this can lead to the collapse or near-collapse of the resource system itself, generating 'crisis' in the social, political, and economic system as well. The resilience of social and ecological systems is therefore linked and co-evolutionary in nature (Gunderson, Holling, and Light, 1995; Perrings *et al.*, 1995; Holling and Sanderson, 1996; Berkes and Folke, 1998; Levin, 1999).

The failures of conventional management have led to a widespread search for new approaches able to anticipate and cope with multiscale demands and stresses while maintaining ecological and social resilience. Increasingly, the search for new approaches has been manifested by a broadening of management paradigms – beyond expert-based, control-oriented management and instrumentalist, reductionist science – to include greater emphasis on community-based, participatory approaches to management (Webler, Kastenholz, and Renn, 1995; Knight and Meffe, 1997; VanNijnatten, 1999; Pimbert *et al.*, 2000; Wondolleck and Yaffee, 2000), and re-examination and revaluation of the diverse spectrum of resource systems based on local and traditional ecological knowledge (Anderson and Grove, 1987; Kloppenburg, 1991; Holmberg, 1992;

Pimbert, 1993; Reichhardt *et al.*, 1994; Rocheleau, Tomas-Slayter, and Wangari 1996; Sarin, 1996; Berkes *et al.*, 1998; Posey, 1999). Increasingly, strategies for incorporating such practices into agency resource management are being pursued within modern pluralist democracies in an attempt to manage conflicts between competing users, negotiate through and out of social and ecological crisis situations, and avoid or pre-empt future conflicts and crises (Wondolleck and Yaffee, 2000).

Many authors have argued that human individuals and groups appear to do the majority of 'out-of-the-box learning,' or breakthrough thinking, in response to crisis (Kingdon, 1984; Holling, 1986; Lee, 1993; Light and Dineen, 1994). In the adaptive cycle heuristic developed by Holling (1986), 'crisis' can serve as a source of renewal, the Schumpeterian 'creative destruction' which allows re-ordering and reorganization of system 'capital.' Such 'capital' may be present, for example, in the form of the resource base (ecological systems or natural capital), knowledge, relationships, and values (social systems), or available financial capital (economic systems). The organizational or ecological response to crisis depends both on the 'capital' present in the existing system as well as on the unfolding of events that lead to reorganization (Kingdon, 1984; Lee, 1993; Westley, 1995). In the relatively short periods of rapid change which follow in the wake of creative destruction, reorganization of component relationships can occur such that the new system that emerges is fundamentally different from the old one (Holling, 1986). At this stage, individuals, small influences, and/or random events – 'novelty' – can have a major impact on the configuration of the new system that emerges. Control strategies and management skills that are effective in traditional bureaucracies and agencies may be inappropriate or counterproductive in this period from creative destruction to reorganization, or 'the backloop.' Facilitating radical reorientation in resource management, therefore, may require development of skill sets and management principles that differ from those that have served conventional resource management.

As in traditional resource management systems, which have been shown to avoid over-harvest by codifying management 'rules of thumb' in social and religious belief systems (Gadgil, Hemam, and Reddy, 1998), modern resource management practitioners may develop 'tacit understanding' – intuitive, context-specific understanding based on practical experience and observation over a career or a lifetime. Such knowledge contributes in particular to resilience because (a) it is based on a long-term, qualitative understanding of the system and therefore incorporates understanding of long-term change, or 'slow variables,' (b) it includes insight derived from experiences with rare events, or surprise, and thus may aid recognition of thresholds in order to avoid flips, and (c) it complements quantitative monitoring by helping recognize when a system has shifted from being driven by key processes which are essentially linear (such as those occurring from exploitation to conservation) to being driven by nonlinear processes (disturbance, release, and renewal).

In practice, 'experts' do not operate by deriving general rules from case-by-case experience. Rather, they may begin by applying rules, but they gradually begin to rely on intuitive knowledge without applying explicit rules as their experience grows (Dreyfus and Dreyfus, 1986, quoted in Capra, 1996: 278). When asked to articulate such understanding, however, experts translate their knowledge as heuristics and abstract 'rules,' using language, which is in itself an abstraction. These 'rules,' therefore, are presented not as substitutes for experience, but rather as guidelines for experimentation, inquiry, and dialogue. A major characteristic of adaptive, participatory, and indigenous resource management systems is a focus on learning-by-doing (Walters and Holling, 1990; Bawden, 1992; Allen *et al.*, 1998; Berkes, 1998; Borrini-Feyerabend *et al.*, 2000). Experiential learning and reflection are essential because the ecosystems under management are constantly changing. There are no 'cookie cutter' approaches that will work for more than one system or for more than brief periods of time.

Furthermore, the observation that local-level organizations often develop the capability to respond to feedbacks faster than do centralized agencies implies a need for greater decentralization of management learning and decision making (Westley, 1995; Berkes and Folke, 1998). Multiple, modest experiments may yield more new learning about a problem than one general design applied widely (Brunner and Clark, 1997). At the same time, the cross-scale nature of many social, ecological, and economic problems, particularly in the modern global economy, creates a need to find effective strategies to address linkages across scales (Holling, 1986; Grumbine, 1994; Gunderson *et al.*, 1995; Folke *et al.*, 1998; Woodhill and Röling, 1998). For this reason, we have selected a set of diverse cases representing local and regional experiments in watershed management, forestry, and fisheries within the Minnesota Department of Natural Resources (DNR) for a comparative approach to the examination of management practices applied to specific complex problems at nested scales.

9.1.2 Objectives and methods

The goals of this study were (a) to identify the key elements contributing to adaptive or novel responses to natural resource crises in a set of spatially and temporally nested local and regional examples, and (b) to identify and investigate management practices and principles deriving from resource practitioners' direct experience in facilitating organizational renewal in the Minnesota DNR. The DNR is a state agency charged with the management of Minnesota's wildlife, fisheries, water, mineral, forest, and recreational resources, and, in more recent language, its ecosystems and ecological services. The case studies are based on interviews conducted with the practitioners who were primarily responsible for implementation on-the-ground. The case studies were selected on the basis of testimony by peers and participants that consistently suggested a major shift in approach or understanding had been achieved through the project. Interviews were conducted one-on-one over a 2-month period during the fall of 1998. The interviews were centered around 16 questions developed to investigate how practitioners identified and implemented innovative strategies in cases where traditional strategies were no longer working. Each interview lasted 2–3 hours and all were taped and transcribed. Case study research was based on methods outlined in Yin (1994). Follow-up interviews with two practitioners were conducted for the Forest Creek[2] case study in the spring of 1999 to obtain additional factual information pertaining to the case. Five of the six managers interviewed for the cases were trained as scientists in the fields of biology, fisheries and wildlife, watershed management, and forestry; one was trained as an educator. Five were men, one a woman. All six worked for the state natural resource management agency. Years of experience ranged from 5 to 31. Notably, practitioners interviewed were widely respected and noted for their passion and commitment to the natural resource itself.

The case studies are profiled in Box 9.1. In interpreting interviews with practitioners, we focus in particular on practices developed explicitly to deal with productively negotiating through crisis and change, the release and reorganization phases, or the 'backloop' of the adaptive cycle. We evaluated interviews qualitatively to identify themes that were consistently emphasized (Babbie, 1992; Miles and Huberman, 1994). We then located these 'rules of thumb' in a matrix based on the release and reorganization phases of the Holling adaptive cycle (Table 9.1). We address several hypotheses proposed in the challenge of understanding dynamics between ecosystems and institutions: (a) that there are ecological and management practices that contribute to resilience and adaptive response in linked social–ecological systems; (b) that such practices serve to 'put the brakes on release' or 'conserve memory and opportunity for renewal' during reorganization; (c) that processes are nested at multiple scales; (d) that self-organization plays a critical role during renewal; and (e) that qualitative knowledge complements conventional quantitative data in helping to assess the status of systems and to determine appropriate context-contingent responses during reorganization and renewal.

Box 9.1 Case studies for practitioner interviews

Forest Creek

The issue Angling groups wanted a popular trout stream in a local state park in the Mississippi blufflands to be managed for trophy fishing, but many other stakeholders were concerned about the potential impacts of habitat improvement projects on other uses and values for the park. Habitat improvement has been used successfully for 30 years to satisfy public demands for quality trout fishing. It serves as a single-use, 'bandaid' approach – stabilizing banks and engineering cover for fish (primarily introduced brown trout) – given managers' limited ability to address the ultimate causes of stream degradation in the watershed. Over the years, however, as uses and values for the park had broadened, concerns had been raised by other users regarding the impact of trout habitat improvement projects. Concerns pertained to the broader local ecology, particularly state and federally listed threatened plants and animals, as well as to cultural values such as archaeological sites and artifacts present in state park lands.
The challenge DNR staff wanted to avoid a repeat of a contentious battle which had occurred a few years earlier over a similar stream. When several trout associations approached DNR with a proposal for a habitat improvement project in the popular state park, managers knew they needed to do things differently. A new temporary acting manager, with the aid of the new regional management team, created a facilitated process that was fair, open, and flexible. A critical change was devolving the authority and accountability for the final decision making to the local managers and to the process.
The outcome After only 8 months, an agreement was reached that satisfied all parties and soothed residual community tensions. 'Memory,' in the form of working relationships and information about system hydrology gained in resource assessments, persisted long enough to defeat a quarrying operation subsequently proposed at a nearby farm which would have severely affected the hydrology and water quality of the stream in question.

Rainy Lake fishing roundtable

The issue Rainy Lake and Rainy River straddle the USA–Canadian border, and present special problems of international coordination and management. The Rainy Lake resort economy on both sides of the border depends on good fishing to attract tourists/anglers. Although regulations in the 1980s had put an end to commercial fishing in the lake, total catch and average size had continued to decline in the years leading up to the roundtable process, displaying the classic signs of an overexploited fishery. Around the same time, DNR had cut back on stocking in natural walleye lakes based on internal research showing annual stocking was a waste. Resort owners and anglers, convinced DNR was the problem, demanded more stocking.

The challenge To develop science-based consensus with stakeholders and resort owners that catch limitations, rather than stocking, were necessary to improve fishing, along with mitigation of water quality and water level impacts caused by the operations of two paper mills.

The outcome DNR built support for experimental catch regulations by involving stakeholders in research, modeling, and experimental design. Involvement in the research and modeling processes and in the final management decisions built broad-based support for the policy. Catch rates and sizes improved after experimental regulations based on roundtable recommendations were implemented. Subsequently, the recreational fishery showed signs of a healthy recovery, and anglers praised the slot limits. Negotiations with stakeholders from both sides of the border over water level issues were ongoing.

The Boundary Waters Canoe Area Wilderness controversy: from extensive harvest to intensive forestry

The crisis The forestry profession and lumber industry were completely in agreement that it was necessary to manage the wilderness area for timber, despite strong public opposition at the state and national levels. They were still building roads, using herbicides, and taking other actions that were inconsistent with wilderness law and public opinion, despite data showing there was abundant timber available outside of the BWCAW.

The challenge How to break through the rigid, conventional forestry paradigm carried by commercial and government foresters to accept their own data showing that a shift from extensive exploitation to intensive forestry practices could provide an adequate volume of timber without the need to harvest timber in the wilderness area.

The outcome Public opposition and convincing testimony from state foresters using the Forest Service's own data countered the federal agency testimony that harvesting in the wilderness was scientifically or economically necessary. A Congressional mandate ended harvesting and road building in the BWCAW. The timber industry and professional foresters began shifting from extensive to more intensive forestry practices, longer rotations, and adoption of best management practices, with minimal job losses and positive economic impacts in forestry and tourism. Several state parks in historically fire-dependent ecosystems have reintroduced fire management. The Forest Resources Council, a 13-member board representing commercial, recreational, scientific, and conservation interests, was created to provide sound management advice to federal, local, state, and county governments.

Agriculture

Heron Lake Watershed Project

The issue Various stakeholders were concerned about the decline of the watershed, historically one of the richest wildlife and waterfowl areas in southwestern Minnesota. Various recreational goals and single-target management strategies, such as fishing, public access, and waterfowl habitat, were perceived to be in conflict.
The challenge To find a way to resolve multiple, sometimes conflicting goals for a limited resource. The project moved over time to embrace a whole system or watershed approach.
The outcome Dozens of public and private groups and individuals, including sportsmen's organizations, farmers, local and national conservation organizations, and local, state, and federal government, formed a watershed project in 1989 to develop a comprehensive watershed plan, restore wetlands, acquire easement lands, improve water level management, and address nonpoint source pollution. DNR voluntarily assisted the watershed project leadership in an advisory role. Ecosystem management efforts were continued.

Preserving prairie remnants in the Glacial Lake Agassiz region

The issue Development of land for agriculture had eliminated 99 percent of the original extent of prairie ecosystems; DNR and other conservationists sought a strategy to protect remaining prairie remnants.
The challenge Conventional agricultural systems in the midwest rely on extensive inputs and extensive hydrological modifications that have severely impacted and permanently altered the original prairie, savanna, stream, and wetland ecosystems. The challenge in the Glacial Lake Agassiz area was how to envision a future which could preserve remaining prairie remnants as well as honor the strong agricultural identity of the region, in a time of social and economic crisis in agriculture and rural agricultural identity, declining agricultural profitability, and low international commodity prices.
The outcome The DNR helped spearhead and then participated in a successful visioning process through a series of facilitated dialogues. Local communities took ownership of their needs and concerns and developed strategies for preserving and stewardship of remaining prairie resources.

Table 9.1 *Matrix of practitioners' 'rules of thumb'*

	Local →	Watershed →		State/regional →	
'Rules of Thumb'	Forest Creek	Heron Lake Watershed Project	Rainy Lake fishing roundtable	Glacial Lake Agassiz citizen forum	BWCA forests and MN DNR ecosystem-based management
Looking outward and inward					
Double-loop learning: identify governing values and paradigms; key driving variables; use of metaphor	•	•	•	•	•
Drawing on memory: remembering the past	•	•	•	•	•
Protecting and nurturing capital					
Thorough resource assessment, inventory, and monitoring	•	•	•		•
Valuing diversity of perspectives and experience	•	•	•	•	•
Fair, open, honest process to build trust	•	•	•	•	•
Sense of place, connection to land Relationship building through shared experience	•	•	•	•	•
Detecting and fostering novelty					
Mobilizing capacity for inquiry Engaging all stakeholders	•	•	•	•	•
Open process and information flow, listening	•	•	•	•	•
Developing shared language and understanding	•	•	•	•	•
Coping with surprise Disturbance, crisis, and conflict as change agent	•	•	•	•	•
Encouraging and amplifying experimentation Devolved decision-making, self-organization	•	•	•	•	•
Dampening barriers to renewal and learning Creating 'safe spaces' for experimentation; tolerance of mistakes	•	•	•	•	•
Minimizing learner's sense of vulnerability; 'respect' for process and individuals	•	•	•	•	•
Proactive negotiated consensus process; avoiding charged, polarized settings	•	•	•	•	•
Speeding the contagion					
Cultivation of networks	•	•	•	•	•
Metaphor, shared language and understanding	•	•	•	•	•
Developing readiness at multiple levels – replication of efforts at local, regional, and state scales	•	•	•	•	•
Cross-scale interactions	•	•	•	•	•
Vision, leadership, passion, and commitment	•	•	•	•	•

9.2 Background and organizational history

What do resource managers manage? Resource management agencies have at best only limited control over the interactions between society and nature (Fig. 9.2). In a complex democratic society with private ownership of land and capital governing the production of public and private goods, interactions between society and local ecosystems are driven by the structure and scale of the social–economic system as a whole. Most of the structuring linkages occur well outside the sphere of government or bureaucratic regulation. Consequently, few 'resource managers' in the USA today actively manage resources, and are instead engaged in managing organizations, staff, and human use. More than ever, as others have observed, resource management is people management (Gerlach and Bengston, 1994; Berkes and Folke, 1998).

In the late 1970s, resource management theory began to shift from control of the resource to regulation of human demand (Gerlach and Bengston, 1994). Changes in the orientation of federal agency programs paralleled changes in state agency programs in many parts of the USA as well as grassroots efforts at the local level to incorporate ecosystem management principles and greater public participation in resource management decision making (Grumbine, 1994). In Minnesota, the shift to ecosystem-based management began officially

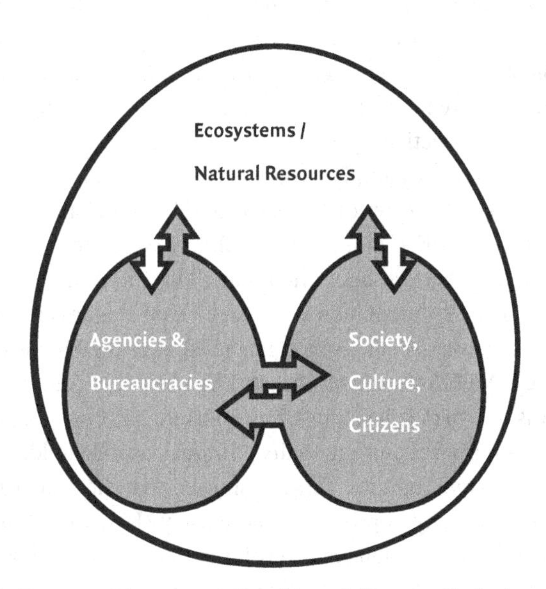

Figure 9.2 The Resource Management Practitioner's Context: limited control or relative influence over resources and public, and two-way flow of linkages between resources, institutions, and society.

in 1995, but the groundwork for such a shift had been in the works for decades. As at the federal level, the impetus for change had come both from outside the agency, in the form of pressure exerted in the political arena by adversaries, and from the internal contradictions that tend to surface in linearly evolving bureaucracies as they seek to implement a growing list of mandates that may be in direct conflict (Peltzman, 1976; Yaffee, 1997).

In Minnesota, state agency resource management had its roots in a long political struggle to end the rapacious commercial and private exploitation that in less than 100 years had resulted in widespread deforestation, driven dozens of fish and game species to commercial or local extinction, and drained and channelized wetlands and streams extensively (Breining, 1981). The Minnesota Department of Conservation was created in 1931 when four units of state government – forestry, game and fish, drainage and waters, and lands and timber – were combined. These early commissions and departments, set up in the late 1800s to deal with rapid settlement and exploitation, had performed unevenly, and had at times exacerbated the destruction of the state's natural resources. The new Department of Conservation was marked by ambitious attempts to stem the tide of resource losses and to foster a growing conservation ethic. In 1971, the name of the agency was changed to better reflect the DNR's broadening responsibilities.

Over the years, each of the various departments of the DNR had evolved close working relationships with their primary constituent groups, resulting in internal fragmentation of DNR's goals and activities and lack of communication among and between the different divisions of the agency. Tight coupling of stakeholder groups with their respective agency counterparts based on shared agendas – section of fisheries with angling groups, wildlife management with sportsmen's organizations, Division of Waters with lake associations, forestry with timber interests, and parks with recreational users – led to poor coordination of messages and activities and often pitted divisions against each other (Yaffee, 1997). This fragmentation was also evident at a larger scale, between the DNR as a whole versus other state agencies such as the Minnesota Pollution Control Agency (MPCA) or the Minnesota Department of Agriculture (MDA). The agency was often a magnet for conflict, both internally between departments working at cross-purposes and externally with various stakeholders (Anderson, 1995). Although public opinion surveys showed that the 'silent majority' of state residents essentially approved of the agency (Kelly and Sushak, 1996), the DNR suffered from a serious erosion of trust and downright hostility among specific constituent groups, particularly farmers and loggers in the rural areas of the state.

In the 1970s and 1980s, agency initiatives had focused on resource assessments and planning initiatives, in addition to the traditional single-target management activities. In the late 1980s, managers and other individuals interested in planning approaches began experimenting with roundtables on various resource issues and the first wave of public and private watershed partnerships. It was not until the early 1990s, under a new administration, that the agency began seriously to study alternatives for fundamental reorganization. The agency developed a hierarchical system for classifying the state's ecosystems and ecological communities, and initiated strategic planning efforts based on these ecological units. In 1995, DNR outlined a plan to adopt the ecosystem approach as a way to redesign its basic organizational structure and operating principles. 'Ecosystem-based management' encompassed a set of strategies for developing integrated planning processes and building teamwork at all organizational levels, greater budget flexibility to foster shared responsibility for common goals, and increased use of partnerships to foster interdisciplinary collaboration within the agency, between the DNR and other state agencies, and with citizens and communities (MNDNR, 1996). The agency initiated pilot projects for multiple-use, ecosystem-based natural resource management in two of the ecologically defined regions; consolidated integrated planning and budgeting activities and ecological support services; developed statewide natural resource forums to convene citizens, agencies, and other organizations for sustainability dialogues organized around forestry and agriculture; and created Regional Environmental Assessment Teams to facilitate coordination and collaborative decision making earlier in project planning to develop better working relationships with local units of government (MNDNR, 2000). The agency also continued and expanded a roundtable process it had used to resolve contentious issues as they arose, including experimental game and fish regulations, de-listing the wolf as an endangered species, and old-growth forest management. The roundtable process brought stakeholder representatives from all sides of an issue together at regular intervals for facilitated meetings to review the science of an issue, discuss policy, and develop consensus or compromise recommendations for working through the issue.

These changes form the context for the case studies outlined in Box 9.1. They represent forestry, fishery, and agricultural issues and cover a range of scales and ecological areas of the state. At the state and federal level, strong leadership by state foresters helped to resolve the controversy over logging and road building in the Boundary Waters Canoe Area Wilderness (BWCAW) in the late 1970s. The resulting changes in forest management and forestry paradigms, from extensive harvest to intensive forest management, have been accelerated through a series of public-input processes which led to improvements in forest management, attracted economic investment in forestry and tourism, and stemmed the tide of conflict. At regional scales, the Rainy Lake fishing roundtable and the Heron Lake Watershed-Project highlighted the ability of diverse groups of constituents to reach innovative, science-based decisions for resource management based on negotiated consensus. Because such decisions involved a broad base of stakeholder participation and perspectives, they proved to be more resilient in the face of attacks in the political arena that had frequently derailed decisions made by DNR scientists in isolation. At the local level, the Forest Creek trout habitat improvement project highlights how a facilitated public-input process that is committed to being open, fair, and respectful can resolve long-standing community conflicts, develop consensus on a detailed resource management plan, and generate a shift in focus from single-use, single-species management to balanced multiple-use management.

9.3 Practitioners' rules of thumb

Despite practitioners' emphasis on the contextual, improvisational nature of managing resource crises and change, many principles or guidelines emerged repeatedly in interviews. In contrast with conventional paradigms characterized by instrumentalist methodologies for targeted problem solving, practitioners presented insights as loose guidelines for managing an organic process while not being overly directive. Direct control in such systems is, in fact, not possible given the multitude of interacting, independent agents and the role of chance events. 'Rules' are geared more at maintaining the parameters and conditions for learning – the conditions for meaningful dialogue, communication, and innovation – than at producing particular outcomes (Table 9.1). We categorized the principles in relation to the backloop of the adaptive cycle as follows:

1. Looking outward and inward for understanding.
2. Protecting social and natural capital.
3. Detecting and fostering novelty.
4. Speeding the contagion.

9.3.1 Looking outward and inward for understanding

Practitioners identified being sensitive to initial conditions, focusing on slow variables, and facilitating learning as important practices in leading change.

Such practices, drawing both from memory and visioning, served to stimulate reflection on the internal and external sources of current dilemmas and to expand the temporal frame of reference beyond the immediate present. Rather than directing a process, practitioners focused on creating safe spaces for dialogue among diverse players to stimulate learning at multiple levels – both single-loop and double-loop learning. Single-loop learning generally refers to learning within a framework or paradigm. It is often viewed as basic 'how-to' knowledge acquisition, for example learning about rules and regulations to achieve set goals (Pimbert *et al.*, 2000). Double-loop learning involves learning at a deeper, more fundamental level. It involves examining assumptions, identifying not just the immediate, proximate mechanisms behind current structures and problems, but also the governing values, underlying frameworks, and paradigms that structured the previous problem-solving response (Argyris, 1990). Double-loop learning is about reorganizing conceptual models and behavior based on a revised understanding of the system or problem. It concerns the diagnosis and treatment of causes, rather than the symptoms.

At Forest Creek, the need to be 'sensitive to initial conditions' was provided by the lingering conflict over previous habitat improvement projects. DNR staff wanted to avoid a repeat of a contentious battle over another state park stream, Trout Creek, that had occurred a few years earlier. That battle had led to significant and lingering controversy, negative publicity for the DNR, internal conflict within the agency, enormous costs in terms of time and attention, and an outcome not widely perceived to have been positive. The Trout Creek experience had set much of the tone. Relationships – some positive, but many negative – had already been formed among individuals in the various agency and stakeholder groups. Initial decisions about the process – internal communication and meetings, setting ground rules for respect in the public meetings, solicitation of informal, one-on-one feedback from key players in advance of meetings, and an open-door policy of DNR managers toward inquiries from the public and participants – were all based on this understanding.

'Slow' variables, or variables that change over longer time scales, such as cultural history, governing values and paradigms, and economic structures, were identified both as sources of memory and as barriers constraining bureaucratic options and social imagination. The agrarian traditions of northern and central European settlers who largely settled Minnesota conditioned attitudes toward resources. Conservation philosophies and local knowledge passed down from small farmers of the Depression era blended elements of Christian stewardship ethics with elements reminiscent of Leopold's land ethic (1949). Many second-generation and third-generation farmers recall being raised with strong stewardship values, and many continue specific farming practices linked to that ethic, such as the maintenance of fencerows, woodlots, and wetlands as refuges for on-farm biological diversity (Blann *et al.*, 1998). At the same time, these traditions structured rigid paradigms for the management of forestry, fish, and game, combining an agrarian view of human management and control and an entitlement view of the rights of Western culture to fully exploit natural resources wherever and whenever they could (R. Sando, personal interview, October, 1998, former Commissioner of Minnesota DNR). The agrarian view saw fish stocking and wildlife feeding programs as a DNR mandate. Monitoring data showing such programs to be ineffective at best were to be ignored or mistrusted. The pro-development view saw the efficient liquidation of natural capital on both public and private lands as an economic and moral imperative to sustain growth that was being thwarted by naïve preservationists.

At Glacial Lake Agassiz, dialogues began with facilitated discussions drawing the historical picture of commodity agriculture in the local and global economy. The paradigms, economic incentive structures, information flows, and federal policies which influenced conventional agricultural decision making were seen as the factors driving prairie loss, as well as farm consolidation, rural depopulation, and political pressure on agriculture from urban environmentalists. The state natural resource agency's long-term focus on hiring technically and scientifically trained staff to fill primarily technical job descriptions, despite a need for communication skills and significant public interaction in many of these jobs, has continued to influence agency 'capital' in terms of its relationships with its public 'customers.' A long history of extensive timber exploitation in Minnesota, leading to tight economic and social coupling of state and federal agency foresters with the timber industry, yielded fierce resistance to the wilderness movement's attempt to bring an end to timber exploitation in the BWCAW. Such observations lend support to the caricature of the perversely resilient bureaucracy (Gunderson, 1999) or the Kuhnian notion of nonlinear cycles in dominant scientific paradigms. During the exploitation phase of the adaptive cycle, the success of linear thinking and control strategies leads to stable configurations of paradigms, management rules, and relationships. Scientists, managers, and other players who have experienced success in this system are likely to resist risking fundamentally new approaches, and to systematically ignore accumulating evidence of impending crisis or failure of the current policy or approach. Past success becomes the motivation for persisting in conservative behaviors. The rigidity of this configuration serves to thwart efforts at renewal, and underscores the observation that the impetus for reorganization generally comes from the 'fringe' (Kuhn, 1970; Holling, 1995).

9.3.2 Protecting social and natural capital

In the adaptive cycle, the period from reorganization to renewal is rapid, chaotic, and subject to chance events. Events occurring in this phase often lay the foundation for the order that emerges, while innovation emerges from novel combinations of existing social and natural capital. It is critical in this phase to protect and retain the 'raw ingredients,' or social and natural capital – whether in the form of soil fertility or experience-based knowledge and wisdom. Managers expressed a conscious framework for protecting such 'capital', with practices such as:

- thorough resource assessment, inventory, and monitoring;
- valuing diversity of input, perspectives, and experience;
- developing trust through fair, open, honest process;
- building on sense of place: developing local knowledge, emotional commitment, shared experience, and relationship.

One of the most significant examples of identifying and protecting capital – conserving memory and opportunity – is provided in the form of a threat to the resource at Forest Creek that did not materialize. Once the habitat improvement plan was completed and approved, the decision-making platform at Forest Creek had achieved its purpose and had therefore been dissolved – freeing capital while retaining memory in the form of skills, relationships, and knowledge. In the fall of that year, a quarry operation was proposed on a property near the creek. Due to the resource survey that had been conducted, it was known that that land was part of the recharge area for the creek. A quarry operation on the land would have undoubtedly posed a threat to the spring and to the entire resource through alterations of groundwater hydrology or contamination of water quality from the quarrying process. Individuals responded quickly through the informal communication network that the Forest Creek project had spawned. They managed to get the property designated fairly rapidly as an important 'Scientific and Natural Area', a state land acquisition and management program. In a memo to a colleague, the practitioner involved in the Forest Creek event commented:

> The value of a facilitated stakeholder involvement process is sometimes found in the costly things that don't happen, such as the time and energy that is not spent because the

situation does not spin out of control, or the extra meetings that do not happen because the initial processes work effectively. However, it is hard to measure things that don't happen.

Capital is often unrecognized until it is eroded. When facilitation is working, it is often invisible. Likewise, when community capital is in place – e.g., civil society or community institutions are able to perform essential social services and cope with small-scale shocks and disturbances – avoided crises, or crises that never materialize, may never be perceived at all. A linear problem-solving approach is unlikely to recognize those crises that do not materialize. Because we lack measures and means of identifying when capital is functioning to maintain resilience, a problem-solving orientation is likely to prevail, identifying problems and proposing radical solutions to small-scale, fast-variable problems that may result in cures that are worse than the disease. A focus on what is working, and what resources are in place and functioning, may play a key role in generating solutions to perceived problems.

9.3.2.1 Thorough resource assessment, inventory, and monitoring

Practitioners emphasized the importance of having as thorough as possible a baseline understanding of the system. Scientific surveys played a key role in inventorying existing natural capital. Thorough inventory and assessment of ecological resources as well as careful monitoring of the fit between agency activities, financial outlays, and outcomes were key to the success of the reorganization efforts in the DNR at all levels. Statewide, federal and state forestry inventory data provided the justification needed to cease the harvest of old-growth timber in the BWCAW. At Rainy Lake, long-term data on exploitation rates and catch sizes provided input to models assessing the impact of catch regulations on the fishery. The Minnesota County Biological Survey (MCBS), a statewide natural resource inventory which had been initiated in 1987 in order to identify, map, and facilitate land acquisition to protect rare ecological features and communities, helped practitioners to communicate with citizens at Forest Creek and in other outreach efforts.

At Forest Creek, project leaders augmented MCBS data with detailed resource inventory and mapping of the archaeological and biological resources of the riparian area. An interdisciplinary team of DNR scientists was assembled and worked closely to develop a Geographic Information Systems (GIS) depicting the precise locations of each valued resource. These maps were then used to develop the proposed locations for habitat improvement so as to be unlikely to undermine other resource values for the creek. A key benefit of the resource surveys conducted as part of the Forest Creek project was not just the expanded ecological knowledge base, but the working relationships forged between the biologists in different disciplines, the stakeholder groups, and the agency managers. Fisheries staff organized a series of walking tours for the benefit of any and all interested stakeholders, as a new component of all habitat improvement projects. The park supervisor played a key coordination role throughout the project, working closely with interested parties and participating in walking tours. The walking tours formed the basis for developing a shared understanding of the diverse values people held for the creek. On one tour, for example, a botanist led the group away from the creek in search of a rare fern that grew only in small patches characterized by a cool microclimate. For the regional fisheries manager, experiencing the collective search for this rare, little-known fern in its unique microhabitat helped to enlarge his own appreciation for the diversity of ecological conditions and values for the park.

9.3.2.2 Valuing diversity of input, perspectives, and experiences

Creating adaptive capacity requires valuing diversity and individual experiences for what each can contribute to the process in the form of knowledge capital. By honoring each individual's right to participate and creating a climate of respect for differing perspectives and opinions, positive-feedback loops are created in which participants move from tolerating diversity to valuing and enjoying it for the role it plays in their own learning process. While many resource managers may go through the motions of facilitating public participation, these managers expressed a strong belief that working with stakeholders as partners was vital to addressing the resource issue. As the manager of the Heron Lake Watershed Project stated: '... the plan they asked me to write ... it was their ideas I had listened to ... I carried those plans everywhere with me. Anybody could have a copy of the plan. It was very available.' The manager of the Glacial Lake Agassiz described his own epiphany: 'We had been approaching [the problem] as saving the prairie *from* the farmer instead of *with* the farmer.'

Protecting capital was embodied in a focus during Forest Creek on helping people to perform at the peak of their individual abilities. Part of what had eroded the public's lack of trust in the agency during the Trout Creek controversy was the obvious conflict and disagreement, both about facts and uncertainties, present among agency staff – the non-game biologists versus the fisheries staff in particular. In order for each biologist and manager to present his or her best work at public meetings, it was necessary to develop consensus within the agency about what was known, what was unknown, and what were the likely uncertainties, as well as to anticipate as much as possible the likely criticisms and concerns that would arise in the meetings. Public meetings regarding the

Forest Creek habitat improvement plan were characterized by a well-planned agenda that laid out careful time limits, ground rules for respectful interaction, a coordinated and consistent message for the state resource agency, and an inclusive process for assessment, decision making, and accountability. Agency agreement with respect to the science and the process had been carefully crafted through a series of internal formal and informal meetings and one-on-one conversations carried out in advance of the public meetings. Careful attention to process in advance allowed the process to become invisible. The knowledge and skills of the agency scientists could then emerge, while the knowledge and concerns of the various nonagency participants would also be honored, heard, and addressed without major damage to relationships.

9.3.2.3 Fair, open, honest process to establish trust

Woven through all the discussions about protecting capital was an emphasis on the necessity of building trust by maintaining a consistently fair, open, and honest process. In complex systems, relationships form the basis for all communication, motivation, and action. Trust is critical (Nelson, 1994; Faast and Simon-Brown, 1999). Practitioners emphasized creating opportunities for people involved in an issue to meet and interact socially in the resource environment, sharing food, stories, experience, collective learning, and work. They stressed the need to allow time and space for relationships and ideas to incubate, and to resist assuming that just because they were not doing anything, nothing was happening. Once the conditions for learning and dialogue were in place, their role was to step back and allow events to unfold.

9.3.2.4 Building on sense of place: developing local knowledge, emotional commitment, shared experience, and relationships

Without exception, practitioners stressed the need for decision makers to have regular and direct experience of the resource. Such experiences served to build context-specific knowledge and passion. The role of science was to serve this process of building connections and local knowledge, rather than the reverse. It was not the knowledge *per se* that developed commitment to stewardship, but the sense of ownership behind that knowledge, the connection to place embodied in the range of knowledge acquired about a place. Knowledge about the uniqueness of a resource or a place helped individuals to enlarge their appreciation for ecological knowledge in general, by helping them both to recognize and value diversity and to connect with others' feelings of stewardship and pride for unique resources in other places.

9.3.3 Detecting and fostering novelty

As the introduction to this volume suggests, self-organization plays an important role both in ordering ecological knowledge and in emergent organizations. Ideas and heuristics emerge over time through dialectic between individuals in a group as well as through the group's interactions with the ecological system. Individuals within stakeholder groups 'co-evolve' with the ecological system they have formed to address – 'self-organization through mutual entrainment' (Folke *et al.*, 1998). Stakeholders and planners reframe their understanding of a situation through a kind of dialogue (like those practitioners described), but may be unaware that this is happening (Innes, 1998). Self-organization may also play a role in the emergence of 'novelty' in terms of non linear shifts in understanding. Numerous practitioners suggested that their own ideas as well as those of others had evolved as a result of a collaborative process.

Strategies for creating the conditions for novelty to emerge in the ecological or social system included mobilizing capacity for inquiry, anticipating and capitalizing on surprise and uncertainty, encouraging and amplifying experimentation, and dampening barriers to renewal and learning.

9.3.3.1 Mobilizing capacity for inquiry

Mobilizing capacity for inquiry is a process of identifying and freeing capital. In each case, this involved engaging all stakeholders in an inclusive process and developing shared language and understanding. Inclusive dialogue was seen as bringing more capacity for innovation to the process. According to one member of a regional DNR interdisciplinary team: 'Ecosystems are not only more complex than we think, they're more complex than we *can* think. So the more brains you have working on the problem, the better chance you have of coming out with something that's acceptable to everyone or successful.'

Developing shared language to communicate system metaphors, ecological processes, and community vision is a gradual process that occurs hand in hand with developing relationships and understanding. Collective science-based dialogue, experiential learning, and modeling were used to develop a shared knowledge base. Practitioners stressed the need to be aware of and avoid the use of overly narrow language, such as scientific jargon, especially in the initial stages. At Rainy Lake, a fishing roundtable composed of lay persons, resort owners, and fishermen gradually developed a firm grasp of the terminology and theory used by fisheries scientists. According to the fisheries manager on the project, the roundtable's final report issuing management recommendations for Rainy Lake 'could have been written by a biologist.'

9.3.3.2 Coping with surprise and uncertainty

Practitioners and participants tended to characterize the 'time being ripe' in terms of the existence of a threshold level of concern or frustration and the willingness of key people to begin actively searching for new approaches to dealing with a problem. 'Crisis' was frequently cited as playing a necessary or sufficient role in spurring action, but was rarely defined. At what scale must crisis occur in order for meaningful learning to occur? Crises shared elements of being both real and perceived or socially constructed, both social and ecological. Glacial Lake Agassiz highlighted economic crisis in agriculture as well as loss of remaining prairie remnants. At Heron Lake, concerns focused on the loss of waterfowl habitat and degraded water quality. At Rainy Lake, concerns focused on the deterioration of the walleye fishery and the impact this might have on the tourism economy. Forest Creek was primarily characterized by a perception of crisis in the social arena, where desire to avoid repetition of a difficult and negative process at Trout Creek led to internal reorganization and rethinking of the agency's approach to habitat improvement in state parks. In other cases, concerned groups responded to the perception of impending crisis. The impetus for reorganization of the DNR at the state level emerged from the convergence of small, medium, and large-scale contradictions and conflicts, none of which individually might have precipitated action.

The differing nature and role of 'crisis' may lie in the structure of land tenure and management as well as in existing social capital. Where such capital exists in the form of strong local or grassroots community ties, an individual motivated to act is more likely to find and 'plug in' to an active support network, or a 'shadow network' working at multiple scales. By contrast, where people's lives are relatively isolated at the level of the nuclear family, decoupled from place and community, as was the case in several of the case studies related to a specific issue such as recreation, more immediate crises are required to spur collective action and generate novelty. The impetus for such collective efforts – the perception of crisis – may arise out of fears about being subject to larger uncontrollable forces of globalization and rapid change, out of a historical and cultural context of values and political empowerment, via the enabling institutions which support directly or respond to civic action, and/or via the articulation of vision by practitioners involved in leading change regarding the inadequacy of traditional approaches.

Practitioners were generally philosophical about the role of surprise and uncertainty, recognizing that not everything could be anticipated. A diversity of perspectives was credited with conferring a level of insurance against the type of unintended consequences generated by narrow thinking: 'Keep your eyes open and watch for the unexpected . . . you just have to be careful about linear tracking on all of this stuff. Having the bigger group . . . really helps to avoid a lot of the problems.' The conflicts that arose in the case of the earlier situation at Trout Creek had come as a complete surprise to the DNR fisheries and park managers. The defensive reaction to surprise allowed the issue to spiral out of control. In broadening input, practitioners were able to better anticipate surprise and take it in stride when it materialized. In this sense, commitment to honoring diverse perspectives serves both to break open renewal cycles as well as to 'put the brakes on' release.

In several cases, surprise provided the opportunity for re-evaluating views. At Forest Creek, an unusually severe summer flash flood occurred at a time when resource assessors, in the process of thoroughly mapping the 1.5-mile stream, had begun to get bogged down in specific, single-purpose plans for each segment of the stream. The flood literally washed away several areas of contention. In so doing, it created a more flexible climate for negotiation, and served as a reminder of the pitfalls of micromanagement.

9.3.3.3 Encouraging and amplifying experimentation

Practitioners saw the involvement of the public as both a challenge and an opportunity. Reorganization and renewal in particular require a faster pace of learning and organizational change than agencies typically achieve on their own. Mistrust of government, lack of scientific literacy and/or of a common language for speaking about ecology, and poor communication between agencies and the public serve as obstacles to the implementation of management plans. Narrow interest groups of citizens sometimes serve to co-opt and corrupt processes of public involvement. Agencies often feel that they are being hijacked and hamstrung by political controversy. However, loose networks of activist citizens and non-governmental organizations (NGOs) can play the role of change agent by regularly lighting fires under the slower-moving, task-oriented bureaucracies. Thus, ironically, 'slowing down' to broaden the process of planning input, if done correctly, serves to facilitate more rapid fundamental change.

Several practitioners experimented with devolution of the actual decision-making authority, or 'leadership from behind' as it has been termed within the agency, but continued to pay close attention to the dynamics of helping people to work through the process collectively. Devolution of decision making explicitly recognizes the inability of an agency or a manager to truly 'manage' a complex system itself. The manager cannot direct change or control change, and therefore must focus instead on creating the conditions for learning and for self-organized contagion. Practitioners often felt the need to articulate a vision or a set of ground rules, but then stepped back to 'let self-directed discussions nourish themselves.' At Forest Creek, the regional manager supervised the process at a distance, but devolved decision making and clear accountability for the final decision on the outcome to the park manager. The park manager, in turn, coordinated communication between biologists involved in the resource mapping and surveys as well as with the public. The practitioner involved most intimately with the Heron Lake Watershed Project gave the following prescription: (1) listen, (2) contribute information, (3) sit back and be patient, and (4) expect your partners to make good decisions. He added 'I can remember thinking: "I'm the resource person, I have the training and the experience, *I'm* the only one who is going to be qualified to make the correct decision" . . . [but] actually, if you trust in the process, your partners will make better decisions than you will.' Prescribed burning to reintroduce historic disturbance in one of Minnesota's largest and most beloved state parks was planned and carried out entirely at the local level in a public process led by the park superintendent. In each case of devolved authority, practitioners admitted to having made a leap of faith – and to feeling pleasantly surprised by the quality and the scientific soundness of the decision reached.

Practitioners emphasized the improvisational style, the lack of a 'master plan,' as a way to encourage experimentation. The primary focus was on 'discovery learning,' engaging people in dialogue, observation, and in some cases monitoring in such a way as to allow them to draw their own conclusions.

9.3.3.4 Dampening barriers to renewal and learning

Because learning and experimenting with new ways entail risk, an important strategy for dampening barriers to renewal is to create space where participants are free to experiment, the learners' sense of vulnerability is minimized, and mistakes are actively tolerated or even rewarded. At Forest Creek, an active 'open-door policy' for communication between the agency and the various stakeholder groups sought to 'create safe places where conflict could be managed and learning could take place.' 'Safety' in this case refers to regular opportunities for informal or one-on-one communication with process participants via phone, office visits, and field visits. Agency decision makers regularly contacted key leaders of stakeholder organizations to discuss concerns and convey information regarding the status of the process.

Many practitioners focused on the importance of 'active listening,' or listening for ideas that emerge from the different ways in which people frame problems. Some brought up the paradoxical value of extremists. While extremists can be barriers in the sense of creating strained relationships, gridlock, and barriers to communication, they can also be sources of novelty as the sparks for inducing 'crisis' (as opportunity), as the parameters for enlarging the range of options considered, as a source of passion for energizing a process, or even as a common 'enemy' helping to unite others. The Trout Creek experience, for example, had taken such a toll in terms of stress, mental energy, and broken relationships that individuals explicitly expressed a 'readiness to do things a different way' at the outset.

A major concern of practitioners was how to handle individuals who were particularly charismatic, powerful, and/or disruptive to the collaborative process. Several members of stakeholder groups at Heron Lake and Forest Creek were known for their tendency to be vocal and disruptive at public meetings, to be manipulative of the agenda, to interrupt when other individuals were speaking, and to go outside of the process to achieve their ends. Such negative tactics were dampened via a kind of 'Tao' of facilitation: individuals who tended to disrupt discussion were encouraged to speak up within their allotted period. Ground rules agreed upon at the beginning of each public meeting helped to diffuse the tension and anxiety experienced by participants who tended to hijack the conversation out of fear that their concerns might not be aired otherwise. Skilled facilitation helped to maintain ground rules, to ease anxieties about the fairness of the process, and at times to transform the passionate but negative energy these individuals brought to the table into a force for constructive change.Practitioners confirmed that an open process that respectfully honors diverse input will be self-censoring. The group itself becomes the arbiter of the rules, and the group will sanction individuals who consistently violate the established norms of respectful input.

9.3.4 Speeding the contagion

Practitioners worked to ensure that processes of change occurred at multiple scales, from local to regional to organizational. They recognized the need to operate at multiple readiness levels – not just scaling up from the local or imposing top-down processes from the state level. Lessons learned from DNR's early experiments with ecosystem-based management were shared across scales through internal and external publications within the agency, links with the state university, media coverage, informal communication, and formal exchanges at events and conferences. Organizational changes in the DNR have occurred to capture the memories from Forest Creek and Trout Creek and to scale up the learning (Fig. 9.3). The regional environmental review teams and process are in place to replicate the planning process that occurred at Forest Creek in any future settings. During the Forest Creek planning process, the regional

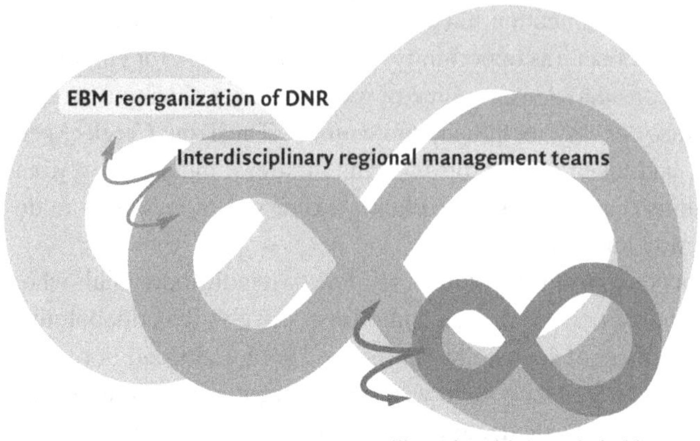

Figure 9.3 Nested interactions at Forest Creek.

coordinator and the regional managers met informally several times to review the process.

Several practitioners knew each other or were familiar with one another's published work and drew parallels in their own work. Novelty in the social system emerges as an organic property of collective work, and is facilitated by lateral diffusion of successful models through informal networks. As one practitioner phrased it 'you can't say that one idea came from here and then it was picked up over here . . . it's more like a cloud taking shape. Everyone's ideas are like little nudges, they're constantly nudging themselves and others. Some nudges made a bigger difference than others.'

The commitment and passion of specific key individuals, especially, play a critical role in generating enthusiasm, maintaining momentum, and speeding the contagion of a process. Practitioners recognized an opportunity to act on behalf of the resource and capitalized upon it. Their own passion and commitment to a successful outcome were evident in interviews. Many emphasized the need for patience and understanding of the lag times that follow legwork that has been in the works for decades. One practitioner said: 'You can only go as fast as local people want to go or can go. If you try to go faster it doesn't work. We're almost 20 years later [now] and we really are talking about wetland restoration, riparian buffers, feeder streams, and erosion control' (having moved from a focus on in-lake restoration). This may have problematic implications in terms of measuring and reporting on outcomes, especially from the point of view of a task-oriented bureaucracy. Nonprofit organizations, which can afford to be more patient with these kinds of informal development processes, play a valuable support role.

The process of negotiating resource use conflicts is often intrinsically rewarding in itself. Participants regularly express a profound sense of satisfaction resulting from learning about the environment or others and the relationships they developed both with others and with the natural world, particularly relationships that evolved from adversarial to cooperative. At Forest Creek, participants were extremely positive about the learning embodied in the respectful process. Handling small steps well, especially in the initial stages, laid the groundwork for success. With Trout Creek in mind, participants at the initial, well-coordinated public meeting about Forest Creek went away pleasantly surprised by how smoothly it had been conducted. Early small successes created a snowball effect, helping to reinforce the positive feelings participants retained after the plan was developed.

9.4 Synthesis: adaptive practices for navigating through the 'backloop'

9.4.1 Against prescriptions: resource management as jazz

The management of complex social–ecological systems is highly context specific. There are no formulas for 'technology transfer' that can be bottled and applied to other resource management problems with assurances of success. Management, especially the management of change, is as much an art as a science, and requires continuous re-evaluation and monitoring. Fostering change requires 'institutional leaders' who have particular personal qualities and abilities as well as opportunities for influence (Stein, 1997). The 'new' practitioners, as facilitators of learning and change, need different skill sets, including the ability to articulate vision and metaphor for double-loop learning and to create safe, open, and respectful platforms for dialogue, learning, relationship building, and experimentation. Looking outward and inward to understand the roots of crisis, protecting and conserving human and natural capital – the 'memory' of the system – through release and reorganization, detecting and nurturing novelty to generate renewal, and speeding the contagion by which adaptive capacity can be replicated and transferred across scales, lead to new configurations of social and ecological capital (Fig. 9.4).

Facilitating adaptive learning and renewal involves an explicit recognition of a fundamentally adaptive, iterative paradigm, rather than one of a linear planning and implementation process geared toward efficient reaching of targets specified at the outset. Practitioners do not subscribe to an instrumental view of public participation as a means to broaden the base of public support to counter challenges to bureaucratic power and to liberate themselves from demands of special interests. They view the construction of alternative processes for

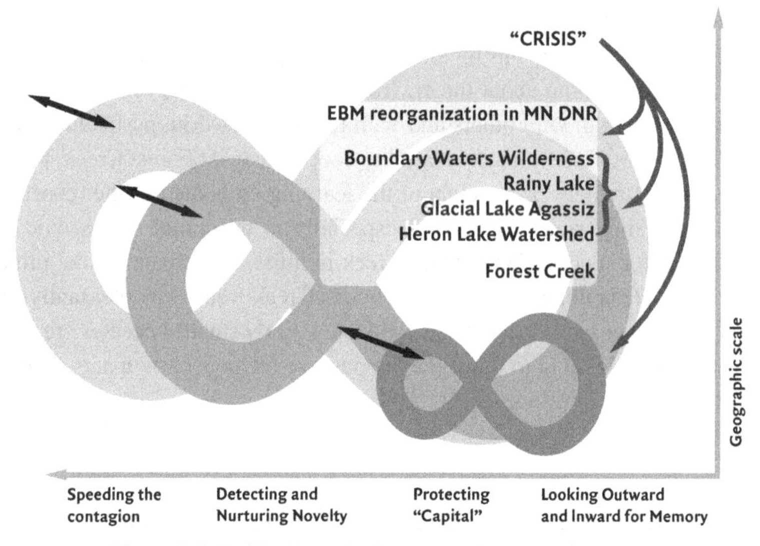

Figure 9.4 Facilitating adaptive renewal at multiple scales.

devolved or shared decision making as a process valuable as an end in itself insomuch as it promotes learning and experimentation. They stress the importance of orchestrating diverse, interdisciplinary working groups of scientists, field staff, managers, landowners, and other stakeholders to build platforms for learning about specific issues that can be scaled up to broader problems. Such platforms are a necessary alternative to the formalized processes by which most public agencies make decisions. As temporary learning systems, they retain flexibility without additional bureaucratic costs. They generate self-organized adaptive capacity, allowing diverse communities of interest and place to strengthen or renew social, economic, and ecological resilience. Practitioners recognized the role of these broad networks as antidotes to the pattern of increasing conservatism that develops in permanent, specialized, fragmented communities.

Facilitating adaptive renewal also requires complementing quantitative knowledge with qualitative understanding of social and ecological dynamics. It requires balancing soft systems with the hard systems; scientific understanding with human values; the instrumentally rational, goal-oriented, problem-solving approach with the organic, emergent, self-organized process of facilitated learning and human resource development. Managers focused as much on the human dynamic as on the ecological in seeking to produce adaptive capacity. According to one, 'We understand adaptive management as a way of looking at ecology, but the other factors for how we proceed are driven by social, political, and economic forces that need to be looked at in an adaptive way.' In each of these cases, scientific resource data were being obtained and balanced day by day, side by side, hand in hand with a platform for negotiating resource use outcomes contingent upon values, beliefs, and learning. Thorough resource assessment helped to lay the groundwork for science-based discussion of options, trends, and key driving forces shaping the ecological, economic, and social system at local and regional scales. Open communication, information flow, and information quality played a key role in building solid working relationships and mutual trust. By establishing a safe, open climate for dialogue, practitioners were able to facilitate double-loop learning and to begin building capacity for making long-term, fundamental change. The development of networks operating at multiple levels of readiness enabled cross-scale transfer of knowledge and learning. Together, these principles encompass strategies for navigating through crisis in ways that lead to renewal and resilience.

Notes

1. The cost of this study was underwritten by a grant from the Resilience Network.
2. Real name changed.

References

Allen, W.J., Bosch, O.J., Gibson, R.G., and Jopp, A.J. 1998. Co-learning our way to sustainability: an integrated and community-based research approach to support natural resource management decisionmaking. In *Multiple Objective Decision Making for Land, Water and Environmental Management*, pp. 51–9, ed. S.A. El-Swaify and D.S. Yakowitz. Boston: Lewis Publishers.

Anderson, D. 1995. Hunters, anglers need to question DNR plan. *Minneapolis Star Tribune* January 20, 1995.

Anderson, D. and Grove, R. 1987. *Conservation in Africa: People, Policies, Practices.* Cambridge: Cambridge University Press.

Argyris, C. 1990. *Overcoming Organizational Defenses: Facilitating Organizational Learning.* Boston: Allyn and Bacon.

Babbie, E. 1992. *The Practice of Social Research.* Belmont: Wadsworth.

Bawden, R. 1992. Creating learning systems: a metaphor for institutional reform for development. Paper for joint IIED/IDS Beyond Farmer First: Rural People's Knowledge, Agricultural Research and Extension Practice Conference, 27–29 October, Institute of Development Studies, University of Sussex, UK. London: IIED.

Berkes, F. 1998. Indigenous knowledge and resource management systems in the Canadian subarctic. In *Linking Social and Ecological Systems. Management Practices and Social Mechanisms for Building Resilience*, pp. 98–128, ed. F. Berkes and C. Folke. Cambridge: Cambridge University Press.

Berkes, F. and Folke, C., eds. 1998. *Linking Social and Ecological Systems. Management Practices and Social Mechanisms for Building Resilience.* Cambridge: Cambridge University Press.

Berkes, F., Kislalioglu, M., Folke, C., and Gadgil, M. 1998. Exploring the basic ecological unit: ecosystem-like concepts in traditional societies. *Ecosystems* 1(5): 409–15.

Blann, K., Light, S., Barton, K., Carlson, E., Fagrelius, S., and Stenquist, B. 1998. *Citizens, Science, Watershed Partnerships, and Sustainability: the Case in Minnesota.* St Paul, MN: Minnesota Department of Natural Resources.

Borrini-Feyerabend, G., Farvar, M.T., Nguinguiri, J.C., and Ndangang, V.A. 2000. *Co-management of Natural Resources: Organising, Negotiating and Learning-by-Doing.* Heidelberg: GTZ and IUCN, Kasparek Verlag.

Breining, G. 1981. *Managing Minnesota's Natural Resources: the DNR's First 50 Years, 1931–1981.* St Paul, MN: Minnesota Department of Natural Resources, Bureau of Information and Education.

Brunner, R.D. and Clark, T.W. 1997. Practice-based ecosystem management. *Conservation Biology* 11(1): 48–58.

Capra, F. 1996. *The Web of Life: a New Scientific Understanding of Living Systems.* New York: Anchor Books.

Dreyfus, H. and Dreyfus, S. 1986. *Mind Over Machine.* New York: Free Press.

Faast, T. and Simon-Brown, V. 1999. A social ethic for fish and wildlife management. *Human Dimensions of Wildlife* 4(3): 86–92.

Folke, C., Pritchard, L. Jr, Berkes, F., Colding, J., and Svedin, U. 1998. *The Problem of Fit Between Ecosystems and Institutions.* International Human Dimensions Project Working Paper No. 2. International Human Dimensions Program on Global Environmental Change, Washington DC. Cambridge: Cambridge University Press.

Gadgil, M., Hemam, N.S., and Reddy, B.M. 1998. People, refugia, and resilience. In *Linking Social and Ecological Systems. Management Practices and Social Mechanisms for Building Resilience*, pp. 30–47, ed. F. Berkes and C. Folke. Cambridge: Cambridge University Press.

Gerlach, L. and Bengston, D. 1994. If ecosystem management is the solution, what's the problem? *Journal of Forestry* 92(8): 18–21.

Grumbine, R.E. 1994. What is ecosystem management? *Conservation Biology* 8(1): 27–38.

Gunderson, L.H. 1999. Antidotes to spurious certitude? *Conservation Ecology* 3(1): 7. [online] URL: http://www.consecol.org/vol3/iss1/art7

Gunderson, L., Holling, C.S., and Light, S.S., eds. 1995. *Barriers and Bridges to the Renewal of Ecosystems and Institutions.* New York: Columbia University Press.

Holling, C.S. 1986. The resilience of terrestrial ecosystems: local surprise and global change. In *Sustainable Development of the Biosphere*, pp. 292–317, ed. W.C. Clark and R.E. Munn. Cambridge: Cambridge University Press.

Holling, C.S. 1995. What barriers? What bridges? In *Barriers and Bridges to the Renewal of Ecosystems and Institutions*, pp. 13–14, ed. L.H. Gunderson, C.S. Holling, and S.S. Light. New York: Columbia University Press.

Holling, C.S. and Sanderson, S. 1996. Dynamics of (dis)harmony in ecological and social systems. In *Rights to Nature: Ecological, Economic, Cultural, and Political Principles of Institutions for the Environment*, pp. 57–85, ed. S. Hanna, C. Folke, and C.G. Maler. Washington DC: Island Press.

Holmberg, J., ed. 1992. *Policies for a Small Planet.* London: Earthscan.

Innes, J.E. 1998. Information in communicative planning. *Journal of the American Planning Association* 64(1): 52–63.

Kelly, T. and Sushak, R. 1996. Using surveys as input to comprehensive watershed management: a case study from Minnesota. General Technical Report NC-181.

St Paul, MN: United States Department of Agriculture, Forest Service, North Central Forest Experiment Station.

Kingdon, J. 1984. *Agendas, Alternatives, and Public Policies*. Boston: Little, Brown, and Company.

Kloppenburg, J. 1991. Social theory and the de/reconstruction of agricultural science: local knowledge for an alternative agriculture. *Rural Sociology* 56(4): 519–48.

Knight, R.L. and Meffe, G.K. 1997. Ecosystem management: agency liberation from command and control. *Wildlife Society Bulletin* 25(3): 676–8.

Kuhn, T. 1970. *Structure of Scientific Revolutions*. Chicago: University of Chicago Press.

Lee, K.N. 1993. *Compass and Gyroscope: Integrating Science and Politics for the Environment*. Washington DC: Island Press.

Leopold, A. 1949. *Sand County Almanac and Sketches Here and There*. New York: Oxford University Press.

Levin, S.A. 1999. *Fragile Dominion: Complexity and the Commons*. Reading, MA: Perseus Books.

Light, S.S. and Dineen, J.W. 1994. Water control in the Everglades: a historical perspective. In *The Everglades: the Ecosystem and its Restoration*, pp. 47–84, ed. S. Davis and J. Ogden. Delray Beach, FL: St Lucie Press.

Miles, M.B. and Huberman, A.M. 1994. *Qualitative Data Analysis: an Expanded Sourcebook*. Thousand Oaks, CA: Sage Publications.

MNDNR 1996. DNR Regions IV and V plan for managing ecosystems. February 15, 1996. Internal report. St Paul, MN: MNDNR.

MNDNR 2000. Organization reinvention efforts. St Paul, MN: MNDNR. URL: http://www.dnr.state.mn.us/ebm/activities/orginvent.html

Nelson, L. 1994. Leading from behind to solve natural resource controversies. *Transactions of the 59th North American Wildlife and Natural Resources Conference*. Washington DC: Wildlife Management Institute.

Peltzman, S. 1976. Towards a more general theory of regulation. *Journal of Law and Economics* 19: 211–40.

Perrings, C.A., Maler, K-G., Folke, C., Holling, C.S., and Jansson, B-O., eds. 1995. *Biodiversity Loss: Economic and Ecological Issues*. Cambridge: Cambridge University Press.

Pimbert, M.P. 1993. IPM options for Asia – explorations for a sustainable future. *Journal of Asian Farming Systems Association* 1: 537–55.

Pimbert, M.P., Bainbridge, V., Foerster, S., Pasteur, K., Pratt, G., and Arroyo, I.Y. 2000. Transforming bureaucracies: institutionalising participation and people centred processes in natural resource management. An annotated bibliography. International Institute for Environment and Development. Sustainable Agriculture and Rural Livelihoods. URL: *http://www.iied.org/agri*.

Posey, D.A., ed. 1999. *Cultural and Spiritual Values of Biodiversity*. London: Intermediate Technology Publications.

Reichhardt, K.L., Mellink, E., Nabham, G.P., and Rea, A. 1994. Habitat heterogeneity and biodiversity associated with indigenous agriculture in the Sonoran Desert. *Ethnoecologica* 2(3): 21–33.

Rocheleau, D., Tomas-Slayter, B., and Wangari, E. 1996. *Feminist Political Ecology. Global Issues and Local Experiences*. London: Routledge.

Sarin, M. 1996. From conflict to collaboration: institutional issues in community management. In *Village Voices, Forest Choices: Joint Forest Management in India*, pp. 165–209, ed. M. Poffenberger and B. McGean. Oxford: Oxford University Press.

Stein, J. 1997. How institutions learn: a socio-cognitive perspective. *Journal of Economic Issues* 31(3): 729–40.

VanNijnatten, D.L. 1999. Participation and environmental policy in Canada and the United States: trends over time. *Policy Studies Journal* 27(2): 267–87.

Walters, C.J. and Holling, C.S. 1990. Large scale management experiments and learning by doing. *Ecology* 71(6): 2060–8.

Webler, T., Kastenholz, H., and Renn, O. 1995. Public participation in impact assessment: a social learning perspective. *Environmental Impact Assessment Review* 15: 443–63.

Westley, F. 1995. Governing design: the management of social systems and ecosystems management. In *Barriers and Bridges to the Renewal of Ecosystems and Institutions*, pp. 391–428, ed. C.S. Holling, L.H. Gunderson, and S.S. Light. New York: Columbia University Press.

Wondolleck, J.M. and Yaffee, S.L. 2000. *Making Collaboration Work: Lessons from Innovation in Natural Resource Management.* Washington DC: Island Press.

Woodhill, J. and Röling, N.G. 1998. The second wing of the eagle. In *Facilitating Sustainable Agriculture: Participatory Learning and Adaptive Management in Times of Environmental Uncertainty*, pp. 283–306, ed. N.G. Röling and M.A.E. Wagemakers. Cambridge: Cambridge University Press.

Yaffee, S.L. 1997. Why environmental policy nightmares recur. *Conservation Biology* 11(2): 328–37.

Yin, R.K. 1994. *Case Study Research: Design and Methods.* Applied Social Research Methods Series, Vol. 5. Thousand Oaks, CA: Sage Publications.

10

Caribou co-management in northern Canada: fostering multiple ways of knowing

ANNE KENDRICK

10.1 Introduction

The links between social and ecological systems are represented by diverse ways of looking at human–environment relations. The continuing exchange between different ways of knowing may be crucial to integrative thought about social–ecological linkages. For many indigenous societies, the separation of social and ecological systems does not make sense. A 'human–environment' divide is especially absent from many arctic and subarctic cultures. How does this fundamental ideological difference play out in resource management systems that incorporate stakeholders both from 'the West' (Euro-American) and from indigenous cultures for whom a human–environment or social–ecological divide is a relatively new and foreign concept?

This chapter looks at the differences that exist in the perceptions of indigenous caribou-using communities, caribou managers, and scientists in co-management processes in arctic and subarctic North America. It is contended that these differences represent potentials to expand how we think about human–*Rangifer* (caribou) systems as much as they represent obstacles to caribou research, monitoring, and management decision-making. The process of negotiating cross-cultural differences in the co-management of caribou herds indicates the potential for the growth of alternative resource management systems capable of accommodating varied ways of knowing and learning.

The question of how humans learn to respect other ways of knowing is represented here as an examination of humility, a respect for diverse realities. There are multiple epistemologies outlining ethical positions of human–environment relations and human perceptions of nature (Folke, Berkes, and Colding, 1998). Attempts to develop resource management systems in balance with resource dynamics must address the suite of alternatives available to 'navigate nature's dynamics.' However, a fundamental issue in social sciences with respect to resource management is the mistrust that can occur among stakeholders. This mistrust may stem from one thought system's domination or outright dismissal of alternative ways of knowing. An appreciation of coexisting but different ways of knowing may improve the chances of developing sustainable resource management systems. At the same time, the options at hand for interpreting and adapting to ecological change are broadened.

Increased conceptual diversity alone does not lead to the increased resilience of social–ecological systems. However, building the adaptive capacity for change may hinge on the existence of varied tools for change. This chapter is an attempt to promote discussion of the mechanisms supporting conceptual diversity that may develop within co-management regimes. It is postulated that the trust, respect, and feedback internal to co-management regimes play a role in building the capacity to deal with change in social–ecological systems. Co-management is defined here as '[a] blending of [indigenous and government[1]] systems of management in such a way that the advantages of both are optimized and the domination of one over the other is avoided' (Royal Commission on Aboriginal Peoples, 1996: 665–6).

There are discrepancies between the attitudes and beliefs of government caribou managers, biologists, and traditional caribou users within co-management regimes (Kruse *et al.*, 1998). Do these differences represent fundamental obstacles to resource management decision-making or the respect for multiple ways of knowing? What are the differences in how and what caribou managers, biologists, and users learn and think about caribou? The lingering differences between the beliefs and attitudes of indigenous resource users and government managers may reveal much about humility. For instance, continued differences in perceptions of caribou population dynamics (Kruse *et al.*, 1998) represent a significant epistemological problem. How can different ways of thinking about social–ecological realities be reconciled within resource management systems? Epistemologically speaking, co-management may contain clues about how to overcome the human deficit of *what* we are able to know and think ecologically (Bateson, 1991), and an increasing tendency to homogenize *how* we are able to know and think about ecological systems. The role of narrative and the larger potential to understand the mismatch between human behavior and ecological processes may be best reconciled in co-management settings. As one caribou biologist phrases it, through involvement in co-management, he has:

> . . . come to realize that there is a very different method of storing this knowledge and of examining what's going on with the caribou that's not related in the numerical sense or in written words
>
> *(Kruse* et al., *1998).*

Dominant global ideologies emphasize the 'one-sided divorce, not only from nature but also from our own biology, and thus of course from our very selves' (Livingston, 1981: 82) leading to a mismatch between human behavior and natural processes. This mismatch has strong implications for the human capacity to think about living processes and to act on our knowledge of living systems. In other words, *how* we learn about social–ecological linkages is as important as *what* we learn about these links. The sustained recognition that there exist discrepancies between the thought and belief systems of traditional caribou users and government caribou managers may lead to significant integrative and complex learning about human–environment relations.

10.1.1 Objective and method

This chapter attempts to broadly answer the question: where attempts have been made to create a dialogue between government scientists/managers and indigenous caribou-hunting communities, what is the nature of the cross-cultural information exchange? Findings are based on conversations with caribou users (the term 'users' is employed here to encompass not only those individuals who hunt caribou, but those who process and consume the animals as well) in three traditional caribou-hunting communities, and an analysis of caribou co-management systems in North America. First, the issue of the trust necessary for the maintenance of diverse conceptual constructs is discussed along with the limitations of knowledge systems. The re-awakening of thought about the integrative nature of belief systems that may shape the relationship between sustainable resource use and human values is then explored. Finally, the chapter looks at the suggestion that the learning occurring in cross-cultural co-management settings may lead to the conditions necessary for the formulation of resource management systems encompassing alternative and diverse ways of knowing.

10.2 Caribou co-management in the Canadian North

Caribou co-management systems provide unique opportunities to explore the learning necessary for the respect of diverse systems of thought. A Man and Biosphere research project completed an extensive comparative study of the influence of caribou management history and beliefs on current caribou management systems (Kruse *et al.*, 1998). The two management systems examined were those established for the Beverly-Qamanirjuaq (of northern Canada, hereafter the Canadian system) and western Arctic (of Alaska, hereafter the

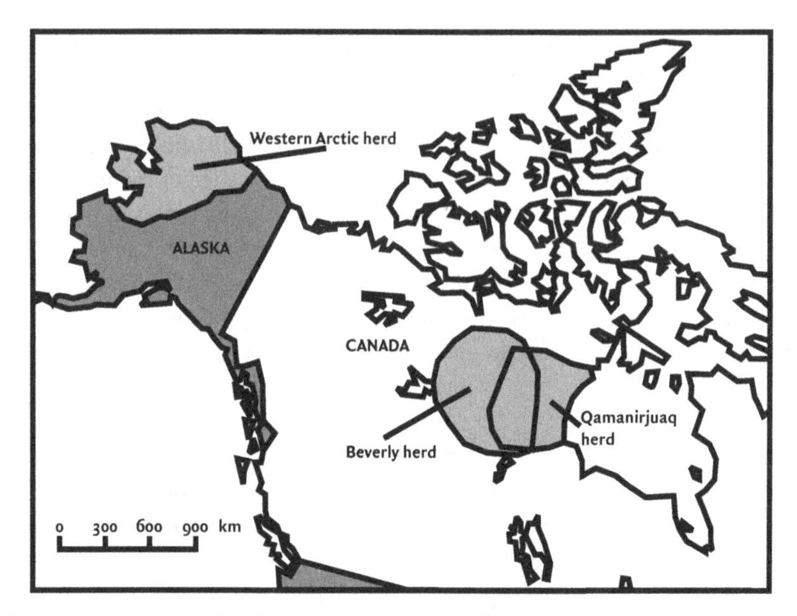

Figure 10.1 Ranges of the three Alaskan and Canadian caribou herds mentioned in the chapter.

Alaskan system) caribou herds (Fig. 10.1). These management bodies were initiated because the caribou populations involved seemed to be encountering population declines in the 1970s. These herds were subsequently shown to have recovered quickly or not to have declined to the extent first assumed following the perceived 'crisis' episodes of the 1970s (Kruse *et al.*, 1998: 449).

There is a high level of uncertainty regarding barren-ground caribou (*Rangifer tarandus*) population dynamics and the interpretation of aerial survey data for population estimations. A 1993 census survey of the Beverly herd suggested that the population had fallen well below the co-management board's ascribed critical threshold. However, a census completed the following year showed three times as many animals (Anon, 1995). The increase in numbers between the 1993 and 1994 surveys could not be explained by population growth alone. The complexity and variability of caribou population dynamics are vast (Klein, 1991). Such variation and survey data uncertainty exacerbates the questions of who interprets changes in caribou behavior and dynamics and subsequently makes management decisions, and how this is done.

The legacy of caribou population 'crises' continues to reverberate in the Canadian and Alaskan management systems (Freeman, 1989). It is extremely difficult to separate differences in the belief structures between and among communities and government. In addition, there are schisms between government and indigenous communities, created by the uncertainties of population survey techniques (Usher, 2000). It is in this history that the complexity of the issues surrounding the 'who, what, and why' of the definition of resource crises can be illustrated. The history of the Canadian system is a unique example of an ongoing conversation about resource crises and their real or socially constructed nature. This history also illustrates the enigma of assessing co-management achievements. For instance, co-management institutions are examples of social interactions where 'what you don't see is as important as what you do see.' The lack of protracted legal battles over monitoring and enforcement methods and increased environmental literacy enabled by 'cross-checking' knowledge of herd dynamics (including population numbers, behaviors, and health) among co-management participants is as important to the definition of co-management achievements as any listing of the 'outcomes' of the process (Singleton, 1998).

A survey of acceptable harvest practices reveals that in both the Alaskan and the Canadian caribou management cases, consensus between caribou users and managers exists for only half of the harvest practices discussed (Kruse *et al.*, 1998: 453). There is much less agreement on acceptable population monitoring practices (Klein *et al.*, 1999). The comparative Alaskan and Canadian study shows that whereas a majority of caribou managers find aerial cow-calf counts and radio collaring of caribou acceptable, only a minority of indigenous caribou users find these practices acceptable (Fig. 10.2). Moreover, there are considerable differences between the cultural beliefs held by the various user communities represented by the Canadian and Alaskan systems (Kendrick, 1994; Kruse *et al.*, 1998).

Despite co-management efforts, managers believe that their knowledge is clearer to users than it was during the 'crises' of the 1970s, whereas users do not agree (Fig. 10.3). A majority of government managers in Alaska and Canada believe indigenous caribou users are more likely to trust their knowledge now compared to the situation in the 1970s. In contrast, only a minority of indigenous caribou users find the knowledge of government caribou biologists believable. What are the reasons for this discrepancy? One suggestion is that indigenous caribou users interpret changes in the prevalence of caribou as a matter of location, not changes in herd size (Ferguson and Messier, 1997). Other suggestions include misunderstandings about changes in survey methods as they were refined from early efforts (in the late 1950s) to present-day techniques, and changes in information needs complicated by changing herd dynamics as well as management goals. An example of such changes includes the switch from visual aerial surveys (caribou counts made by human observers flying overhead) to photographic aerial surveys of the herds, first carried out in the

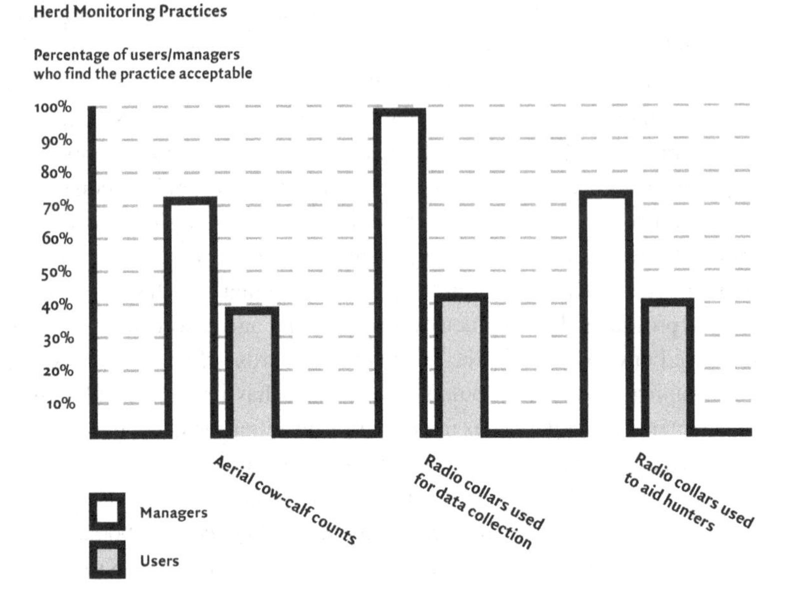

Figure 10.2 Attitudes of caribou managers and users toward herd-monitoring practices Adapted from Klein *et al.* (1999: 495).

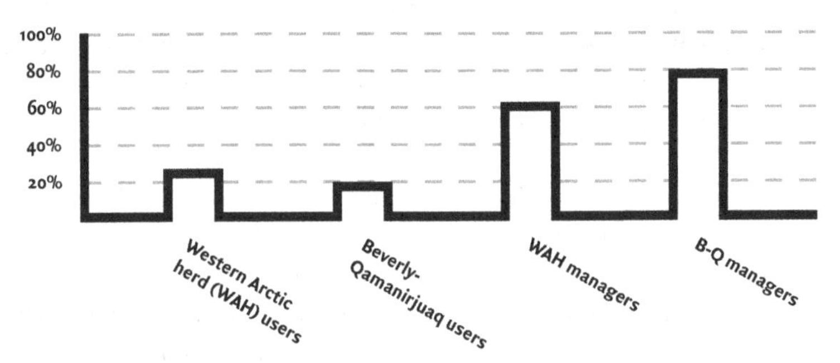

Figure 10.3 Proportion of caribou users more likely to believe biologists now than in the 1970s. From Kruse *et al.* (1998: 454).

early 1980s (caribou counts estimated from photographs of the caribou range taken from the air). Results of a survey of the Beverly caribou herd released in 1984 showed that the new aerial photographic technique revealed *twice* as many animals as the visual aerial survey technique.

The explanation and meaning given to population changes often differ between indigenous caribou users and government managers. The connection between Chipewyan observations of caribou population declines in the 1950s and 1960s, and some of the first caribou population surveys by government biologists and Chipewyan traditional beliefs is illustrated below:

> A wide-spread tradition holds that caribou never die, unless killed, but if one is captured or mistreated his spirit will go to the others and warn them to remain away from the area . . . [t]he decline in caribou numbers in the 1950s and 1960s coincided with the onset of serious caribou studies by the Canadian Wildlife Service. The Chipewyan attributed the decrease in caribou in this area to the capture and tagging, which caused the caribou to avoid the area, rather than to any real decline in numbers
>
> *(Smith, 1978: 72).*

This is not to imply that indigenous belief systems exist in isolation from quantitative observations of change. The contribution of indigenous caribou-hunting communities to empirical knowledge of caribou population dynamics cannot be dismissed. It is not unusual to find, especially in the north where research costs are expensive, that biologists collect data on relatively few variables within specific geographical areas for short periods of time. However, it is very difficult to generalize findings at short time and small spatial scales to the variable and fluctuating environments of boreal ecosystems (Ferguson, Williamson, and Messier, 1998). An increased appreciation of the uncertainty involved in understanding fluctuations in caribou populations is also playing a role in increasing academic interest in the contributions of local knowledge to understandings of caribou ecology (Klein *et al.*, 1999). Similarly, the field of ecology does not solely focus on the description of reductionist quantitative patterns to the exclusion of integrative and qualitative thought (Holling, 1998). The differences between indigenous and government knowledge systems are not black and white. However, it is interesting to contemplate whether traditional caribou-hunting peoples, caribou biologists, and managers gather, interpret, and take action on their knowledge of caribou in fundamentally different ways.

10.3 Human thought and ecological processes

Not only are there cultural differences in human perceptions of nature; there are mismatches between natural processes and human thought in many societies (Bateson and Bateson, 1987). Gregory Bateson, a leading thinker on this issue, was concerned that the materialist framework of knowledge dominating ecological science leads to interpretive error, and as a result helps to deepen ecological crises. Attempting to correct for such interpretive error, he partially developed a theory of an integrative biological dimension of experience.

Bateson used a model of 'mental process' (where nature has a mentality – similar in some ways to animism) to describe the interaction of structure and process by abduction, a widespread phenomenon of human thought. Abduction is a term adopted from philosophy to describe a qualitative method of knowledge construction. It is evident in metaphor, dream, parable, allegory, comparative anatomy, etc. (Bateson, 1979: 142). Abduction permits a 'lateral extension of abstract components of description,' allowing formal comparisons through 'contrasts, ratios, divergences of form, and convergences' (Harries-Jones, 1995: 177). Unlike deduction or induction, abduction is a process of modeling information characteristic of both humans and other living organisms in their own environments. 'Mental process' is a model Bateson created in part as a tool for comparative study, bridging the gap between epistemology and ethics, and in part because he felt that occidental (Western) languages do not lend themselves easily to the discussion of process versus structure. Bateson metaphorically described 'mental process' as very large mental systems of ecological size or larger. He saw the mentality of a single human being as a subsystem characterized by constraints in the transmission of information (information is defined as communicated knowledge or news of a difference from one state to another) between the parts of the larger mental system (Bateson and Bateson, 1987).

This body of thinking acknowledges that every individual and every cultural, religious, and scientific system has particular habits governing knowledge creation. However, Bateson contended that most epistemologies confuse 'map' (the domain of distinctions and differences) with 'territory' (the physical domain that we can never perceive in its entirety). Local epistemologies usually assume that the rules for drawing maps (receiving information) are inherent in the nature of that which is being represented in the map (Bateson and Bateson, 1987). This epistemological confusion of map with territory is the equivalent of believing that the 'name is the thing named.' However, while we cannot 'know' an individual 'thing,' we can know something about the *relations* between things.

For Bateson, *metaphor, not classification, is the logic upon which the biological world is built*. The logic of metaphor identifies and connects all living processes. In contrast, classical logic is ultimately limited because of its dependence on language, unavoidably structured by the discontinuous nature of description or 'naming.' One of the first steps to new ways of thinking about nature is to look at the limitations of any act of description (Bateson and Bateson, 1987).

Consistent with Bateson, the Dene concept of *inkonze* (Ridington, 1990; Sharp, 1997; Smith, 1998), loosely translated as 'little bit know something,' emphasizes the inferiority of human knowledge and power in comparison to nature. The Dene are the indigenous Athabascan peoples of the Canadian subarctic. It is the Gwich'in, Dogrib, Slavey, and Chipewyan subgroups of the Dene that hunt the caribou populations of the Canadian subarctic and arctic. *Inkonze* is a complex concept that is echoed in Bateson's thinking about the difficulties of human attempts to describe and understand living processes. *Inkonze* emphasizes the experiential nature of life. Living and learning are intertwined and nature is the source of knowledge and power. As expressed by Tuan (1979), 'knowing is an engagement with the world, rather than a reflection of the world.'

Description is a 'spinoff of our perceptions and thought' and the name is never the thing named. The *Ding an sich*, or 'the thing itself,' is equivalent to an 'infinitude of details' we can never fully describe or comprehend (Bateson and Bateson, 1987: 164). Our descriptions of the world around us will always be marked by 'gaps' so that descriptions, 'form,' or 'structure' are human constructs that are discontinuous or digital in character, whereas process or the world of flux is continuous or analogic. The Dene concept of *inkonze* emphasizes the limitations and uncertainty involved in human understandings of reality. The knowledge of traditional caribou users 'tends to reserve a place for phenomena which are basically and fundamentally unsuitable for research and the unknowable core gives strength to the knowledge system' (Roots, 1998).

Description is obviously necessary if humans are to communicate their knowledge of living systems. Extended metaphor (such as that of a vast 'mental process') is a way to classify statements of description consistently without denying the primary nature of process (Bateson's 'territory:' Bateson and Bateson, 1987: 193). By studying human descriptions, and human nature as information-processing creatures, we may learn much about the mismatch between human actions and natural processes and therefore about sustainable resource use. The connections between language and human reliance on learning and teaching are adaptive mechanisms of crucial significance in human efforts to avoid ecologically disastrous behavior (Bateson and Bateson, 1987). Until we understand the necessary limits of language and, by extension, the limits of science, we will continue to ignore unavoidable epistemological problems (Bateson and Bateson, 1987). Regarding the limitations to conscious human knowledge, Bateson firmly believed that the correctives for errors implicit in human language and science lie in metaphor or narrative. Bateson was frustrated by so-called solutions to 'environmental problems' that fail to understand the limitations of description and fail to use metaphor to achieve integrative and complex thought. Humankind's resource and environmental problems path may be based on a systemic, epistemological problem because of:

> . . . a destructive mismatch between human behavior and the characteristics of the biosphere within which human beings live and on which we depend. This is a mismatch rooted, not in the mistakes of particular chemists or the wastefulness of hunters or farmers, but in the human capacity to think about natural systems and act on that knowledge *(Bateson, 1991:* x*).*

10.4 Co-management: the potential for institutional transformation?

Co-management may be an arena where the human capacity to think about natural systems is remembered and innovated. This can be achieved by respecting the metaphors or beliefs of indigenous communities that inform the technical aspects of traditional knowledge and practice. Caribou co-management institutions constantly negotiate fundamental resource management concepts, including those summarized in Table 10.1. Terms like 'conservation' and 'management effectiveness' are culturally derived. Indigenous caribou-using communities are often not homogeneous collective units, depending upon the manner in which people's seasonal movements on the land changed to permanent year-round settlement less than 50 years ago. The needs of community representation and resource use and allocation are constantly shifting as human and caribou occupation of the land changes through time. This constant negotiation of meaning makes caribou co-management a dynamic process. Co-management is a discourse in which interactive and mutual learning takes place. For example,

Table 10.1 *Caribou co-management: areas of negotiation*

Concept	Contrast of meaning
Conservation	Caribou as a depletable bank of resources versus a 'partner' in a reciprocal relationship feeding not only economic needs, but spiritual, cultural, and intellectual life and playing fundamental roles in social organization, kinship relations, and cultural transmission
Management effectiveness	Marked by a disparity between traditional caribou users and state management's perceptions of acceptable monitoring, harvesting practices, and notions of expert knowledge
Caribou-using community	A homogeneous collective voice versus heterogeneous settlement of people that moved seasonally; social and ecological contexts changed throughout the year until less than 50 years ago
Resource crisis	Who defines a caribou crisis? How is it defined? Why are crises defined?
Community representation	What is the adequacy of representation? What is the shifting nature of representation needs as caribou herd ranges expand and shrink?
Resource use/allocation	If all uses are not given equal priority, how are commercial, recreational, and subsistence uses prioritized so that uses recognize that often subsistence harvesting depends upon the wages earned through commercial uses?

in many resource management settings, there are long-standing questions not only about who has the authority to define when a resource crisis has been reached, but also about how resource crises are defined, and the equity behind the decisions made to pronounce a perceived resource crisis.

Co-management embodies a tension between government-mandated wildlife management regimes, the expertise of indigenous resource users, and the 'social and political questions about how practices of recognition occur in contexts of power, dominance and resistance' (Feit, 1998: 124). Most of the co-management literature establishes and discusses the power dynamics that are central to the linking of government and indigenous knowledge systems (Berkes, George, and Preston, 1991; Campbell, 1996). The problems of power imbalances are well illustrated in the 'us–other' cycle where the knowledge of 'other' marginalized societies is compared and measured against 'our' mainstream thinking as illustrated in Figure 10.4. The 'other' (indigenous communities)

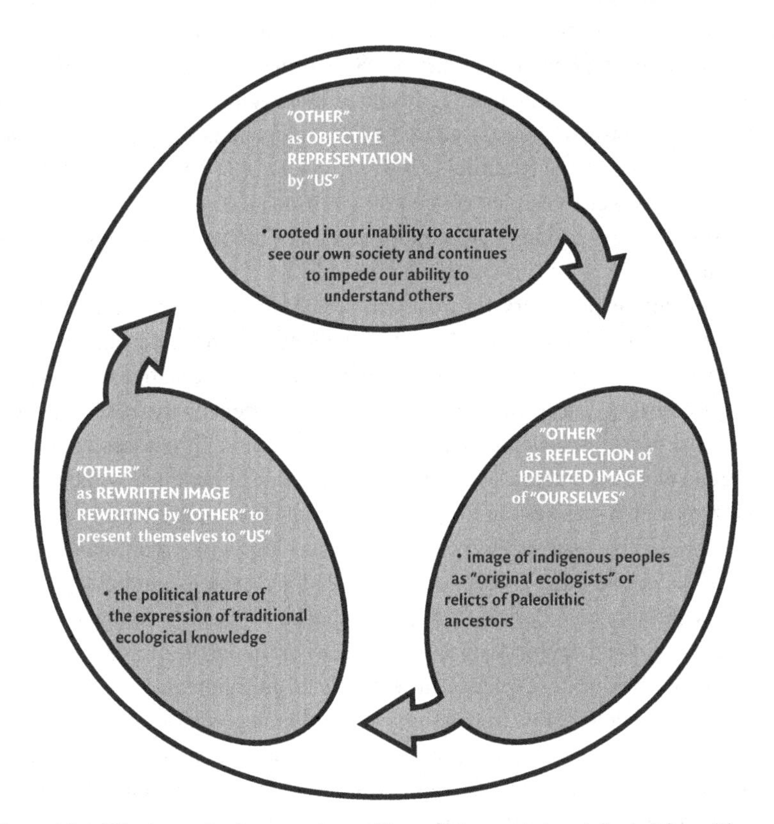

Figure 10.4 The 'us–other' comparison of knowledge systems. Adapted from Fienup-Riordan (1990).

are often represented as idealized images of 'ourselves' (mainstream society), forcing the politicization of this image in order to allow indigenous communities to represent themselves to us in a more realistic and empowered fashion. Fienup-Riordan (1990) speculates this cycle of image making stems from 'our' inability to understand our own society, leading to objective representations of 'other' societies that are essentially images mirroring ourselves.

Perhaps because of this 'us–other' dynamic, co-management is pictured in a number of different manners in the literature. Some describe it as a process of management devolution from government to indigenous responsibility. Others see it as one of convergence between government and indigenous management systems. Co-management is also depicted as an act of compromise and, finally, especially in the case of caribou co-management, as a model of community burden where the risks to communities of participating in co-management processes can be high. These burdens include community concerns that indigenous rights and titles may be undermined through participation in co-management (Kofinas, 1998). Co-management is not a formula for the resolution of long-standing resource management conflicts. There are considerable transaction costs endured by the indigenous societies that participate in these processes (Caulfield, 1997; Kofinas, 1998). However, while not denying that power dynamics are central to co-management decision-making processes, this discussion is an attempt to expand upon the concept of 'trust between actors.'

The conditions necessary for trust will be unfolded here by examining evidence of *learning*. Cross-cultural co-management learning is dependent on the mutual recognition of the belief systems, metaphor, and alternative narratives of the parties involved in co-management. Co-management analyses have largely ignored the potential development of innovative learning processes within co-management arrangements. Most analyses have almost completely focused on political power dynamics. In an analogous manner, ecology largely focused on the competitive aspects of ecological relationships while marginalizing the study of the cooperative aspects of living relations, such as mutualisms, until relatively recently (see Berkes, 1989; Rybczynski, 1997). *Informal* and *flexible* conditions allowing double-loop learning ('learning about the ultimate or underlying factors behind a response, in addition to the immediate or proximate causes;' see Chapter 9) or frame-shifts where the negotiation of the meaning of resource management can occur are possibly critical components of caribou co-management dynamics. The trust necessary for co-management involves the mutual recognition of the learning patterns of the cultures participating in caribou co-management processes.

10.5 Resource management, belief systems, and societal aspirations

There are dangers laden in any analysis of resource management systems that does not acknowledge differences in cultural values and societal aspirations. Northern indigenous groups in Canada are negotiating self-government in some form or another. It is often politically risky for indigenous communities who have not completed these negotiations to participate in resource management processes that may later undermine their efforts to achieve political autonomy. For instance, co-management can become a process reflecting the conformity of traditional caribou-using communities to existing government management systems, rather than a shared perception of fairness and respect of diverse cultural values among participants.

The majority of Alaskan and Canadian indigenous caribou users involved in the aforementioned Man and Biosphere research project see formal government management regimes as mechanisms to control hunting in the event of caribou herds experiencing a decline (Kruse *et al.*, 1998). However, in the face of herd declines, it is difficult to imagine that conventional contingency plans will be workable. The communication occurring between government and indigenous institutions can be fraught with difficulties, including: (1) historical conflicts between caribou users and managers, (2) differences in cross-cultural ecological knowledge, (3) jurisdictional differences between communities, and (4) low levels of community identification with government institutions (Kendrick, 2000).

There is also a lack of recognition of the 'customary practices' of caribou-using communities among government institutions. This lack of understanding of customary practices may be couched in a lack of understanding of the changing history of human–caribou relationships. Just as wildlife population dynamics change, the human–caribou dynamic has shifted dramatically in the last century. In a comparison of human–*Rangifer* (the genus *Rangifer* includes the caribou of North America and the reindeer of Eurasia) relationships across the circumpolar North, Anderson (2000: 169) asks, 'when we know that the history of human communities has changed, is it not reasonable to assume that the identity and quality of *Rangifer* populations has also changed?'

There are problems inherent in caribou research that attempts to understand the very complex and unpredictable ecology of caribou through methodologies that simplify human–*Rangifer* relations. In a description of historical Dene movements, Smith relates that:

> Visiting between local and regional bands was common . . . information on shifting herd movements would be widely known. The distribution of hunting groups may be viewed in terms of the anticipated dispersal of caribou . . . The apparent breakdown of

the network about 1950 was a consequence of a new phase in the late contact-traditional era . . . indicative of a greater degree of sedentism

(Smith, 1978: 82).

Communities are very conscious of the dramatic changes in their relationship with caribou. The caribou of today are not the same as the caribou at the turn of the twentieth century, any more than today's caribou hunters are the same as the hunters of the early twentieth century (Anderson, 2000). There is an intimate and long-standing shared history between caribou and traditional caribou-hunting communities. External forces can profoundly influence the relationships between these communities and the caribou populations they hunt.

The connection between indigenous self-determination and resource management in the Canadian North (Caulfield, 1997; Nuttall, 1998) cannot be over-emphasized. Harvest controls are potentially as devastating to community identity and self-determination as any other process of acculturation (Ames, 1979; Bussidor and Bilgen-Reinart, 1997).

In the light of the discussion above, how do indigenous communities perceive co-management arrangements? Among Canadian managers, 87 percent feel that co-management has increased users' sense of control, whereas *only 27 percent* of users have an increased sense of control. Two-thirds of Canadian managers think that user involvement is as great as it needs to be, whereas *less than one-third* of users feel the same. In the Alaskan case, almost all (94 percent) managers and a majority of users (two-thirds) would like to see increased user involvement (Kruse *et al.*, 1998). Evidently, users and managers have different perceptions of the appropriate level of caribou user participation in co-management processes.

Neither the Canadian nor the Alaskan systems have found effective mechanisms for incorporating user and manager observations in management decision-making (Kruse *et al.*, 1998). Managers comment that user observations are often difficult to interpret. Such communication difficulties are not uncommon in many other efforts to include traditional ecological knowledge (TEK) in resource management decision-making. The crux of this problem is described by one indigenous scholar in the following manner:

By reducing processes into factual data, much of the power of Indigenous Knowledge is lost. The dominant society is willing to use Indigenous generated factual data in co-management agreements, but they are not willing to use the *process* of Indigenous management. Instead of strengthening and using Indigenous processes, the dominant society inserts factual knowledge into its own processes, models and management plans. The ability of Aboriginal [indigenous] peoples to affect change in environmental management then becomes greatly reduced

(Simpson, 1999: 74).

Despite the formal co-management arrangement, Canadian government managers do not seem to be placing a higher value on indigenous knowledge than Alaskan managers who are not party to a co-management regime. In another twist, when queried, a higher proportion of Alaskan users said they would cooperate with managers compared to the Canadian case. This may be related to the amount of time that Alaskan users and managers spend together on a day-to-day basis (Kruse *et al.*, 1998). Is it the informal social relations between government and community that lead to real, but informal institutional change that includes processes of indigenous management?

Fienup-Riordan's (1999) work with Yup'ik communities in Alaska confirms the importance of the social connections between government and community. Without the development of personal connections, collaborative work between communities and government remains limited in scope. As expressed by one Yup'ik elder (Fienup-Riordan, 1999: 19): 'There are different kinds of biologists. Some stick with what they know, they don't try to expand their knowledge. There are the others who want to learn more and expand their knowledge to help us.'

It is important to note, however, that in the comparison of the Alaskan and Canadian systems, Canadian indigenous users have achieved governmental recognition of indigenous rights to resource use and management that Alaskan users do not hold to the same extent. In contrast, Alaskan indigenous users are recognized as 'rural people' under a federal law that has provided the means for rural subsistence harvesters to protect their use rights from commercial interests through allocation preferences (Wolfe, 1998). However, the 1980 Alaskan Subsistence Law is out of step with the 1971 Alaska Native Claims Settlement Act that requires protection of the subsistence needs of *indigenous* Alaskans. As a result, an uneasy tension exists between federal and state laws where the federal government regulates subsistence on federal lands (60 percent of Alaska) and Alaska maintains authority over the majority of the rest of state lands (Thornton, 1998).

It is not as incumbent on Canadian users to cooperate with managers as it is in the Alaskan case. In Alaska, managers *and* users do recognize the *potential* of reformed management regimes. However, government managers and traditional users may *recognize different potentials*. The clash of perceptions of users and government managers during the caribou 'crises' of the 1970s may be more fundamental than we have even begun to realize. *Formal* co-management institutions still treat community-based thinking as something so unfamiliar as to be essentially nonexistent in formal decision-making. For example, users may see co-management as a place to secure a voice in government resource management decision-making. Managers may recognize that both the enforcement of hunting restrictions in the face of a decline in caribou populations and the monitoring of animals are impossible without community involvement.

10.6 Cognitive commitments

Little work exists documenting indigenous cognition of caribou population dynamics. Attempts include work with Inuit, James Bay Cree, and Gwich'in communities (Gunn, Arlooktoo, and Kaomayok, 1988; Berkes, 1995; Kofinas, 1998). When scientific survey results conflict with community observations, conventional management measures (i.e., limitations placed on harvest activities) may be achieved only at the expense of significant short-term and long-term social costs (for example community distrust and noncompliance with harvest quotas). Very little has been documented about the ways that caribou users share (and shared before year-round settlement patterns) information about the location and movements of caribou (Smith 1978; Speiss, 1979; Burch, 1991). Pre-contact indigenous harvesting levels were probably *not* limited *solely* by low population numbers and inefficient technologies. There were local rules, behavioral norms, beliefs, tribal territories, and other social mechanisms of harvest control (Csonka, 1991; Berkes, 1999). It is possible that the hesitancy of communities to accept the methods and technology of caribou population surveys is also related to the ways that such research may threaten the authority of indigenous knowledge and management systems. For example, Dene elders have expressed concerns about past caribou programs by biologists that targeted animals at river crossings, perhaps causing a change in their movements (Smith, 1978; Kendrick, 1994).

Hunters in some settlements use citizen's band (CB) and high-frequency (HF) radios to communicate with people on the land or in communities hundreds of miles away about wildlife movements (Nakashima, 1991; Kendrick, 1994). It could be argued that the way knowledge of caribou movements is now spread by people with home-operated, two-way radios helps to replicate the manner that information was exchanged in the past. Before settling year-round in one location, people moved on the land, visiting neighboring hunting camps and occasionally gathering in large numbers for caribou group hunts in which they shared knowledge of caribou movements.

There are remarkable parallels in caribou-hunting techniques across the circumpolar region, including mass traps such as corrals with drive fences and drives of animals into water bodies (Speiss, 1979). These techniques indicate that the indigenous knowledge of caribou movements and distribution of societies even marginally dependent on caribou was extensive. Ethnohistorical records show that 'regional bands' of between 200 and 400 people (of 'Caribou Eater Chipewyan' and Yellowknife Dene) gathered for the sole purpose of communal hunts at drive fences in the fall and winter (Speiss, 1979: 115). Even brief consideration of the complexity involved in coordinating these hunts should give us pause about the level of knowledge, communication, and cooperation involved.

What kind of relations, relevant to increasing trust and respect, are driven by co-management efforts? State managers increase their awareness of the sharing and kinship relations involved in hunting and food distribution. Co-management also increases awareness of the knowledge exchange occurring at the local level. Notions of sharing, equity, and reciprocity are very different in traditional caribou-hunting societies versus government bureaucracies. This is especially apparent when these principles are applied to the negotiation of resource use rights. For instance, in the 1980s, the Beverly-Qamanirjuaq caribou co-management board, with its majority of indigenous members, recommended the removal of barriers to the shipment of meat to community members hospitalized in southern cities. Government regulations had prohibited the shipment of wild meat to cities. In a further example, user representatives on the same board initiated research that documented community-identified critical caribou habitat that subsequently led to changes in one political jurisdiction's fire-fighting policy. Canada's Northwest Territories have a fire-management plan that now includes caribou habitat protection as well as commercial timber areas for fire control.

The examples listed in Table 10.2 further illustrate the depth of the concepts currently negotiated in co-management contexts. Biologists, government managers, and indigenous communities may not only define caribou herds and land differently, they may also relate to caribou and the land differently. There are therefore difficulties inherent in translating across these linguistic as well as conceptual barriers. In some Dene communities, it is said that time and space join in ways understandable only from the standpoint of a five-dimensional or six-dimensional model (Sharp, 1997: 97). The reality of *time* created in such a conceptual model is more similar to Western concepts of *space*. Time is a variable rather than a fixed point, so that 'history is not past, history is; future is not maybe, future is; both are equally real' (Sharp, 1997: 97). This concept of time is in sharp contrast to Western thinking in which time is organized in a linear sequence and causality is an implied part of this sequence (the classical definition of logic). Time moves in a direction from a determined past to an undetermined future so that only 'now' is real (Sharp, 1997: 97). It is obvious that if Dene conceptions of time are more closely related to Western ideas of space, there are huge differences between these conceptions.

Table 10.2 *Blurred conceptual realities: Dene caribou users, managers, biologists, and caribou*

On one hand...	Concept	On the other hand...
A discrete population of animals displaying long-term fidelity to a definable calving area and subject to abundance and scarcity	A caribou herd...	Groupings of animals whose behavior and discreteness are responsive to human–caribou interactions or a relationship of collaborative reciprocity
A resource that is a biophysical component of habitat that supports populations	Land...	Nde – a Dogrib (Dene) term describing land as a living entity encompassing a holistic notion of living; a landscape complete with animals, plants, and other living processes with spiritual lives
Time is linear/space is three-dimensional and impersonal	Time and space...	For the Bearlake Dene, time and space join in a five-dimensional or six-dimensional model (Sharp 1997: 97) Time is a variable rather than a fixed point so that the past, present, and future are current
Caribou as a 'resource' that can be managed and belief in control and rational explanation; belief that population dynamics can be manipulated	Management...	The word is nonexistent in Dene languages; however, *inkonze* illustrates the fundamental differences between the scientific human–nature split and *inkonze*'s recognition of human dependence and place 'in' nature Human–nature relationship is one characterized by ethics, and reciprocity, but not by 'management;' acceptance of uncertainty, uncontrollability, and unknowability

The realization of such differences makes cross-cultural caribou management a huge challenge, especially when the word 'management' itself is a word that has no equivalent in Dene languages and characterizes a human–environment divide that is antithetical to Dene cultural values. Moreover, it may be detrimental to communities to fully engage themselves with government institutions or to try to achieve a synthesis of knowledge and concepts (Kofinas, 1998). Mainstream influences may decontextualize community structures and knowledge to such an extent that they are no longer meaningful or viable (Weinstein, 1996).

10.7 Learning to recognize diverse knowledge systems

In recent years, there have been efforts to complement the TEK of resource users with the science of resource managers. Comparisons are often made between Western science and TEK. There are dangers inherent in this comparison because it is easy to forget that 'what we know' is framed by 'how we know.' Comparing TEK as knowledge associated with human values and ethics to 'Western scientific' knowledge – seemingly differentiated from human value systems – is an almost self-defeating task when the multidimensional nature of knowledge is forgotten.

Since Cartesian times, Western society has marginalized the science of the integration of the parts in favor of the prescriptiveness offered by the science of the parts (Bateson, 1991; Capra, 1996; Holling, 1998). The science of the parts gives an incomplete and misleading picture of social–ecological systems (Chapter 1). The '... inherent unknowability, as well as unpredictability, concerning ecosystems and the societies with which they are linked ... [and the] inherent unknowability and unpredictability to sustaining the foundations for functioning systems of people and nature [is forgotten]' (Holling, 1998).

However, other scientists point beyond the *imbalances* in the *application* of the 'two cultures of biological ecology.' Bateson (1991: 199–201) outlined the *misleading orthodoxies* of the natural sciences. Namely, (1) the artificial isolation of the observer from the object observed, (2) the false sense that time is independent of process, when in fact time is a consequence of process, and (3) the misapplied logical typing that makes structure primary and process secondary.

Many thinkers have outlined the pitfalls of seeing Western science as an objective, value-free practice (Longino, 1990; Latour, 1999). Comparisons made between TEK and Western science tend to construct TEK as a sounding board for Western science, true to the 'us–other' comparison described in Figure 10.4. The troublesome aspect of such comparisons is that diverse knowledge systems boxed as TEK or indigenous knowledge are heavily generalized or stereotyped, often in the effort to redress the imbalances, misleading orthodoxies, or forgotten presuppositions of Western science (Agrawal, 1995; Cruikshank, 1998). There is a responsibility to understand the cultural context of TEK *and* Western knowledge in order to avoid untenable comparisons:

> Some of the people using TEK do so on the false basis of a comparison between a selected part of Western societies' knowledge, i.e. 'science,' and the whole of a culture which is regarded as knowledge ... Such opinions are stated when it is quite clear that science without intuition would be nothing; that many scientists were strictly moral and religious; that the discoveries of Einstein would not have been made had Einstein not believed in a principle of divine harmony; that Indigenous people have hierarchical classification;

and that the same people always, and I say always from my own experience, count the number of fish or whatever they catch in a season, and that they are not merely qualifiers of nature but also quantifiers

(Clément, 1998: 12).

In this way, the TEK–Western science dichotomy can be an illusion that co-management regimes may or may not identify and avoid.

10.8 How something is known is as important as what is known

Real innovation and progress in thought may be discovered in resource management approaches that concentrate on recognizing cultural differences in learning patterns. The manner in which knowledge is learned is as important as the ways it is shared. Animism and shamanism, 'described as among the most significant characteristics of northern cultures . . . [and the] least analyzed . . .' (Yamada and Irimoto, 1997), may be viewed as systems of thought and practice articulating the associations between human will and environmental potential (Ridington, 1990: 96; Irimoto, 1997). Anthropologists working in northern Canada have explored this idea (Bielawski, 1992; Cruikshank, 1998). Mythic beliefs and practices are forms of technology – a system of knowledge representing living practices. They are the means for sharing and interpreting knowledge and often provocative tools for problem solving (Cruikshank, 1998). Myths and stories are powerful agents for thinking, not simply for entertainment. Separating the technical knowledge of a culture from its place within a system of belief can lead to misleading conclusions about the relevance and reliability of that knowledge.

Mythic thinking, story telling, dreaming, and ceremonies are activities that members of Western society might refer to as artistic or ritualistic. However, indigenous story tellers have used narratives to raise significant epistemological issues about Western classificatory practice and contemporary theoretical constructions (Cruikshank, 1998). For example, government resource management regulations in one part of the north allow the hunting of swans for subsistence purposes only. The subsistence classification prompted a Dene elder to ask, 'So, if "subsistence" means "food," and "nonsubsistence" means "culture," how do you get a swan bone for a ceremony?' (Cruikshank, 1998: 17). Fundamental questions are being asked: how do Dene continue to observe hunting practices that respect the reciprocity inherent in human–environment relations? Western resource management defines acceptable harvesting practices in a utilitarian manner: hunting for food is legitimate use, but hunting to maintain a relationship of mutual obligation between the hunter and the animal is another matter.

Stories and ceremonies play a role in linking human history (of human–environment relations) to a sense of place, so that when '[o]ld people they tell you a story, you've got to listen. When you don't listen, you're going to be crazy. You're going to be crazy, and you're not going to live long' (Cruikshank 1998: 19). Should we be concerned that mainstream resource management ignores and marginalizes the integrative learning offered by narrative and metaphor? By narrowing the kind of knowledge and learning that is relevant to resource management decision-making, are we helping to sever the feedback between human actions and the environment?

10.9 Conceptual diversity: the wisdom to respect what we may not understand?

The cross-cultural translation of knowledge, regardless of the rationale behind human actions, is a big task. What happens, for instance within caribou co-management regimes, when the policies of northern governments require the inclusion of TEK within management decision-making and monitoring activities? A biologist or social scientist speaking to indigenous elders about their knowledge of wildlife may sideline the dances, songs, or stories accompanying such knowledge sharing. An academically trained scientist may not have the knowledge of the metaphor or context that these expressions represent. Aspects of learning or of respecting another way of knowing are therefore lost or at the very least forgotten. Western cultures often ignore the 'background' or context that is in fact the 'operator' in an interaction, and selectively place the 'operands' in the foreground, believing that these parts can be understood distinct from their contexts (Bateson, 1991: 66).

In a series of films recording the thoughts of Inuit hunters and caribou biologists in the Keewatin region (west coast of Hudson Bay) in the late 1970s, one hunter stated:

> I think that they [Inuit hunters, government managers, and biologists] would stop disagreeing with each other if they both started showing things in a way that doesn't make the other person look bad but so that the other person or party understands what the other is trying to do about the caribou
>
> *(National Film Board, 1982, video tape #8[1]).*

In this statement there lies a plea to recognize the difference in the contexts from which caribou hunters and government managers speak.

10.10 Conclusions

Not only are we more aware that space and time are not conceptualized by all cultures in the same way, but we are beginning to understand that language plays a large role in the way we learn about the world around us. This chapter looked at the discrepancies between the attitudes and belief systems of caribou-using communities and government caribou managers and the ways these differences can frustrate caribou co-management efforts. No doubt, cooperative management *action* is difficult because of the continued differences between government managers and caribou users despite almost two decades of direct dialogue in Canada. However, it is here that encouraging conditions for the learning of resilient thought processes lie. Co-management may be the path to an expanded capacity to recognize the diversity of what we can know about *ourselves* as biological beings, as well as *how* we can know and understand natural processes. The creation of resource management systems that include stakeholders with fundamental perceptual differences – such as caribou co-management boards – may lead to profound and important insights into human–environment relations.

The fact that caribou-using communities have significantly different notions of caribou population phenomena from caribou biologists has the potential to stimulate rather than stall co-management decision-making. Caribou biologists are still coming to grips with issues such as herd discreteness, range use, the periodicity of population cycles, and the effects of human disturbance activities on caribou behavior and viability. Inuit and Dene elders question the reality of population 'crises' and the necessity of handling wild animals outside of the respect and reciprocity of harvesting relationships. Are we seeing evidence of a consciousness that emphasizes the primariness of relations over structure or something similar to Bateson's 'mental process'?

Caribou co-management has enabled an exchange of ideas between resource users and government managers and biologists about research approaches (Urquhart, 1996; Ferguson *et al.*, 1998; Kofinas, 1998). Co-management as a mutual education process involves a fundamental reform of attitudes, both of government and community representatives (Thomas and Schaefer, 1991). As a consequence, co-management is a dynamic process and its outcomes are difficult to describe. There is evidence that slow learning is taking place in caribou co-management settings, but not yet necessarily in a way that ensures that both community and government equally share the costs of this process. It is suggested that indigenous communities run the risk of undermining their aspirations for political autonomy by participating in co-management arrangements that represent partnerships with outside government institutions (Caulfield, 1997; Feit, 1998; Kofinas, 1998). Indigenous communities also diminish their capacity to participate in other land-use planning processes when they take on the enormous burden of translating co-management discussions (conceptually and linguistically) to community members. There is a large amount of time and effort involved in enabling collective community participation in co-management decision-making.

Caribou co-management may represent an emerging dialectic of conceptual diversity in practice. However, the trust and humility involved are complex. These elements are fostered in arenas beyond the rigid frameworks of formal co-management board meetings. The differences in the knowledge of caribou biologists, managers, and users may involve more than the spatial or temporal contexts of the knowledge (i.e., synchronic versus diachronic observations). The resilience of human abilities to think about natural processes may lie in learning how to challenge mainstream orthodoxies. To do this we must better understand social (human) and ecological linkages. Without mechanisms allowing the respect and support of knowledge systems based upon precepts that are fundamentally different from mainstream resource management thinking, future resource management systems may not only distort or ignore viable ways of thinking, but altogether destroy them.

Learning to respect differences does not lie in the codification of knowledge (Cruikshank, 1998). If we are to think and learn in an adaptive manner, then world views or metaphors that add to the range of human integrative and complex thinking need to be supported rather than ignored. By focusing *only* on the codification of marginalized knowledge, we risk eliminating the resilience of the human capacity to know in integrative and complex ways.

This chapter looks at caribou co-management as a case study of the accommodation of different views of human aims and perceptions of the environment, and illustrates the potential co-management has to provide the space for intellectual discourse between mainstream thinking and marginalized indigenous thought on human–environment relations. Resource management systems tend to fragment the meaning and values inherent in indigenous knowledge in the search for technical explanations of resource dynamics that fit into current ecological models and theories. Co-management may be the place where these problems will be overcome. There is also a relatively unexplored role for indigenous narratives within mainstream resource management systems. These narratives represent ways of looking at social and ecological systems and the links between them that could both complement and challenge established thinking. Co-management is fundamentally a process of joint problem solving with positive outcomes for all parties. All stakeholders lose when knowledge of 'the other' is mystified. Narrative may play a more fundamental role in the way human beings learn and promote resilient, integrative, and complex social–ecological thought than mainstream resource management yet recognizes.

Acknowledgements

The author would like to acknowledge the kind support of Fikret Berkes in the writing of this chapter, and the stimulating comments of the reviewers. The communities of Tadoule Lake, Manitoba; Arviat, Nunavut; Churchill, Manitoba; and Lutsel K'e, Northwest Territories (all northern Canadian settlements) have taught me many lessons about tolerance and sharing. This research has been supported over the years by grants from the Northern Scientific Training Program, Canadian Northern Studies Trust Caribou Bursary, Beverly-Qamanirjuaq Caribou Management Board Scholarship, The Walter and Duncan Gordon Foundation, University of Manitoba Graduate Fellowship, and Social Science and Humanities Research Council, and, of course, the love and patience of my family.

Note

1. The use of the term 'government' in this chapter denotes nonindigenous government structures. Many indigenous groups in Canada consider their negotiations with Canadian government structures as a nation-to-nation or government-to-government process.

References

Agrawal, A. 1995. Indigenous and scientific knowledge: some critical comments. *Indigenous Knowledge and Development Monitor* 3(3): 3–6.

Ames, R. 1979. *Social, Economic and Legal Problems of Hunting in Northern Labrador.* Nain: Labrador Inuit Association.

Anderson, D.G. 2000. Rangifer and human interests. *Rangifer* 20(2–3): 153–74.

Anon. 1995. Two herds total almost 800,000: 1994 surveys. *Caribou News* March 1995, 15(1): 1–2.

Bateson, G. 1979. *Mind and Nature: a Necessary Unity.* New York: Dutton.

Bateson, G. and Bateson, M.C. 1987. *Angels Fear – Towards an Epistemology of the Sacred.* New York: Macmillan.

Bateson, M.C. 1991. *Our Own Metaphor – a Personal Account of a Conference on the Effects of Conscious Purpose on Human Adaptation.* Washington: Smithsonian Institution Press.

Berkes, F., ed. 1989. *Common Property Resources: Ecology and Community-based Sustainable Development.* London: Belhaven Press.

Berkes, F. 1995. Indigenous knowledge and resource management systems: a native case study from James Bay. In *Property Rights in a Social and Ecological Context: Case Studies and Design Applications*, pp. 99–109, ed. S. Hanna and M. Munasinghe. Washington DC: Beijer International Institute of Ecological Economics and the World Bank.

Berkes, F. 1999. *Sacred Ecology – Traditional Ecological Knowledge and Management Resources.* Philadelphia: Taylor and Francis.

Berkes, F., George, P., and Preston, R.J. 1991. Co-management. *Alternatives* 18(2): 12–18.

Bielawski, E. 1992. Inuit indigenous knowledge and science in the Arctic. *Northern Perspectives.* 20(1): 5–8.
Burch, E. 1991. Herd following reconsidered. *Current Anthropology* 32(4): 439–45.
Bussidor, I. and Bilgen-Reinart, U. 1997. *Night Spirits – the Relocation of the Sayisi Dene.* Winnipeg: University of Manitoba Press.
Campbell, T. 1996. Co-management of aboriginal resources. *Information North* 22(1): 1–6.
Capra, F. 1996. *The Web of Life.* New York: Anchor Books.
Caulfield, R.A. 1997. *Greenlanders, Whales and Whaling – Sustainability and Self-Determination in the Arctic.* Hanover: University Press of New England.
Clément, D. 1998. Evolution of concepts in ethnobiological studies. In *Aboriginal Environmental Knowledge in the North*, pp. 7–19, ed. L-J. Dorais, M. Nagy, and L. Müller-Wille. Québec: GÉTIC, Université Laval.
Cruikshank, J. 1998. *The Social Life of Stories: Narrative and Knowledge in the Yukon Territory.* Lincoln: University of Nebraska Press, and Vancouver: UBC Press.
Csonka, Y. 1991. Les Ahiarmiut (1920–1950) dans la perspective de l'histoire des Inuit caribous. PhD Thesis, Université Laval, Québec.
Feit, H.A. 1998. Reflections on local knowledge and institutionalized resource management: differences, dominance and decentralization. In *Aboriginal Environmental Knowledge in the North*, pp. 123–48, ed. L-J. Dorais, M. Nagy, and L. Müller-Wille. Québec: GÉTIC, Université Laval.
Ferguson, M.A.D. and Messier, F. 1997. Collection and analysis of traditional ecological knowledge about a population of Arctic Tundra caribou. *Arctic.* 50(1): 17–28.
Ferguson, M.A.D., Williamson, R.G., and Messier, F. 1998. Inuit knowledge of long-term changes in a population of Arctic Tundra caribou. *Arctic* 51(3): 201–19.
Fienup-Riordan, A. 1990. *Eskimo Essays: Yupik Lives and How We See Them.* New Brunswick: Rutgers University Press.
Fienup-Riordan, A. 1999. *Yaqulet Qaillun Pilartat* (what the birds do): Yup'ik Eskimo understanding of geese and those who study them. *Arctic* 52(1): 1–22.
Folke, C., Berkes, F., and Colding, J. 1998. Ecological practices and social mechanisms for building resilience and sustainability. In *Linking Social and Ecological Systems. Management Practices and Social Mechanisms for Building Resilience,* pp. 414–36, ed. F. Berkes and C. Folke. Cambridge: Cambridge University Press.
Freeman, M.M.R. 1989. Graphs and gaffs: a cautionary tale in the common property resource debate. In *Common Property Resources – Ecology and Community-based Sustainable Development*, pp. 92–109, ed. F. Berkes. London: Belhaven Press.
Gunn, A., Arlooktoo, G., and Kaomayok, D. 1988. The contribution of the ecological knowledge of Inuit to wildlife management in the NWT. In *Traditional Knowledge and Renewable Resource Management in Northern Regions,* pp. 22–30, ed. M.M.R. Freeman, and L.N. Carbyn. Edmonton, Alberta: IUCN Commission on Ecology and the Boreal Institute for Northern Studies.
Harries-Jones, P. 1995. *A Recursive Vision: Ecological Understanding and Gregory Bateson.* Toronto: University of Toronto Press.
Holling, C.S. 1998. Two cultures of ecology. *Conservation Ecology* 2(2): 4. Available from the Internet. URL: http://www.consecol.org/vol2/iss2/art4
Irimoto, T. 1997. Ainu shamanism: oral tradition, healing, and dramas. In *Circumpolar Animism and Shamanism*, pp. 21–46, ed. T. Yamada and T. Irimoto. Sapporo, Japan: Hokkaido University Press.

Kendrick, A. 1994. Community perspectives, caribou user participation and the Beverly-Qamanirjuaq Caribou Management Board in northcentral Canada. Unpublished MA thesis, McGill University, Montreal.

Kendrick, A. 2000. Community perceptions of the Beverly-Qamanirjuaq Caribou Management Board. *The Canadian Journal of Native Studies.* 20(1): 1–33.

Klein, D.R. 1991. Caribou in the changing North. *Applied Animal Behaviour Science* 29: 279–91.

Klein, D.R., Moorehead, L., Kruse, J., and Braund, S.R. 1999. Contrasts in use and perceptions of biological data for caribou management. *The Wildlife Society Bulletin* 27(2): 488–98.

Kofinas, G.P. 1998. The costs of power sharing: communities in porcupine caribou herd co-management. Unpublished PhD thesis, University of British Columbia, Vancouver.

Kruse, J., Klein, D., Braund, S., Moorehead L., and Simeone, B. 1998. Co-management of natural resources: a comparison of two caribou management systems. *Human Organization* 57(4): 447–58.

Latour, B. 1999. *Pandora's Hope: Essays on the Reality of Science Studies.* Cambridge, MA: Harvard University Press.

Livingston, J.A. 1981. *The Fallacy of Wildlife Conservation.* Toronto: McClelland and Stewart Ltd.

Longino, H.E. 1990. *Science as Social Knowledge – Values and Objectivity in Scientific Inquiry.* Princeton, NJ: Princeton University Press.

Nakashima, D.J. 1991. The ecological knowledge of Belcher Island Inuit: a traditional basis for contemporary wildlife co-management. Unpublished PhD thesis, McGill University, Montréal.

National Film Board 1982. *The Kaminuriak Herd Film/Videotape Project.* (Film crew: Snowden, D., Kusagak, L. and Macleod, P.)

Nuttall, M. 1998. Critical reflections on knowledge gathering in the Arctic. In *Aboriginal Environmental Knowledge in the North*, pp. 21–35, ed. L-J. Dorais, M. Nagy, and L. Müller-Wille. Québec: GÉTIC, Université Laval.

Ridington, R. 1990. *Little Bit Know Something – Stories in a Language of Anthropology*. Vancouver: Douglas and McIntyre Ltd.

Roots, F. 1998. Inclusion of different knowledge systems in research. *Terra Borealis* 1: 42–9.

Royal Commission on Aboriginal Peoples 1996. *Restructuring the Relationship*, Vol. 2, Part 2, pp. 665–774. Ottawa: The Commission.

Rybczynski, N. 1997. Predators and cosmologies. *The Canadian Journal of Native Studies* 17(1): 103–13.

Sharp, H.S. 1997. Non-directional time and the Dene life-cycle. In *Circumpolar Animism and Shamanism*, pp. 93–104, ed. T. Yamada and T. Irimoto. Sapporo, Japan: Hokkaido University Press.

Simpson, L.R. 1999. The construction of traditional ecological knowledge: issues, implications and insights. Unpublished PhD thesis, University of Manitoba, Winnipeg.

Singleton, S. 1998. *Constructing Cooperation: the Evolution of Institutions of Co-management.* Ann Arbor: The University of Michigan Press.

Smith, D.M. 1998. World as event: aspects of Chipewyan ontology. In *Circumpolar Animism and Shamanism*, pp. 67–91, ed. T. Yamada and T. Irimoto. Sapporo, Japan: Hokkaido University Press.

Smith, J.G.E. 1978. Economic uncertainty in an 'original affluent society': caribou and caribou eater Chipewyan adaptive strategies. *Arctic Anthropology* 15(1): 68–88.

Speiss, A.E. 1979. *Reindeer and Caribou Hunters – an Archaeological Study.* New York: Academic Press.
Thomas, D.C. and Schaefer, J. 1991. Wildlife co-management defined: the Beverly and Kaminuriak Caribou Management Board. *Rangifer*, Special Issue No. 7: 73–89.
Thornton, T.F. 1998. Alaska native subsistence: a matter of cultural survival. *Cultural Survival Quarterly* 22(3): 29–34.
Tuan, Y.-F. 1979. Thought and landscape: the eye and the mind's eye. In *The Interpretation of Landscape: a Cultural Geography*, pp. 89–102, ed. D.W. Meinig. New York: Greenwood Press.
Urquhart, D. 1996. Caribou co-management needs from research. Simple questions – tricky answers. *Rangifer* Special Issue No. 9: 263–71.
Usher, P.J. 2000. *Caribou Crisis or Administration Crisis? Wildlife and Aboriginal Policies on the Barren Grounds of Canada, 1947–1960.* Paper presented at the 12th Inuit Studies Conference, University of Aberdeen, Aberdeen, Scotland, 26 August 2000.
Weinstein, M. 1996. *Traditional Knowledge, Impact Assessment, and Environmental Planning.* Canadian Environmental Assessment Agency, BHP Diamond Mine Environmental Assessment Panel. Ottawa: Minister of Public Works and Government Services.
Wolfe, R.J. 1998. Subsistence economies in rural Alaska. *Cultural Survival Quarterly* 22(3): 49–50.
Yamada, T. and Irimoto T., eds. 1997. *Circumpolar Animism and Shamanism.* Sapporo, Japan: Hokkaido University Press.

Part IV

Cross-scale institutional response to change

Introduction

The chapters in Part II of this volume explored how resilience thinking helps ask questions regarding the adaptive capacity of institutions to deal with change. The chapters in Part III added the dimension of learning in adaptive management and co-management. The chapters in Part IV now turn to the topic of cross-scale interactions. In most of the cases in this volume, and in most cases in real life, there are external drivers, factors that impact local management systems. In an age of globalization, governance has become cross-scale. There is a need to analyze management institutions at more than one level, with attention to interactions across scale from the local level up. What used to be local management now has regional, national, and often international dimensions, leading to the emergence of new players with new power relationships. How can we approach the understanding of cross-scale relationships, and how do these relationships relate to resilience and sustainability?

Systems theory reminds us that a key factor for response is the presence of effective and tight feedback mechanisms or a coupling of stimulus and response in space and time. For example, it is relatively easy to get a neighborhood association to act on a problem. But as problems become broader in scale (e.g., the global greenhouse effect), the feedback loops become looser and the motivation to act becomes weaker. Incentives can be created by tightening cost/benefit feedback loops, for example by assigning property rights. In some cases where the market can work properly and social costs are taken into account, privatization may be an effective measure. In other cases, the transfer of communal property rights to local groups can be effective.

Part IV considers three cases in which commons institutions are linked both horizontally (across space) and vertically (across levels of organization). The cases come from different parts of the world – Brazil (Chapter 11), Indonesia (Chapter 12), and the United States (Chapter 13) – and explore the relations among commons management, governance, and institutions in a cross-scale context.

11

Dynamics of social–ecological changes in a lagoon fishery in southern Brazil

CRISTIANA S. SEIXAS AND FIKRET BERKES

11.1 Introduction

Any resource management system has two interrelated dimensions: the social system and the ecological system. These dimensions are often treated separately. In the last decades, considering the failure of many conventional resource management systems (Ludwig, Hilborn, and Walters, 1993), some researchers have started investigating the dynamics of integrated social and ecological systems (henceforth social–ecological systems) in order to improve resource management (Gunderson, Holling, and Light, 1995; Berkes and Folke, 1998). To analyze the dynamics of social–ecological systems, we use common-property theory and adaptive management.

The development of common-property theory (McCay and Acheson, 1987; Berkes, 1989; Ostrom, 1990; Bromley, 1992) has provided key tools for the understanding of the social dimension of management systems. A common-property (or common-pool) resource (defined as a class of resources for which exclusion is difficult and joint use involves subtractability) can be managed under four 'pure' property rights regimes: communal property (community-based management), state property, private property, or open access (lack of a property rights regime). In reality, many resources are managed under various mixes of these regimes, as in co-management characterized by a sharing of responsibility between the government and user groups for resource management. The degree of participation of government agencies and user groups in the decision-making process may vary greatly from one co-management case to another (McCay and Jentoft, 1996; Pomeroy and Berkes, 1997). Co-management is a promising regime in developing adaptive management systems because it allows for cross-scale interactions (combining the local level and higher levels) and for feedback learning enhanced by the existence of these cross-scale institutions. We define institutions as any formal constraints (rules, laws, and constitutions) or informal constraints (norms of behavior, conventions, and self-imposed codes of conduct) that mold interactions in a society (North, 1994).

In the field of ecosystem dynamics and adaptive management, the model of an adaptive renewal cycle and the use of the idea of resilience have provided management insights (Holling, 1986, 1995). The adaptive renewal cycle encompasses four stages: exploitation, conservation, release, and renewal. Typically, an ecosystem proceeds from exploitation slowly to conservation, then rapidly to release, and again rapidly to renewal, before returning to the exploitation phase. The resilience of an ecosystem is its capacity to absorb disturbances while maintaining its main behavioral processes and structure. It can be defined as the capacity to buffer perturbations, to self-organize, and to learn and adapt (Chapter 1).

As ecosystems are hierarchically structured into a number of levels, many adaptive renewal cycles are linked through time and space, termed panarchy by Gunderson and Holling (2002). At least two features of panarchy (or cross-scale interaction) may contribute to understanding resilience: (1) disturbance in the small-scale system can cascade to the broader scale (Carpenter and Kitchell, 1993), and (2) a large-scale system can provide resources (by 'remembering' or carrying over elements through its release phase) for the renewal phase of the smaller-scale system.

In this chapter, we combine the common-property approach with the ecosystem resilience approach to navigate the dynamics of social–ecological systems. Our purpose is to identify some of the key factors that build social–ecological resilience in resource management, and some key factors that threaten it. To this end, we analyze the case of the Ibiraquera Lagoon, which has experienced several drastic changes (flips) of the social–ecological system in the last four decades. In particular, we investigate changes in the socioeconomic system, the management practices used, local and government institutions, and lagoon ecosystem dynamics. We also investigate fishers' local ecological knowledge behind their fishing practices and institutions.

11.2 The case study

The Ibiraquera Lagoon is located in the municipality of Imbituba (population 33 000 in 1991), Santa Catarina State, in the southern part of the Brazil coast. The lagoon is seasonally connected to the Atlantic Ocean. The main fishing activities are the pink shrimp (*Farfantepenaeus paulensis* and *F. brasiliensis*) fishery year round and the mullet (*Mugil platanus, Mugil spp.*) fishery in winter (from May to July). Fish and shrimp are men's activities; crab fishing, especially in the hot months, is a family activity including women and children.

Legally, any Brazilian who has a professional fishing license can fish in the lagoon. Those with sport fishing licenses cannot. Professional fishing licenses, in law, are supposed to be issued only to those who make their main living from fishing. However, in reality, they are issued to almost anyone who requests them. The main requirement for a professional license is the testimony of two professional fishers that the requester makes his living from fishing. Thus, there is no effective legal access restriction to the lagoon. There are about 350 professional fishers living in seven communities around the Ibiraquera Lagoon: Ibiraquera (also known as Teixeira), Barra da Ibiraquera, Arroio, Alto Arroio, Araçatuba, Campo D'Una, and Grama (or Ibiraquera de Garopaba). Many of the fishers are descendants of immigrants from the Azores Islands, who arrived in this part of the country about 200 to 250 years ago. Until the 1960s, most communities were quite isolated, living on subsistence agriculture and fishing.

To understand the interactions over time between the social and ecological dimensions of the lagoon management system, we investigated the socio-economic and ecological history of this area in the last four decades, divided into four periods according to the occurrence of major changes affecting the management system: (1) the 1960s as a baseline; (2) from 1970 to 1981, a period of several socio-economic changes that culminated in a crisis in the management system; (3) from 1981 to 1994, a period of major changes in fishing regulations that resulted in the recovery of the management system; (4) from 1994 to 2000, when the enforcement of fishery management broke down and a new crisis started to emerge.

The Ibiraquera Lagoon was chosen for the study of the dynamics of social–ecological systems because it was known as an area in which resource collapse and recovery cycles had occurred. In the following sections, first we provide a brief overview of the socio-economic history of the local communities, to be expanded later by period. Second, we describe lagoon ecosystem dynamics. Third, we explore the linkages between the lagoon dynamics and the 'traditional' fishery management. We call it 'traditional' because it represents the pre-commercial, pre-modern system; it is the baseline against which we assessed changes over time. Fourth, we describe the management system in each period. Finally, we analyze the interaction of the lagoon management institutions and the dynamics of the lagoon over these periods.

11.2.1 Methods

Fieldwork was carried out between June 1999 and May 2000. Research methods included interviews, archival research, and participant observation. Interviews were carried out in several formats – structured interviews, semi-structured interviews with key informants and small groups, and ethnomapping – to elucidate fishers' knowledge, fishery activities, and main changes in the fishery management system in the last four decades. Archival research was done to trace changes in fisheries legislation and the local socio-economic system. Participant observation was carried out from October 1999 to May 2000 to monitor fish-catching and shrimp-catching activities and the fishing methods used, and to understand the role of middlemen, buyers, resource managers, fishery association officials, and government officials.

Estimates of fish and shrimp abundance and harvest, fish and shrimp migrations, and seasonal cycles of the lagoon were based on information from key informant interviews, backed up by field observations. There are no fish and shrimp population data or harvest statistics available for this locality, or for most inshore fisheries in Brazil. The major species of fish, shrimp, and crabs were collected in the field and identified by the use of species keys or by taxonomic specialists. Of about 39 species identified by fishers, 24 were recorded in the field, and 17 were identified biologically (Seixas, 2002).

11.2.2 Socio-economic background

In the early 1960s, there were relatively few families living in the communities around the lagoon. Four of the seven communities had no road access, none had electricity, five of the seven had no general store, and none had a fish store. The general stores did sometimes sell fish. Fish and shrimp were not marketed outside the area, except for what might be sold by the road in the three villages with road connection, and what could be carried on one's back along the beach to Imbituba. The local economy was based mainly on household-level agriculture (manioc flour and sugar as the main products), and fishing was done mainly for subsistence. There were no job opportunities for young people who often migrated to big cities for work.

From 1970 to 1981, roads were constructed and electricity became available in most communities. With the roads came the tourists. Tourism-related activities created local job opportunities, and precipitated the return of villagers who had migrated to the big cities; they saw new job opportunities in the area, and some brought capital. More markets and retail outlets were created, including fish stores. Fish and shrimp started to be exported to regional markets. The importance of household-level agriculture in the economy of most communities declined.

From 1981 to 1994, the resident as well as the tourist populations increased. All communities now had road access and electricity. Telephone services became available in most communities, and several summer cottage developments, guesthouses, and restaurants were built around the lagoon. In addition, even more retail stores were created, including fish stores. By this period, tourism-related activities had come to dominate the economy of most communities. In most communities, household-level agriculture was reduced to a minor activity to supplement the diet.

Since 1994, community growth has continued at an accelerated pace, as has the tourism-based economy. Although population numbers by village are not available from the census data, a population estimate can be made from the data on households and the number of people per household. For the seven villages in the lagoon area, this estimate comes to about 5000 people. Judging by the number of summer cottages, which is twice the number of resident households, the population of the area is estimated to reach about 15 000 people in the peak tourism season. The area surrounding the lagoon and the nearby beaches on the ocean are hot summer spots for tourists, mainly from Porto Alegre, which is the largest city to the southwest.

11.2.3 The lagoon ecosystem

The Ibiraquera Lagoon is an assembly of four interconnected small basins, Lagoa de Cima, Lagoa do Meio, Lagoa de Baixo, and Lagoa do Saco ('Upper Lagoon', 'Middle Lagoon', 'Lower Lagoon', and 'Saco Lagoon'), with a total area of approximately 900 ha (Fig. 11.1). This is a brackish water, shallow lagoon with a sandy bottom and most of its area is between 0.20 m and 2.0 m deep. Fish and shrimp migration takes place in channels running through the lagoon, with a few points reaching about 4 m deep. The salinity ranges from 7 parts per 1000 in the rainy season to 30 parts per 1000 or more when the channel is open to the sea (unpublished data, Universidade Federal de Santa Catarina). There are no major river-water sources. Freshwater input is mainly through springs, which feed the lagoon at nine or more points. A freshwater fish fauna (as opposed to marine or brackish water fauna) is found in several places around the lagoon system, but especially in the upper lagoon.

The water level in the lagoon system rises as the season progresses. Throughout most of the year, there is a sandbar between the lagoon and the Atlantic Ocean. When sufficient water pressure builds up, a channel bursts through the sandbar. Lagoon water pressure increases with rainfall, which helps the lagoon 'explode' into the ocean. The channel eventually closes through sand deposition by ocean currents and tides, which in turn once again allows the increase of the lagoon water level.

Almost all fish resources in the lagoon come from the ocean when the channel is open. Most fish enter the lagoon in their juvenile stages by actively

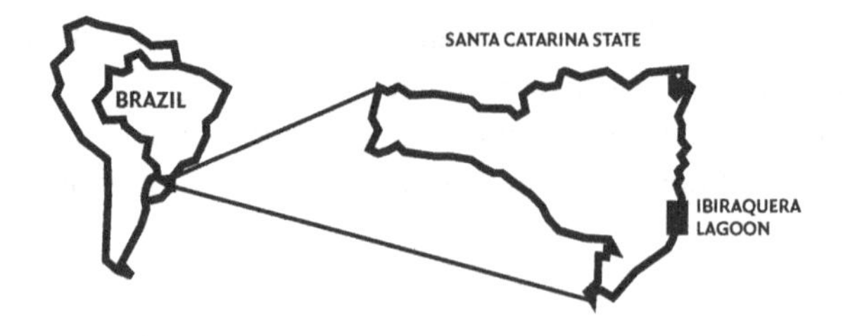

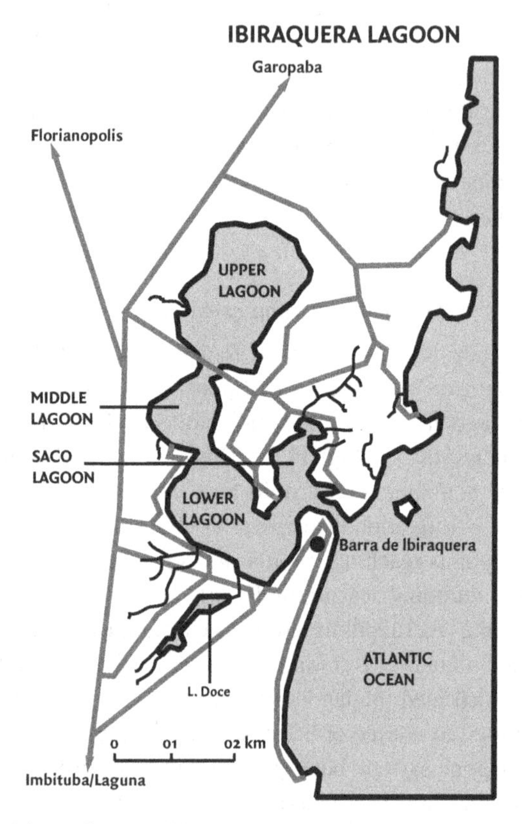

Figure 11.1 The study area: Ibiraquera Lagoon, Santa Catarina State, southern Brazil.

swimming in. By contrast, shrimp are carried in by the tides. Post-larval stage shrimp may be observed in large numbers as they enter the lagoon. Larval stage shrimp, which are planktonic and difficult to observe, are also carried in by the tide. Thus, the lagoon fish stock is determined mainly by the seasons in which the channel is open to the fish and shrimp stocks moving through the ocean in front of the channel, and by the length of time that the channel stays open.

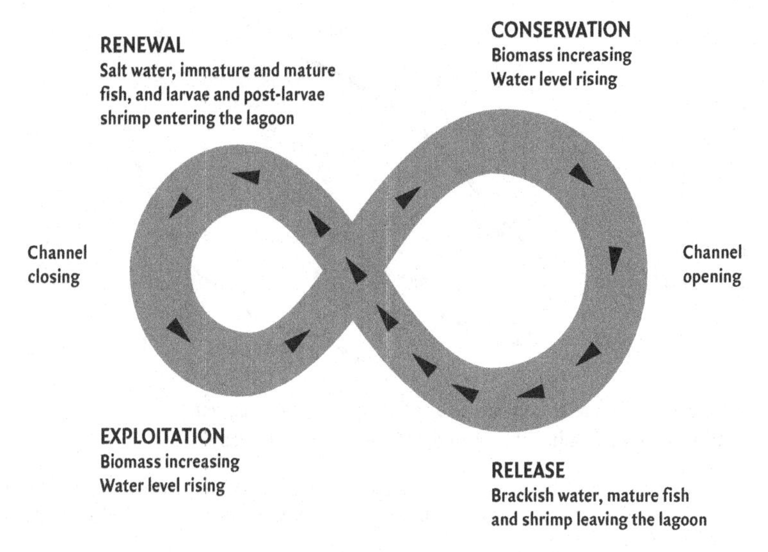

Figure 11.2 Ibiraquera Lagoon ecosystem renewal according to Holling's (1986) adaptive renewal cycle.

Fish and shrimp grow in the lagoon habitat, returning to the ocean as adults in the next channel opening cycle. This represents a 'capital accumulation' in the Holling (1986) sense, or a fish and shrimp biomass increase over the months after the channel is closed.

The Ibiraquera Lagoon is a good example of a small ecosystem going through the adaptive renewal cycle (Fig. 11.2). Following Holling's (1986) adaptive renewal cycle, the release stage is the few hours from the time the channel bursts through the sandbar to the time the lagoon water level matches the water level of the ocean; that is, the period it takes to drain the excess water of the lagoon. The renewal stage is the period that the channel remains open, which can vary from a few days to a few months. In this stage, the lagoon's saltwater and fish and shrimp stocks are renewed. The period encompassing the exploitation and conservation phases, usually the longest one in the Holling model, corresponds to the period when the channel is closed. This period may last from one to several months depending on rainfall. During this time, the lagoon water level rises and the fish and shrimp grow, representing a gradual accumulation of 'capital.' As the system becomes 'overconnected,' the lagoon releases its water and production to the ocean, restarting its renewal cycle.

The Ibiraquera Lagoon system also illustrates how adaptive renewal cycles may be nested in one another over time and space scales, the panarchy (Gunderson and Holling, 2002; Fig. 11.3). During the release phase, the lagoon (the smaller ecosystem) liberates adult fish and shrimp into the Atlantic Ocean

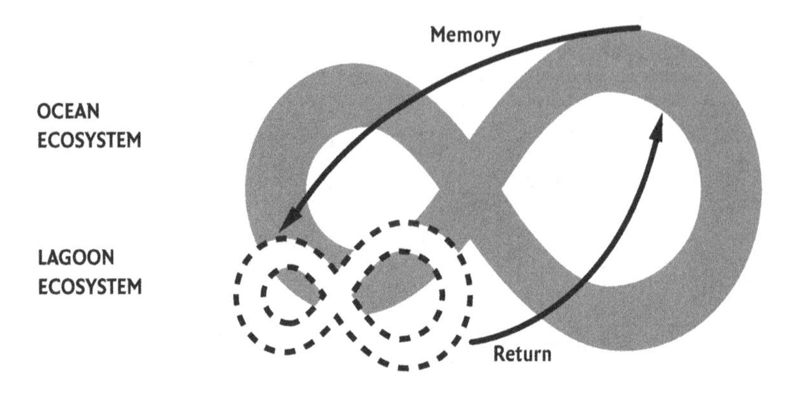

Figure 11.3 Panarchy: the nested relationship between the small ecosystem (lagoon) and the larger ecosystem (ocean). Panarchy idea after Gunderson and Holling (2002).

(the larger ecosystem) where these species reproduce; the lagoon is a source of renewal for the ocean fish and shrimp stocks. In turn, the ocean is the source of saltwater and juvenile and adult fish, and larval and post-larval shrimp during the lagoon renewal stage. Also worth noting, but outside our scope of study, is that changes in the ocean ecosystem such as an oil spill or decisions such as over-exploiting coastal stocks of shrimp and mullet would affect the lagoon ecosystem.

11.2.4 Lagoon ecological dynamics and 'traditional' management

The renewal of the lagoon fishery depends on the season when the channel is open. That is, it depends on the availability of fish and shrimp post-larval stocks moving through the ocean in front of the channel. If nature is left to take its course, the channel does naturally open by the lagoon bursting through the sandbar – but the timing of the opening is highly uncertain. For this reason, local fishers at Ibiraquera traditionally managed the channel opening to coincide with the fish and shrimp season. The fishers dug only a small (2 m wide, 0.5 m deep) channel across the sandbar (about 30–60 m across). The pressure of the water quickly widened it into a channel 1.5–2 m deep and 60–100 m wide.

In a year of normal precipitation, fishers traditionally opened the lagoon at least twice and usually three times. If it were left to nature, the channel opening in most years would probably take place only once, at the end of the rainy season. Thus, the additional openings by the human hand acted to 'put the brakes on release' and served to avoid large disturbances later. Also, they helped avoid the ecological surprise of the lagoon bursting at an unexpected time, and the loss of the lagoon's fish/shrimp stocks into the ocean at a time when no fish or post-larval shrimp stocks were available to renew them.

Fishers opened the channel in the spring (some time around September and October), which is the season of the post-larvae shrimp in the nearby ocean; in late summer (some time between December and February), which is the season of small mullets; and in the late fall (some time between May and June), which is the season when large adult mullets will come into the lagoon. When the channel is opened, several other fish species also enter the lagoon.

Traditionally, fishing activities took place year round in the Ibiraquera Lagoon. Various fishing methods and management practices were used at different stages of the lagoon's adaptive renewal cycle: release and renewal (open channel), exploitation and conservation (closed channel). At the time of the mullet fishing season (May to July), just after the release phase (the period when mullets enter the lagoon, swimming against the brackish water drainage into the ocean), a fence was sometimes built at the lower basin, in order to prevent fish return into the ocean, in case the channel opening lasted for a long period. The use of the fence, made of bamboo and fixed at the two ends by use of two segments of railroad rail, can be viewed as an insurance mechanism.

If there were still many fish or shrimp in the upper lagoon just before the release phase, a bamboo fence was also built in the upper lagoon close to the channel connecting the upper and middle lagoons. During the draining, the fence helped to retain part of the fish stock in the upper lagoon, 'putting the brakes' on the release phase and functioning as an insurance mechanism for maintaining biodiversity. In this case, however, the fence was not only built in the mullet season, but was used in any channel opening. Because the upper lagoon is the shallowest and farthest one from the opening to the sea, it is the one most affected by drainage. If no fence is present, all shrimp and fish stocks may escape the upper lagoon during draining. During the renewal phase, part of the fence (a gate) was taken out to allow fish to come into the upper basin. The gate was opened when a school of fish was trying to come in and closed soon after that, so that fish would not swim back into the other basins and eventually into the ocean if the channel opening lasted for a long period. The same procedure of building bamboo fences just before the release phase was sometimes used at the Saco Lagoon. If there were not many fish and shrimp left in the upper lagoon before the channel opening, a bamboo fence was built only after some fish stocks entered that basin during the renewal phase.

Specialized fishers acted as monitors to check fish movements. In the deeper channels running through the basins, elder fishers sometimes stood watch to assess the amount of fish entering the channels. When a fish monitor decided that enough fish had come through, he gave a sign to close the deep channel using net barriers, allowing the fishers to start fishing. These expert fishers also monitored mullet movements into the lagoon, signaling when to build the bamboo fence (or close it in the case of the upper lagoon), so that the mullet would not swim back out.

Whenever the channel was open (release and renewal stages), gillnets (fixed nets, about 150 m each), surrounding nets (a mobile gear made by tying together several gillnets), and seines (a surrounding net pulled along the bottom) could not be used in the lagoon. This was because these methods relied on slapping the water with poles and hitting the sides of the canoe to drive fish into the nets, thus producing considerable noise in the water. Such disturbance repelled fish back to the ocean. Gillnets, surrounding nets, and seine nets could only be used during the exploitation and conservation phases (closed channel). Cast-nets for both fish and shrimp could be used in all four phases of the cycle, except in the first 2 or 3 days after the channel opening in the mullet season to avoid the problem of fish turning back to the ocean. Cast-nets for fish were not allowed in the channel or on the nearby beach (100 m from each side of the mouth of the channel) when the channel was open. This was to permit fish to enter undisturbed into the lagoon.

The shrimp cast-net fishery was performed only by fishers standing in certain spots close to the channels used by pre-adult shrimp. There was no shrimp fishery along the lagoon shore where the young shrimp come to feed, or from canoes in the middle of the lagoon area, which is the habitat of very small shrimp. There was respect for shrimp fishing locations of the established fishers and for first-comer's rights in other areas. In the fish cast-net fishery, there were certain spots where fishers were allowed to stand and fish. In other spots, even though fish were visible, old fishers advised no fishing because that would cause scattering of the fish school, which could be caught easily in another location. Fish cast-netting from canoes in the middle of the lagoon area was a common practice. Setting gillnets was not allowed in any channel mouth.

11.3 Fishing management by period

11.3.1 Fishing system in the 1960s

In the early 1960s, the fishers' organization, Colônia de Pescadores, already existed but was not responsible for regulating or enforcing fishing rules.[1] One of its responsibilities was to transmit documents so that the government could issue fishing licenses. Although some government fishing regulations existed, these were either unknown or unrecognized by most local people. Most fishing rules were decided locally, and respect for the practices of old fishers was the main measure by which these rules were enforced. State fishery inspectors or police did not normally come to the lagoon area, except when occasionally called on by the Colônia to enforce rules or solve conflicts among fisher groups.

The main fishing gears included the cast-net (25–28 m circumference; mesh size from 4.0 cm to 5.0 cm stretched measure) used for small fish and large shrimp; the cast-net (28–31 m circumference; mesh size of 6.0 cm or larger) used for large fish; the gillnet (180–220 m length; mesh size of 3.5 or 4.0 cm) used for small fish and shrimp; and the gillnet (180–220 m length; mesh size of 5.0 cm or bigger) and seine net (about 300 m length with mesh size of 5.0–9.0 cm, with a cod-end of 15–30 m length with mesh size of 4 or 5 cm) used for large fish. Local fishers used a torch made of dry vegetation to attract shrimp in night fishing. Fishing gears were handmade and so were the boats (dugout canoes). Until the mid-1960s when synthetic fibers were introduced, nets were made of cotton or *tucum* (a fiber made from palm tree). Making nets was costly in both time and money. The cast-net, a small gear, was more affordable to most people than the gillnet or seine net. In fact, it was gear types that separated the user groups. Local fishers were divided into two groups: *tarrafeiros*, those who used cast-nets (most fishers), and *redeiros*, those who used gillnets or seine nets (a smaller group). There were no outside fishers in the system.

The main fishing methods included: (a) an individual fisher using a shrimp cast-net or a fish cast-net while standing in a known fishing spot close to fish and shrimp migration channels; (b) one or two fishers using fish cast-nets from a canoe; (c) an individual fisher setting a gillnet; (d) a group of fishers with two or more canoes encircling a fish school with several attached gillnets, inside which they and others in more canoes threw cast-nets; (e) four or five fishers in a canoe encircling a fish school with a seine net; and (f) two fishers holding an open cast-net used to catch shrimp in the water current when a channel was open. The local management rules, described in the earlier section, were based on respect for the practices and the instructions of the elders.

In the early 1960s, due to low population density, lack of roads, and little market development, supply exceeded demand, and fishers caught a lot of large fish and shrimp. Although there was an abundance of fishing resources all year around, there was a conflict between *tarrafeiros* and *redeiros* for resource access, as *redeiros* caught more fish than *tarrafeiros*. The amount of effort to fish with cast-nets is larger than to fish with gillnets. A person has to throw a cast-net many times to make a living. By contrast, a gillnet set in a fishing spot ‘fishes by itself’ for many hours. The only effort required is to set the net and to take it out. When several attached gillnets are used to encircle a fish school, a few *redeiros* might catch more fish in one short trip than several *tarrafeiros* using cast-nets in an entire day of work. From the point of view of *tarrafeiros*, other factors contributed to this inequity. For example, when *tarrafeiros* fished inside the encircling gillnets, they had to give one-third of their catches to the gillnet owners (*redeiros*).

This conflict between *tarrafeiros* and *redeiros* had existed for decades; however, it intensified in the late 1960s as a result of two technological innovations: (1) the use of synthetic fibers made it easier to produce nets and with a smaller mesh size than before; and (2) *redeiros* started to employ a new fishing method, using several attached gillnets as beach seines, for fish and shrimp along the lagoon coast.

11.3.2 Fishing system from 1970 to 1981

Road access to the communities favored the development of outlets to sell fish from about 1970. At first, local middlemen bought fish and shrimp from the lagoon and sold them in the big cities. A patronage system developed in which a middleman used to give money or fishing gears to a fisher, who in turn had to sell his catch exclusively to the former. After the mid-1970s, when tourism started in the region, opportunities increased for the sale of fish, and particularly shrimp, within the lagoon communities. However, the roads also brought some outside fishers from nearby communities and municipalities, adding another user group to the mix. These outsiders could be either *tarrafeiros* or *redeiros*.

By the 1970s, the Colônia became the decision-making agency for the opening of the channel. Although net fences were legally prohibited, the Colônia informally allowed them. The main fishing gears included all those used in the previous period, plus trap-nets, hoop-nets, and mini-trawl-style pull-nets for shrimp. As the intensity of exploitation increased as a result of market pressures, mesh size started to diminish. Shrimp cast-nets had a mesh size of 2.0 or 2.5 cm and fish gillnets had a mesh size of 3.0 cm. As fishers started making money in fisheries in the early 1970s, they were able to afford kerosene lamps to attract shrimp, which are easier to use and more efficient than torches. In the late 1970s, some fishers started to use butane gas lamps to attract shrimp, which were, in turn, more efficient than the kerosene lamps.

There was an increase of the hand-drawn beach seine fishery, with some nets reportedly as large as 600 m in length. Fishers using canoes started to use shrimp cast-nets all over the lagoon, instead of a few fishing spots. Shrimp trap-nets and hoop-nets (two kinds of anchored nets, one larger and the other smaller) were used in place of cast-nets in the channel. Shrimp pull-nets were drawn along the lagoon shore.

In 1971, an attempt was made to resolve the conflict between *tarrafeiros* and *redeiros*. The arrangement between the two parties, made in the presence of the Colônia president and the director of the State Department for Fishing and Hunting:[2] (a) prohibited the use of gillnets in the upper lagoon and in the Saco Lagoon (gillnets could only be used in the middle lagoon and lower lagoon); (b) prohibited the use of cast-nets and gillnets in the channel whenever it was open; and (c) prohibited sport fishers from selling their catches. A voluntary local fishery inspector who had the support of many *tarrafeiros* in monitoring the lagoon enforced these rules. The first voluntary inspector affirms that the agreement also included the prohibition of cast-nets with a mesh size smaller than 3.0 cm (Lênio Teixeira, personal communication). However, reference to this rule was not made in the Colônia's meeting report. Despite many regulations issued by the Federal Fishery Agency (SUDEPE) in the first half of the 1970s,[3] fishers observed only the local agreement and previous local rules because there was almost no rule enforcement by municipal, state, or federal fishery inspectors.[4] Most of the fishing gears and methods used during this period were in fact legally prohibited.

Initially, the first local voluntary inspector apprehended illegal gears and deposited them in the State Department for Fishing and Hunting office. Fishers were then able to retrieve their gears after paying their fines. As fines were very low, many fishers took the risk of using illegal gears again and again. As a response, the first voluntary inspector started to cut or burn the illegal gears. During this period, there was some recovery in fish and shrimp catches.

In 1974, however, the first voluntary inspector resigned because he got a paid job. The new voluntary inspector was the Colônia president, a *redeiro* himself. According to some informants, he allowed the use of all nets in all basins, and enforced regulations that only favored *redeiros*. Hence, the 1971 agreement failed after the mid-1970s, and the conflict flared up again.

In conclusion, the enforcement of both local rules and government rules broke down. Respect for elders' practices weakened. Profit-oriented fishers now questioned and disrespected old practices. The conflict became worse as differences in fishing income magnified the economic differences between *redeiros* and *tarrafeiros*. Using big gears, *redeiros* caught more and more fish and made more money than *tarrafeiros*; moreover, they bought more material to make even more nets.

By the late 1970s, all fish and shrimp stocks in the lagoon were caught within about 2 months of channel closure. This was the result of the smaller mesh sizes used, and particularly of the intensive use of the hand-drawn beach seine. This meant that there was almost no production in the lagoon for several months before the next opening. The pressure on fishing resources and the conflict between the two user groups triggered the crisis and a 'revolt' of the *tarrafeiros*.

11.3.3 Fishing system from 1981 to 1994

Facing resource overexploitation and the ongoing conflict between *tarrafeiros* and *redeiros*, the *tarrafeiros* (the larger group) organized themselves and in 1981 elected a new Colônia president (an outsider *tarrafeiro* and ocean fisher), who promised to work toward the restriction of nets other than cast-nets in the Ibiraquera Lagoon. The Colônia, then in the hands of a strong and knowledgeable leader who had support of the majority of fishers, and good political relations with the state government, conducted several regulation changes that helped rebuild lagoon management and ecosystem resilience. As a result of the positive outcome of these changes, the president was re-elected five times.

The first, and perhaps the most important, rule change was the banning of any nets in all lagoon basins, except the cast-net. Local fishers, through the Colônia president, demanded this ban from the Federal Fishery Agency (SUDEPE) and two other state agencies working with the fishery (IPEP, ACARPESC).[5] After a study to evaluate the lagoon management situation, SUDEPE agents elaborated a project upon which local fishers voted and decided to ban all net types except the cast-net. The Federal Government approved the regulation (N-027/81) in October 1981.[6] The new regulation, specific to the Ibiraquera Lagoon, banned the use of all nets but cast-nets, with minimum mesh size of 2.5 cm for shrimp and 5.0 cm for fish. These are standard mesh sizes for multi-species coastal fishery in Brazil. The regulation also prohibited any fishing in the channel and in a small channel connecting the upper and middle basins.

In 1986, use of the gas lamp was banned in the lagoon, allowing only kerosene lamps. Gas lamps were being used with a new fishing gear, the shrimp sucker, which caught small shrimp in their feeding areas in the lagoon margins. Also, their bright lights interfered with other fishing activities at night. Because the gas lamp attracted much more shrimp than did the kerosene lamp and because not all fishers could afford to buy a gas lamp, its use was creating equity problems among the fishers. The change was demanded by a majority of fishers, through the Colônia, and officially approved (N-09/86) by the Federal Fishery Agency (SUDEPE).

Until 1988, the Colônia president decided about the channel opening, after consulting with local fishers. Yet, to implement the decision, he had to have approval from the District Navy Commander who acts as the Port Authority ('Capitão dos Portos'). After 1988, the decision and the approval processes were transferred to the Municipal Government. From 1989 to 1992, the person in charge of the opening had no knowledge of the lagoon ecosystem dynamics. He listened to fishers as well as to others living in the lagoon area. Houses in the area have septic tanks and the water levels in the lagoon affect the discharge of sewage because the water table is too superficial, causing a foul smell inside the houses. From the point of view of sewage disposal, it was desirable to open the channel more frequently to improve flushing and to get rid of foul-smelling water. This decision was sometimes made in favor of the sewage problem rather than to optimize the fishery. This, in turn, created a conflict between fishers and nonfishers, especially tourism interests. In 1993, the Municipal Government returned the decision making on channel openings to the Colônia.

According to the Colônia president, due to channel openings in wrong periods and some weather surprises in 1990 and 1991, the amount of shrimp larvae entering the lagoon diminished, severely affecting shrimp production. The upper lagoon was particularly affected due to landfill in the channel connecting it to the other basins. Moreover, the natural shrimp production had became insufficient to supply the growing number of fishers. In the face of these circumstances, the Colônia president contacted the Federal University of Santa Catarina (UFSC) and a state research agency (EPAGRI)[7] to develop a shrimp-stocking project in the Ibiraquera Lagoon. The project, which consisted of releasing post-larval shrimp in the upper lagoon, started in 1992 and lasted until 1998 (Andreatta *et al.*, 1993, 1996).[8] A net fence (1.0-cm mesh) built in the upper lagoon just before channel openings prevented shrimp from escaping into the ocean.

Now that there was a higher abundance of shrimp, the project coordinators, in agreement with the Colônia, showed the local fishers that an increase of the cast-net mesh size from 2.5 cm to 3.0 cm would actually improve fishers' yields and profits, because larger and higher-value shrimp would be caught. Furthermore, such a measure would help exclude most outside fishers, who usually own only 2.5-cm mesh shrimp cast-nets. Accordingly, in 1993, local fishers, through the Colônia, demanded another regulation change, establishing a minimum of 3.0-cm mesh for shrimp cast-nets. This was officially approved (N-115/93) by the Brazilian Department for the Environment (IBAMA).[9]

These three new regulations (N-027/81, N-09/86, and N-115/93), specific for the Ibiraquera Lagoon, replaced most of the regulations of the previous periods. They worked only because of a strong enforcement structure. Between 1981 and 1994, through an agreement with the Federal Fishery Agency (SUDEPE was replaced by IBAMA), the State Government staffed fishery inspector positions in certain localities, including the municipality of Imbituba. Local lagoon fishers also helped these inspectors, but this help had to be withdrawn later because it was generating conflicts between *tarrafeiros* and those *redeiros* who insisted on fishing with prohibited gears. As a result of all these rule changes and strong enforcement, the lagoon's fish and shrimp stocks recovered in about 2 years after the net banning, according to fishers.

Meanwhile, most local and informal fishing rules either disappeared or became formal. Exceptions included the use of net fences in some lagoon basins during channel openings, respect for fishing spots in areas in which only locals fished, and first comer's rights in the remaining areas. With the above exceptions, respect for elders' practices almost completely disappeared. Fences were now made of netting, with a few poles to keep nets in place, as opposed to all-bamboo fences, which were harder to make.

Tarrafeiros were catching more than they did in the previous period, but many of them had, in effect, become part-timers. Fishers who were previously *redeiros* and now fishing with cast-nets were catching less. As a result of the dominance of tourism-related activities in the economy, most local fishers of both groups were looking for other employment, especially in the construction business.

From 1992 to 1998, overall shrimp production increased considerably as a result of the shrimp-stocking project (Andreatta *et al.*, 1993, 1996). At the same time, however, there were emerging challenges to the lagoon fishery from the increase of tourists, whose sailing and sport fishing interfered with professional fishing. Also, there was an increase of outside fishers and the unregulated growth of summer cottages, guesthouses, and restaurants. Excessive development was destroying vegetation on the lagoon edge, which in turn increased erosion, siltation, and mudslides, filling up the fish migration channels and destroying fish and shrimp feeding habitats.

11.3.4 Fishing system from 1994 to 2000

In 1994, the arrangement between IBAMA and the State Government broke down and the fishery inspector positions were eliminated, probably due to budget constraints. A new arrangement was then made between IBAMA and the State Environmental Police,[10] which had the personnel to do the job. In this new arrangement, a small group of officers had to cover a large area encompassing several municipalities, dealing with all resource and environmental issues, including the fishery. A place such as the Ibiraquera Lagoon was only visited sporadically and usually only when infractions were reported. The weakening of enforcement gave many fishers the opportunity to violate regulations. As a result, by 1996, the depredation of the lagoon system was again evident, and some fishers demanded better service from the Environmental Police to avoid a new crisis. Due to the ineffective action of the Environmental Police and the IBAMA, fishers living close to the upper lagoon decided in 1998 to organize themselves into groups to patrol it. Nonetheless, this activity did not last long because fishers did not have the legal right to patrol the area. On several occasions, they called the IBAMA and the Environmental Police to stop illegal fishing, but to no avail. Moreover, monitoring groups were sometimes threatened with shotguns by those fishers using illegal gears.

In addition to changes in the enforcement structure, the only other rule change regarding lagoon management following the 1994 measures was the prohibition of motorized vessels in the lagoon. In 1994, fishers organized themselves to demand the restriction of jet-skis and any other mechanized boats because their use was affecting fishing and threatening the security of fishers and tourists in the lagoon. In 1995, the Mayor of Imbituba issued a regulation (N-1501) prohibiting any type of engines in the lagoon. Dugout canoes with pole or paddles were the most common vessels. However, jet-skis and motor canoes were still used by some tourists.

As a result of lack of enforcement, all prohibited gears and fishing methods used before 1981 returned to the lagoon. In addition, another destructive gear was introduced: the shrimp trawl (*gerival*), which is a small net dragged by a canoe. Both local fishers and outsiders are using banned gears. Evidently, *redeiros* fishers became a user group again; yet, *tarrafeiros* remain the majority. In fact, probably more fishers have (or can afford to buy) big nets today than in the 1970s; fishers do not have to make their own nets anymore. Because of tourism and local economic growth, most fishers fish part-time. Indeed, there were about 10 fishers in 1999 in the whole area that relied only on fishing for their livelihood; many others supplemented their incomes from tourism-related activities and agriculture.

The lack of strong enforcement has also affected the issue of channel openings. In the last few years, instead of waiting for the Colônia decision to open the channel, some fishers are opening it whenever they think it is appropriate. People living in the community closest to the channel prefer it to be opened in December and January – the high season for tourism – so that the lagoon water is flushed constantly to minimize smells caused by sewage. But for other fishers, the important issue is that channel openings at the wrong time have been affecting lagoon fish and shrimp production.

Futhermore, the use of banned gears has been affecting lagoon production. On the one hand, the use of small mesh sizes decreases potential production during the months that the channel is closed. On the other hand, the use of illegal shrimp trap-nets when the channel opens increases shrimp catches. Since 1998, when the shrimp-stocking project ended, there has been a reduction of shrimp production. Lagoon production in the 1990s was mainly sold in the area; there was no excess to sell to big cities. Fishers either sold their product to middlemen or directly to local restaurants and tourists. The patronage system, in which middlemen lent money to the fishers, who in turn had to sell their catches exclusively to them, has disappeared. A large proportion of fishers, however, do not sell their catch; they fish for their own family consumption.

The problems affecting the lagoon have multiplied again since the end of rule enforcement. Some of these new problems are aggravations of old ones and include: (a) the use of illegal gears and fishing methods; (b) the use of motor vessels and windsurf boards interfering with fisheries; (c) lagoon pollution due to the increase of tourists; (d) the increase of tourists' houses draining sewage into the lagoon from poorly constructed septic tanks; (e) illegal constructions inside the lagoon and on its margins; (f) channel openings in wrong periods; (g) renewed conflict between *tarrafeiros* and *redeiros*; (h) increased conflict between local fishers and outside fishers; and (i) conflict between professional fishers and sport fishers. This scenario shows that a new crisis is emerging in the lagoon's management.

'Why is the Colônia not responding to the emerging crisis?' one may ask. 'Because the Colônia has become a "brittle" organization,' is probably the best answer. The Colônia president has been there for a long time, re-elected several times since 1981. However, the organization appears to have lost its ability to respond flexibly to problems, and has become complacent and too centralized. The Colônia has not responded in a resilient manner to a range of feedbacks such as those summarized in the previous section. The president has, in fact, become 'the organization' itself. Although, the Colônia's board of directors includes other members, they play no real role; all decisions are made by the president, who also acts as secretary and controls the Colônia's budget. Meanwhile, to deal with the impacts of unregulated tourism, three of the seven communities surrounding the lagoon re-activated their community councils[11] in 1999/2000. As of the year 2000, the Ibiraquera Lagoon management system seems to be poised for a new round of institutional renewal.

11.4 Navigating the dynamics of a social–ecological system

The history of the Ibiraquera Lagoon fishery is particularly interesting as it shows the resilient 'traditional' management system of the 1960s transforming into a less resilient and non-viable system in the 1970–81 period; rebuilding resilience after experiencing a crisis (1981–94), but once again transforming into a less resilient system since 1994. We evaluate the resilience of the social–ecological system on the basis of its ability to respond to feedbacks and absorb perturbations, its ability for self-organization, and its capacity to learn and adapt (see Chapter 1). Figure 11.4 summarizes these institutional and ecological flips in the Ibiraquera Lagoon system. The adaptive renewal cycles are used here as a

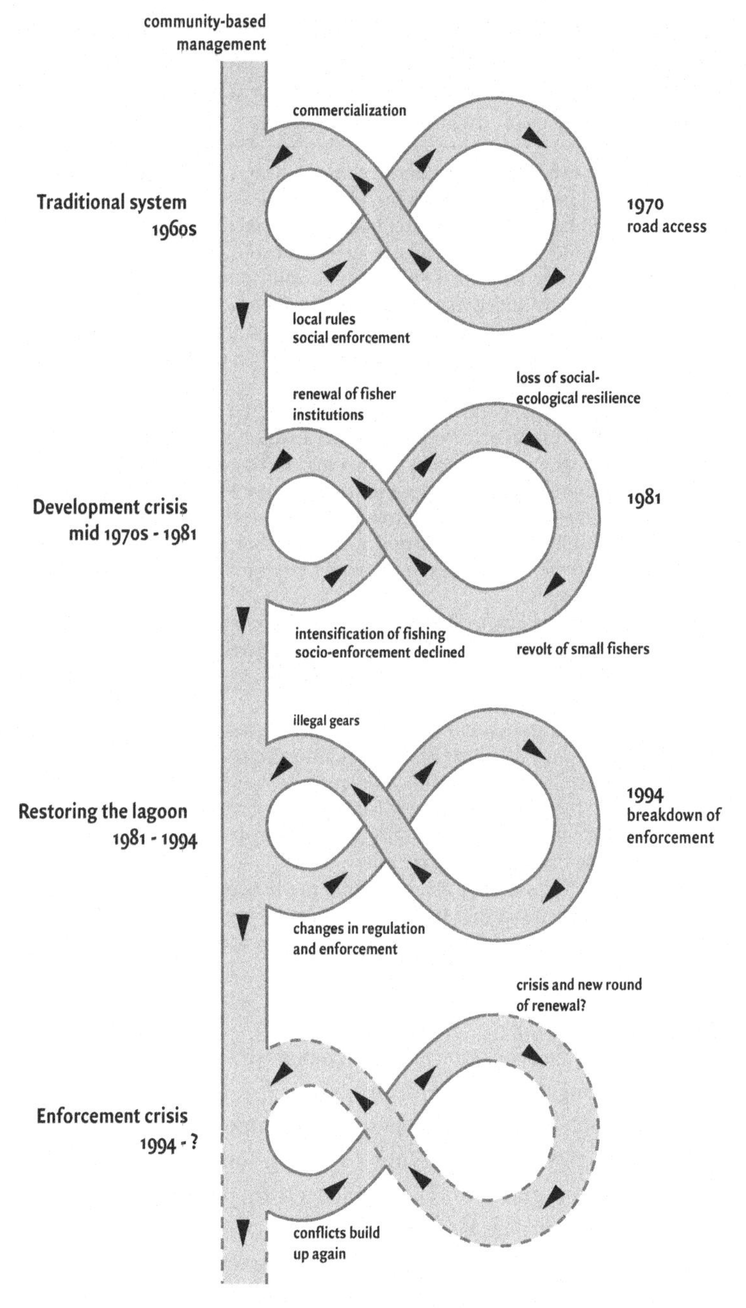

Figure 11.4 Different periods of lagoon and fishery management in Ibiraquera, as represented by successive iterations of the adaptive renewal cycle (Holling, 1986; Gunderson *et al.*, 1995).

Table 11.1 *Traditional management practices until the early 1960s, according to phases of the adaptive management cycle*

Cycle phase	Practices of 'traditional' management
Release	Increasing the frequency of lagoon water discharge into the ocean ('putting the brakes on release')
	Management of channel openings timed to shrimp post-larvae and fish entering into the lagoon to avoid ecological surprises
	Bamboo fence in some lagoon basins to retain part of fishery stock as an insurance mechanism and to 'put the brakes on release'
Renewal	Elders assessing fish migration through the channels connecting the lagoon to the ocean to signal when the bamboo fence should be built
	Fence gate controlling the fish coming in and out of some lagoon basins
	Prohibition of gillnets and other methods that make noise and disrupt fish migration from the sea into the lagoon
	Prohibition of cast-net use in the channel or on the beach near the channel mouth to allow fish to enter undisturbed into the lagoon
	Prohibition of fishing during the first 2 or 3 days after channel openings in the mullet season to avoid the problem of fish turning back to the ocean
Exploitation	Shrimp cast-net fishing allowed only in shrimp migration channels
	Fish cast-net fishing allowed only in particular spots
	Prohibition of setting shrimp/fish gillnets in channel mouth
	Use of large mesh size only
	Elders assessing fish migration in the deep channels of the basins to signal whether fishing should be taking place
Conservation	Shrimp cast-net fishing allowed only in shrimp migration channels
	Fish cast-net fishing allowed only in particular spots
	Prohibition of setting shrimp/fish gillnets in channel mouth
	Use of large mesh size only
	Elders assessing fish migration in the deep channels of the basins to signal whether fishing should be taking place

Holling (1986), Gunderson *et al.* (1995).

heuristic model to understand cycles of change in social–ecological systems. This is by no means a predictive model.

The 'traditional' management practices conferred social–ecological resilience until the 1960s, in addition to a strong informal enforcement based on respect for elders' practices. These management practices mainly applied to the release and renewal phases of the lagoon's ecosystem dynamics (Table 11.1). They triggered critical ecosystem processes, for example opening the channel produced a small-scale disturbance to avoid larger-scale disturbances later. These practices helped to avoid ecological surprises, acting as insurance mechanisms for maintaining biodiversity.

Table 11.2 *Ibiraquera Lagoon ecosystem resilience: management practices that allowed the use of cast-nets and banned other nets contributed to ecosystem resilience and avoidance of overfishing in the face of large market demand for lagoon shrimp and strong rule enforcement*

	1960s	1970–81	1981–94	1994–2000
Phases of the adaptive renewal cycle				
Release				
Lagoon	Cast-net/other nets	Cast-net/other nets	Cast-net	Cast-net/ other nets
Channel	Other nets	Other nets	Not allowed	Other nets
Renewal				
Lagoon	Cast-net	Cast-net/other nets	Cast-net	Cast-net/ other nets
Channel	Not allowed	Not allowed	Not allowed	Cast-net
Exploitation				
Lagoon	Cast-net/other nets	Cast-net/other nets	Cast-net	Cast-net/ other nets
Conservation				
Lagoon	Cast-net/other nets	(No resource left)	Cast-net	Cast-net/ other nets
Market demand[a]	None	Small–medium	Medium–large	Large
Rule enforcement[b]	Strong	Weak	Strong	Weak
Ecosystem resilience[c]	Strong	Weak	Medium–strong	Medium

[a]Scale. None: almost no lagoon shrimp were sold; large: almost all lagoon shrimp were sold.
[b]Scale. Weak: cheating often occurred; strong: cheating hardly occurred.
[c]Scale. Weak: management practices led the lagoon stocks to be overfished before the next release phase (channel opening) – i.e., the lagoon system was not able to absorb disturbances while maintaining its behavioral processes and structure; strong: management practices allowed for part of the stocks to leave the lagoon during the release phase for reproduction in the ocean, while retaining another part to ensure lagoon stock renewal – i.e., the lagoon system was able to absorb disturbances while maintaining its behavioral processes and structure.

From 1970 to 1981, the management system began to lose its social–ecological resilience as fishing effort increased due to changes in the local economy and as social enforcement declined. Although fishing gears and methods used in this period were very similar to those used in the previous period, changes in the socio-economic system affected ecosystem resilience (Table 11.2). Changes in the local economy also reduced the social resilience of the management system. This happened as profit-oriented fishers spurned elders' authority, and as overfishing by *redeiros* magnified the differences in socio-economic status between *redeiros* and *tarrafeiros*. The loss of social–ecological resilience triggered a crisis in the management system.

The rebuilding of social–ecological resilience between 1981 and 1994 depended on a series of changes. The two main responses to the crisis were the election of a Colônia president willing to promote management changes and the implementation of a new enforcement system – a formal one. After 1994, however, the social–ecological resilience of the system was threatened again by the lack of a strong enforcement structure. Also, resilience was threatened by the 'brittle' organization that the Colônia had become. These changes are summarized in Figure 11.4.

Another way to interpret Figure 11.4 is in terms of property rights arrangements. The 'traditional' system in the 1960s is clearly a communal management system.[12] With economic modernization and opening of road access in about 1970, the area entered an open-access competition situation over lagoon resources. Overfishing and conflicts culminated in the revolt of the small fishers, ending the cycle of *laissez-faire* exploitation. The management regime in the

Table 11.3 *The Ibiraquera Lagoon fishing management: changes in property rights over time*

Periods	1960s	1970–81	1981–94	1994–2000
Decision making for fishing rules	Local	Local and national	Local and national	Local and national
Degree of rule enforcement	Strong	Weak	Strong	Weak
Informal organization of fishers	Strong	Weak	Strong	Weak
Formal organization of fishers	Weak	Developing	Strong	Medium
Social–ecological resilience[a]	Strong	Weak	Medium-strong	Weak
Property-rights regimes	Communal management	Open access	Co-management (mix of communal and state)	Mix of state, communal, and open access

[a]Scale. Strong: management practices buffered ecosystem disturbances and allowed for ecosystem renewal, while the social system responded to changes in both social and ecological system; weak: management practices did not buffer ecosystem disturbances or promote ecosystem renewal, and the social system was not able to quickly respond to changes in both the social and ecological systems.

third period in Figure 11.4 may be characterized as co-management. The fishers and their organization successfully lobbied the government to pass a series of regulations, including shared power and responsibility for the management of Ibiraquera Lagoon. However, the fishers did not have the formal power to enforce rules. The restoration success of the 1981–94 period relied on government enforcement of the new regulations. Given the fact that the fishers of the lagoon could not legally exclude outsiders, and given the immense pressure of tourism development, it is not surprising that the system fell apart when the state fishery inspectors were withdrawn in 1994. Table 11.3 summarizes these changes over time with respect to decision making, formal and informal organization, and rule enforcement.

11.4.1 Key factors that affect social–ecological resilience

Having analyzed the case study with respect to historical changes and social–ecological system dynamics, we now turn to the identification of key factors that build or threaten resilience. The case study allows for the identification of both kinds of key factors. The four key factors that weakened resilience included the breakdown of local institutions, rapid technological change, rapid socio-economic change, and institutional instability across political scales (Table 11.4).

Table 11.4 *Key factors that weaken social–ecological resilience*

Key factors	Examples from the Ibiraquera case
1. Breakdown of 'traditional' institutions and authority system	Loss of respect for old fishers' practices and knowledge in the late 1960s and late 1970s, and loss of confidence in the Colônia leadership in the late 1990s
2. Rapid technological changes leading to more efficient resource exploitation	Innovation in fishing gears during the late 1960s led to resource depletion and triggered conflicts among user groups, as the more efficient gears were not affordable by all fishers
3. Rapid changes in the local socio-economic system	Rapid changes in local economy during the 1970s impacted the social system that gives support to management institutions; respect of elders' practices and authority (the enforcement structure) diminished as fishing profits became more and more important
4. Institutional instability in higher political level negatively affecting local management	Changes in the arrangement between state and federal government extinguished fishery inspector positions, leading to an enforcement crisis and management problems

We identified five key factors that strengthed resilience: strong institutions, cross-scale communication, political space for experimentation, equity, and use of ecological knowledge as a source of novelty (Table 11.5). These factors are in fact clusters. For example, the first factor (institutions) includes the elements of robustness, enforcement, and leadership, as detailed in Table 11.5.

Communication appears to be a major factor. The three regulations brought in during the 1981–94 period (banning of all nets but cast-nets, banning of gas lamps, increasing shrimp cast-net mesh size) involved cross-scale cooperation and communication. Four levels of organization were identified: local resource users, the fishers' organization (Colônia), state agencies, and the Federal Fishery Agency. Cross-scale communication was important both during the evaluation of the lagoon fishery and during the decision-making process. The banning of motor vessels in the lagoon also involved the local fishers, the Colônia, and the Municipal Government. Important aspects of cross-scale communication concern the sharing of facts about resource status and threats to management, and the ability of resource users to detect environment modifications and management crises. Another important aspect concerns the co-management of the lagoon using both scientific and local ecological knowledge. Local knowledge concerning the effect of large nets and gas lamps was taken into account by the Federal Fishery Agency. Scientific knowledge about the implications of mesh size on shrimp production was used by local fishers to demand a rule change for larger mesh sizes (Table 11.5).

In the Ibiraquera case, the Federal Fishery Agency allowed local fishers' input for the formulation of regulations, thus creating political space for experimentation. The case shows a multitude of changes, problems, and management responses, including a rich set of fishers' own management measures and fishers' rules incorporated into government management. Given the reality of top-down government management that historically characterized resource management in Brazil, the creation of such political space for experimentation was unusual by the standards of the early 1980s. This was less so in the 1990s (e.g., Barbosa and Hartman, 1997). The positive results of the three major regulations brought in during the 1981–94 period show that the opportunity for local fishers' inputs was used effectively, even though all fishing rules had to be approved by the Federal Fishery Agency. Moreover, the positive result from the first rule modification in 1981 led fishers to demand other modifications in 1986 and in 1993 (i.e., there was positive-feedback learning). These many changes, in fact, add up to adaptive management arising from the creation of political space for experimentation (Table 11.5).

Equity in resource access was the driving force in many of the changes and conflicts observed in the lagoon. The creation of equitable access improved

Table 11.5 *Key factors that strengthen social–ecological resilience*

Key factors	Examples from the Ibiraquera case
1. Strong institutions	
(a) Robust local institutions	Respect of elders' practices in the early 1960s, and a responsive fishers' organization the 1981–94 period
(b) Strong enforcement of rules (local, regional, or national)	Respect of elders' authority (informal enforcement) in the early 1960s, strong informal enforcement (voluntary fishery inspectors) from 1971 to 1974, and strong formal enforcement (fishery inspectors) in the 1981–94 period were central to successful management
(c) Strong leaders with credibility and willingness to promote changes	The Colônia president elected in 1981 was determined to change the fishing rules, whereas the previous Colônia presidents had no such interest, as the old rules suited them well
2. Good cross-scale communication	
(a) Sharing of facts about resource status and threats; ability of resource users to detect environmental modifications and management crises	Fishers detected resource overexploitation at the end of the 1970s, and recognized the threat posed by the use of gas lamps (too efficient); the knowledge generated at local level by qualitatively monitoring the resource was successfully communicated to the federal level
(b) Co-management of the lagoon using both scientific and local ecological knowledge	Two of the three major changes in the 1981–94 period were based primarily on local ecological knowledge, and one primarily on scientific knowledge
3. Political space for experimentation	Three major regulations brought in during the 1981–94 period showed that the Federal Fishery Agency was open to suggestion by fishers
4. Equity in resource access	The banning of the use of large nets in two basins from 1971 to 1974 and in all basins since 1981, as well as the banning of gas lamps in all basins since 1986, led to a more equitable resource allocation among user groups
5. Use of memory and knowledge as source of innovation and novelty	
(a) Innovation in regulations based on past arrangements	The banning of all nets but cast-nets in all four basins in 1981 was inspired by the 1971 agreement
(b) Memory/knowledge of resource monitoring and management practices	Although legally prohibited, the use of net fences in some lagoon basins (traditionally functioning as an insurance mechanism) was informally accepted by the Colônia president and fishery inspectors

system resilience; loss of equity led to conflict and system breakdown, reducing resilience. The banning of overly efficient nets and lamps, as detailed in Table 11.5, contributed to more equitable resource access and allocation. Such equity could not have been achieved by the use of larger nets by all; the experience in the late 1960s and late 1970s proved that the use of larger nets resulted in resource depletion.

Adaptive renewal cycles depend on the existence of memory for the cycle to resume. Further, innovation and novelty allow the reorganization phase of the cycle to respond to changes. In the Ibiraquera case, the use of fishers' memory and ecological knowledge, as a source of innovation and novelty, appeared in two circumstances: (1) in the innovation of regulations based on past arrangements; and (2) in using memory/knowledge for resource monitoring and management (Table 11.5). An example is the banning of large nets in 1981, which was inspired by the first attempts in 1971 to prohibit their use in two lagoon basins. Another example is the regulation prohibiting any fishing activities in the channel and in the small channel between the upper and middle lagoons; these measures were probably based on pre-1970s management practices.

In conclusion, using a combination of common-property theory and Holling's (1986) adaptive renewal cycle, the Ibiraquera Lagoon case clearly demonstrates the feasibility of studying the linked dynamics of social–ecological systems. Although the social dynamics and ecological dynamics in this case have different time scales (decadal in the former and a few months in the latter), the resilience of the management system is analyzed through cycles of change. Over the last four decades, the Ibiraquera Lagoon management system has gone through several cycles of change in property rights regimes. These changes have also had consequences for resource sustainability.

The trend at the end of the 1990s was one of unsustainable resource use. However, the pattern of changes in the last four decades provides the promise that the crisis in the late 1990s may trigger a new round of institutional renewal. The impediments for such renewal include enforcement problems, the brittleness of the local fishers' organization (Colônia), and the fact that the lagoon is increasingly used for a different set of economic benefits (i.e., tourism). In one sense, these are adaptive responses, but at a different time scale (i.e., shorter-term economic gain) from the management of the lagoon's resilience. To the extent that they result in a loss of options, they entail a loss of resilience in the social–ecological system as a whole. However, the experience of the last four decades indicates that the resilience of the Ibiraquera Lagoon management system is not in its maintenance of stable and sustainable resource use. Rather, it is in its ability to turn successive resource crises into opportunities for a new round of institutional renewal.

Acknowledgements

Seixas' research was sponsored by the Conselho Nacional de Desenvolvimento Científico e Tecnológico (CNPq) of Brazil. Berkes' research was supported by the Social Sciences and Humanities research Council of Canada (SSHRC). The authors thank Alpina Begossi, Paulo F. Vieira, Natalia Hanazaki, and Rodrigo Medeiros for help in the development of the project; Joaquim O. Branco, Maurício Hostin-Silva, and Renato A. Silvano for help identifying shrimp, crab, and fish species; and the people of Ibiraquera Lagoon for their hospitality and generosity with their time.

Notes

1. The Colônia de Pescadores was founded in 1952; its members elect its president and board of directors every 2 or 3 years. This organization encompasses fishers from the entire municipality of Imbituba and not only those living around the lagoon. As of 1999, there were about 1500 member fishers, including about 350 fishers living around the lagoon.
2. Departamento Estadual de Caça e Pesca.
3. The Federal Fish Agency, Superintendência para o Desevolvimento da Pesca (SUDEPE), issued several regulations for national or state territory which apply to the Ibiraquera Lagoon, and included the following rules: establishment of minimum mesh size of 2.5 cm for shrimp cast-net (1970), of 5.0 cm for fish cast-net (1972), and of 7.0 cm for fish gillnet (1972); prohibition of setting gillnets longer than one-third of lagoon width (i.e., it also prohibits the use of nets as fences in the lagoon channels) (1972); prohibition of trap-net use in the channel linking the lagoon to the ocean (1972); prohibition of hand-drawn beach seine and seine (1972) and trawling (1975).
4. Those who enforced fishing regulations included members from the local police force and SUDEPE.
5. IPEP: Instituto de Pesquisa e Extensão da Pesca. ACARPESC: Associação de Crédito e Assistência Pesqueira de Santa Catarina.
6. The new regulation (N-027/81) came into force under the Decreto (Decree) 73.632 of 1974, empowering SUDEPE to manage Brazil's fisheries.
7. Empresa de Pesquisa e Difusão Tecnológica do Estado de Santa Catarina (EPAGRI).
8. The shrimp-stocking project was funded by three federal government agencies: Fundação Banco do Brasil (1992–3); Fundo Nacional do Meio Ambiente (1994–6); Programa de Execução Descentralizada do Ministério do Meio Ambiente (1997–8).
9. The Federal Fishery Agency (SUDEPE) was extinguished in 1989 and replaced by the Brazilian Department for the Environment (Instituto Brasileiro do Meio Ambiente e Recursos Naturais Renováveis – IBAMA) in the same year.
10. Companhia de Polícia de Proteção Ambiental (State Police).
11. In 1999, Conselho Comunitário de Ibiraquera and Associação dos Amigos da Praia da Barra da Ibiraquera, and in 2000 Conselho Comunitário de Araçatuba were activated. The Associação de Ibiraquera-Gramense has remained active since 1986.

12. We made no attempt to go back further in history, but suffice it to say that there must have been many changes as different groups succeeded one another in the colonial history of Brazil for this region.

References

Andreatta, E.R., Beltrame, E., Galves, A.O., *et al.* 1996. *Relatório Final: Repovoamento da Lagoa de Ibiraquera com Pós-larvas de camarões*. Report. Universidade Federal de Santa Catarina, Brazil.

Andreatta, E.R., Beltrame, E., Vinatea, L.A., *et al.* 1993. *Relatório Final: Projeto de Repovoamento de Post-larvas das Lagoas Ibiraquera.* Report. Universidade Federal de Santa Catarina, Brazil.

Barbosa, F.I. and Hartman, V.D. 1997. Participatory management of reservoir fisheries in North-Eastern Brazil. In *Inland Fishery Enhancements*, pp. 427–45, ed. T. Pert. Fisheries Technical Paper 374. Toowoomba, Queensland: FAO.

Berkes, F., ed. 1989. *Common Property Resources: Ecology and Community-based Sustainable Development.* London: Belhaven Press.

Berkes, F. and Folke, C., eds. 1998. *Linking Social and Ecological Systems. Management Practices and Social Mechanisms for Building Resilience.* Cambridge: Cambridge University Press.

Bromley, D.W., ed. 1992. *Making the Commons Work: Theory, Practice, and Policy.* San Francisco: ICS Press.

Carpenter, S.R. and Kitchell, J.J., eds. 1993. *The Trophic Cascade in Lakes.* Cambridge: Cambridge University Press.

Gunderson, L.H. and Holling, C.S., eds. 2002. *Panarchy: Understanding Transformations in Systems of Humans and Nature.* Washington DC: Island Press.

Gunderson, L.H., Holling, C.S., and Light, S.S., eds. 1995. *Barriers and Bridges to the Renewal of Ecosystems and Institutions.* New York: Columbia University Press.

Holling, C.S. 1986. The resilience of terrestrial ecosystems: local surprise and global change. In *Sustainable Development of the Biosphere*, pp. 292–317, ed. W.C. Clark and R.E. Munn. Cambridge: Cambridge University Press.

Holling, C.S. 1995. What barriers? What bridges? In *Barriers and Bridges to the Renewal of Ecosystems and Institutions*, pp. 3–34, ed. L.H. Gunderson, C.S. Holling, and S.S. Light. New York: Columbia University Press.

Ludwig, D., Hilborn, R., and Walters, C.J. 1993. Uncertainty, resource exploitation, and conservation: lessons from history. *Science* 260: 17, 36.

McCay, B.J. and Acheson, J.M., eds. 1987. *The Question of the Commons: the Culture and Ecology of Communal Resources.* Tucson: University of Arizona Press.

McCay, B.J. and Jentoft, S. 1996. From the bottom up: participatory issues in fisheries management. *Society and Natural Resources* 9: 237–50.

North, D.C. 1994. Economic performance through time. *American Economic Review* 84: 359–68.

Ostrom, E. 1990. *Governing the Commons. The Evolution of Institutions for Collective Action.* Cambridge: Cambridge University Press.

Pomeroy, R.S. and Berkes, F. 1997. Two to tango: the role of government in fisheries co-management. *Marine Policy* 21: 465–80.

Seixas, C.S. 2002. Social – ecological dynamics in management systems: investigating a coastal lagoon fishery in southern Brazil. PhD dissertation, University of Manitoba, Canada.

12

Keeping ecological resilience afloat in cross-scale turbulence: an indigenous social movement navigates change in Indonesia

JANIS B. ALCORN, JOHN BAMBA, STEFANUS MASIUN, ITA NATALIA, AND ANTOINETTE G. ROYO

12.1 Introduction

A resilient ecosystem is one that will retain the ability to reorganize and renew itself without loss of function or diversity when disturbed, *if disturbance is managed adaptively* (Holling, 1992) – i.e., if management keeps disturbance within certain bounds and/or management is prepared to react to unexpected disturbance in ways that sustain resilience.

Safeguarding resilience requires appropriate management decisions by people using their society's cultural norms and institutions at different (small and large) scales. Conflict between these scales sometimes leads to clashing management decisions, and subsequently an erosion of resilience. Over time, changes in social and political conditions as well as population sizes, technologies, incentives, and values can also result in this erosion unless societies recognize and respond to negative ecological feedback by modifying their management institutions. Changes also create conflict amongst different management levels (national, regional, and local) and/or between social groups; these conflicts can only be resolved through political action. Making responsive management regime adjustments in time to avert damage is not easy, however. Cross-scale conflict between national and local institutions creates turbulence, and so does political change, often holding ecological resilience hostage. A better understanding of effective social movements that promote positive changes in response to ecological feedback is fundamental for understanding and maintaining the resilience of Earth's ecosystems.

Our case study of the Dayak people's social movement in Indonesia offers insights into how the erosion of ecological resilience can be countered by social renewal. We use Holling's model for ecosystem resilience and build on previously published insights regarding the dynamic linkages between social and ecological systems described in Berkes and Folke (1998), and Folke and Berkes

(1998). As Holling and Sanderson (1996) have observed, social system changes related to adaptive management are more complex and less well understood than the natural cycle of ecological renewal described by Holling. Folke and Berkes (1998) have noted that resource crises during the release and reorganization phases of Holling's adaptive renewal cycle can lead to creative destruction – a time when institutions associated with resource management can reorganize themselves in order to prevent a Holling 'flip' into a less valued, less resilient ecosystem.

We add to Holling's framework the observation that if the political system is closed to participants who want to modify institutions in response to negative ecological feedback, then, during crises, ecological resilience will diminish until the system flips. Resilience depends on a vibrant political life in which multiple interests participate. We further posit that social movements can prevent flips if they successfully challenge the political system to accommodate their voices and their concerns about ecological feedback.

Our social movement analysis here is based on McAdam's (1982) 'political process model,' because this model stresses the multiscalar and interactive feedback aspect of social movements, and because it focuses on the elements that undergird the persistent Dayak movement. Like ecological systems (Gunther and Folke, 1993), struggles for social change are autopoetic – creating their own place and direction of evolution within an environment they themselves constantly shape. Social change relies heavily on changing the definitions of what is acceptable and what is not. Meaning is constructed in the confrontations between opposing schemes for evaluating situations. The process of struggle between actors determines whose definition of the situation will prevail (Klandermans, 1992) and, therefore, whose framework of values will be used to plan and judge action.

In our case study, we describe how Dayak self-organized associations work at shaping the cusp of this struggle between different systems of meaning. At the national level, Dayak are, in Gamson's terms, 'challengers' – people who are outside the polity because they 'lack routine access to decisions that affect them' (Gamson, 1975: 140). At the local level, Dayak are innovators and revivers. Their local resource management practices and governance systems have maintained forest resilience at local and regional scales on Borneo for centuries, yet a repressive national-scale government is destroying those same forests. Dayak groups privilege indigenous memory, culture, and institutions, but they have also adapted modern tools for scientific analysis, communication, and networking in order to nurture a movement whose progress proceeds through interactions with the state – the dominant polity that asserts the 'social control response' (McAdam, 1982: 56). Their innovative, indigenous social movement is creating transformative cross-scale communication that is essential for recoupling global, national, and local society with ecological feedback.

Dayak institutions evolved as a good fit (as defined by Folke *et al.*, 1997) for the local ecosystems they manage. This chapter builds on the extensive existing literature on the ecologically adapted management practices of Borneo's people, and extends the analysis to identify some of the key socio-political practices that nurture ecological resilience. Similar to many other indigenous peoples (International Labour Organization, 1996) in Indonesia and elsewhere, the Dayak developed agro-ecosystems adapted to tropical forest ecosystems. The agro-ecosystems, and the behavior of the people who use them, are governed under indigenous institutions – rules created and enforced by community consensus through community-based political processes.

Dayak's indigenous, sustainable resource management faces two problems: modernization, which has altered Dayak lifestyles over the past several decades, and the Indonesian central government's (GOI) attempts to govern access to resources on Dayak lands.

The Dayak problem is typical of the two-pronged quandary found in tropical forests around the world: indigenous peoples struggling to adapt to new technologies and needs while staving off invaders, international investors, or national governments that claim their resources. Dayak are among the world's remaining 'ecosystem people' – people who have adapted to, and depend on, local ecosystems (Dasmann, 1991). Their collective identities, cultural traditions, and management practices have often enabled them to maintain resilient, productive ecosystems (Berkes, 1999). Yet these societies and their management systems are being disrupted by stresses at two different scales: stresses arising from within the local population, and stresses arising from the larger society of the nation-state within which the indigenous society finds itself. The latter, national societies are linked to the global economy; they are 'biosphere people' – they do not depend on local ecosystems, and they are decoupled from ecosystem feedback (Dasmann, 1991).

When the international investment capital of biosphere people funds legal and illegal resource extraction from homelands over which ecosystem people lack formal tenure, ecosystem people protest this social and ecological injustice. Cultural erosion and physical intrusion by state-supported colonists and companies threaten indigenous societies' collective identities and the resource base for their livelihoods. The self-images of ecosystem peoples' societies undergo transformation, from viewing themselves as collective units responsive to ecological feedback into seeing themselves as a disorganized population detached from their local natural environment.

Even when people fight back, their efforts can be swamped by stronger national-scale waves. The dominant society has the power to bless concessionaires' and migrants' land use with legitimizing laws and subsidies, while using coercion and re-education to silence indigenous societies' resistance (Scott, 1990).

Because ecological stresses are often both local and supra-local, the challenge for ecosystem governance is to mobilize society to (a) couple higher-scale governance institutions to local ecological feedback, and (b) renew or create local self-governance responsive to ecological resilience. This chapter analyzes the instructive case of how Dayak society has responded to this challenge.

12.2 A panarchy of institutions

In order to appreciate the applicability of the study's insights, one needs to understand the historical and political constraints of the panarchy under which the Indonesian social movement has evolved. Across Indonesia, local indigenous (*adat*) organizations are nested within nationally defined institutions, which in turn are nested within the global economy. Over the past 30 years, the cross-scale links have intensified through a turbulent process in which the national and local levels clash.

12.2.1 The national level

Indonesia is one of the world's most biologically and culturally diverse countries. It encompasses over 17 000 islands joined by a vast expanse of sea bridging the Australian and Asian continents with their distinct flora and fauna. It is home to 210 million people – over 250 indigenous cultures speaking over 600 languages. Each group has evolved its own ways of sharing space with nature through local laws and political institutions.

In 1949, forests covered 150 million ha, 75 percent of Indonesia. By 1990, this was down to 40 percent (Barber, 1997). These figures reveal how greatly the Indonesian state has disturbed local institutions associated with forest ecosystems – a surprising shock successively suffered in many separate localities with a combined population of 40 million.

The state legitimizes elite interests and reifies their values as pre-eminent in the rhetoric that it uses to gain public acceptance of the *status quo*. A central goal of any social movement is to change this discourse frame. In Indonesia, official discourse about 'development' has valued activities benefiting ruling elites and 'crony capitalists' who surround the president and the military – a process the Indonesian State defines as 'development.' Any resistance to state-initiated resource extraction or labor mobilization is viewed as resistance to 'development,' officially castigated as 'backward' or 'insurgent.'

The Indonesian elite value their personal profits and patronage relationships based on the control of access to the vast mineral and natural resources of the archipelago. They directly affect land-use decisions, which in turn affect ecosystem resilience. The central government's actions to control access to forests and rivers are in direct conflict with *adat* systems controlling access to these same resources. Yet the Indonesian State has finessed this contradiction through rhetorical discourse that extols the virtues of diversity even as it strives to assimilate all its internal nations into one Indonesian national identity and transform them from self-governing landowners into a compliant workforce on plantation estates. The Dayak social movement seeks to replace the state's discourse frame with one that legitimizes the perspective that the state is damaging the environment and society for the benefit of a few. Rather than being seen as subversive, as in the state-framed discourse, the Dayak are bringers of truth, saving the state from the elite.

The history of Indonesia's people is one of stress, change, and resilience. Indigenous systems responded to the shifting constellations of trade networks linked to Asia for thousands of years before Europeans arrived (Potter, 1993; Brookfield, Potter, and Byron,1995; Peluso and Padoch, 1996). In the late 1500s, Europeans added their demands to those of competing Islamic sultans and local kings. During the nineteenth century, the Dutch colonial administration enforced radical transformations of local commercial production systems; local subsistence systems were forced to adapt. Regional wars between colonial powers at the turn of the nineteenth century and the Japanese occupation during World War II caused more disruptions. Yet at the dawn of the Indonesian State, in 1949, the health of Indonesia's extensive forests and productive waters stood as evidence that local institutions had effectively managed their resources through all these disruptions. Thus, the social and cultural elements for renewal were in place when the most recent stresses manifested themselves in the latter half of the 1900s.

At independence, elites from Java took over the colonial administrative apparatus and wrote the Constitution for the neocolonial state. While the Constitution and *Pancasila* (Five Governance Principles) could be interpreted to lay out the basis for a pluralistic society, even possibly a federation of semi-autonomous *adat* units, they have been used instead to legitimize the rule of a strong central government enforced with repressive military force. The Constitution creates a strong president, a weak parliament, and no independent judiciary. The Pancasila supports (1) belief in one God; (2) Indonesian unity; (3) justice and civility among peoples; (4) democracy through deliberation and consensus among representatives; and (5) social justice for all. While *adat* rights and institutions are recognized in the Constitution, agrarian law recognizes *adat* laws, rights, and institutions only if that recognition does not create problems for development. As is typical of the nation states whose borders have surrounded indigenous peoples around the world, Indonesia views indigenous peoples as having rights to development but not to their resources.

The state uses the military (ABRI[1]) and a large internal intelligence force to coerce compliance. ABRI formally enjoys a 'dual role' as military and political power. It created the ruling Golkar Party, which controls legitimizing elections. According to official policy, the people are to be a 'floating mass,' apart from politics. For three decades, political parties have been forbidden to have any political mobilization or other activity in rural areas, except at election time. Golkar, through ABRI, has maintained a constant presence in all villages. Political activity is further restricted by the Anti-Subversion Law under which no more than five people are allowed to meet without informing the police.

The suppression of political feedback and response left the state brittle and vulnerable to social movements. The Reformasi protests of 1998, following the 1997 economic downturn, forced the aging General Suharto to step down. Elections brought a new coalition to power in 1999, with a freer political climate. The state's structure and policies have changed little, however, especially in rural areas. Instead, political adventurers and regional elites are taking advantage of their new freedoms to extract resources – arming militias and orchestrating new violence to maintain control.

12.2.2 The local levels

People have made their homes in Borneo's forests for at least 35 000 years, adapting to changing political and ecological circumstances (King, 1993). The indigenous cultures living in the interior of Borneo are collectively known as Dayak – numbering over 3 million, with an average density of 14 people per km^2 (Cleary and Eaton, 1996). Dayak societies share many features, although there are dozens of subgroups with different languages, social structures, and governance traditions.

For centuries, Dayak have relied to varying degrees on agriculture, fishing, hunting, and gathering products from the forest, shifting their emphasis as situations changed. Dayak territory is rich in natural resources, including watersheds of great rivers and vast forests that contain an extraordinary diversity of fish, over 500 species of birds (many of them endemic species), and a poorly documented flora including over 350 species of the dipterocarps valued for their timber. This richness also includes a rich fauna of rare species, such as orangutans, banteng cattle, sun bears, elephants, and rhinos (Cleary and Eaton, 1992; Brookfield *et al.*, 1995).

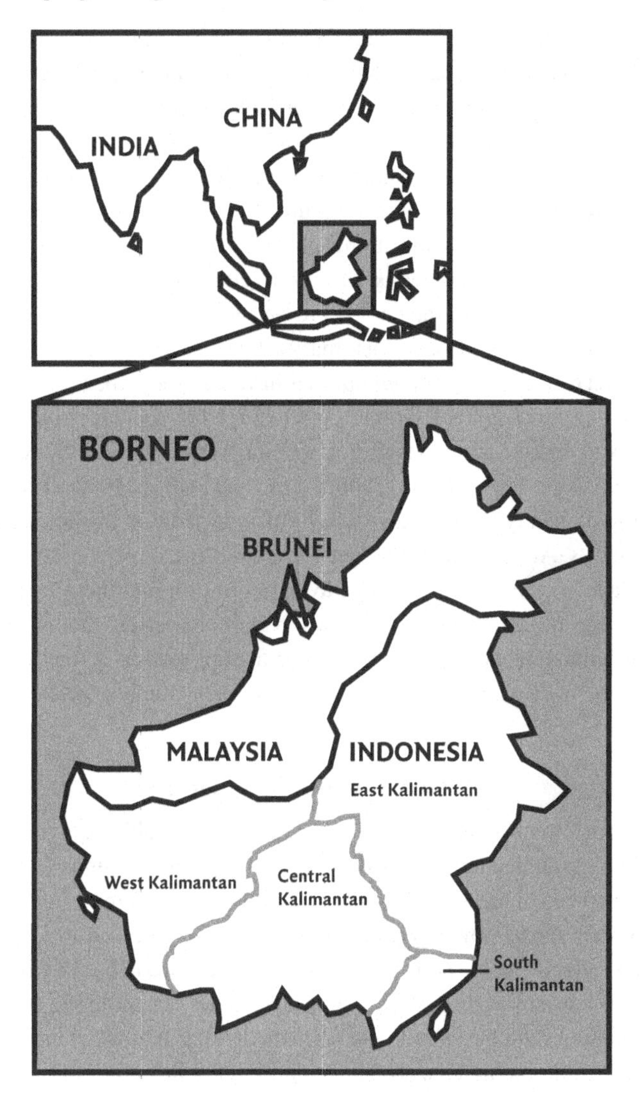

Figure 12.1 Map of Borneo.

Dayak have historically warred amongst themselves. But, as Padoch and Peluso (1996: 3) note, '[t]he dynamism of how Borneans manipulate resources must . . . be understood in the context of a constantly changing forest environment.' Dayak have adapted their resource management institutions to the droughts, famines, fires, and population fluctuations that have been common, if not predictable, challenges to making a living on Borneo.

Borneo is now divided by three states – Indonesia, Malaysia, and Brunei (Fig. 12.1). Indonesian Dayak territory (collectively called Kalimantan) is divided into four provinces – East, South, Central, and West Kalimantan. Political activity was vibrant in the territory through the mid-1900s. After independence from the Dutch, the Dayak Unity Party formed, and won the 1955 elections in West Kalimantan. But in 1959, General Sukarno signed a regulation stipulating that political parties must have branches in seven provinces in order to field candidates. This effectively disrupted the link between Indonesia's diverse 'ecosystem peoples' and national politics, because the political legitimacy of Indonesia's internal nations is based on geographically localized polities, none of which spans more than a few provinces.

While all Dayak subscribe to customary *adat* law and tradition, even within a given district different Dayak communities have different processes for creating and enforcing those rules (Bamba, 2000). *Adat* law includes rules and procedures for allocating tenurial rights to land and trees (Appell, 1992; Cleary and Eaton, 1996; Ngo, 1996; Peluso, 1996; Peluso and Padoch, 1996). The structure of authority typically includes a group of elders in addition to other leaders and their assistants. Anyone, including outside authorities, must obey the *adat* laws of the village in whose territory they travel. In some regions there are small federations of several communities, led by a paramount leader. Shamans are also important village leaders whose ritual and healing powers know no borders; they are free to enforce their laws and perform their rituals anywhere.

12.3 Dayak resource management

When asked, 'What is happiness?,' a typical Jalayi Dayak's response is 'Sasak Behundang, Arai Beikan, Hutan Bejaluq' – 'There are shrimps in the leaves sunk in riverbeds, there are fish in the waters, there are animals in the forests.'

This vision shows how river, land, and forest are essential to Dayak identity, and is reflected in the shifting mosaic land-use pattern that Dayak create in their forest ecosystems (Fig. 12.2). There are patches of natural forest, managed forests, rotating swidden/fallow, and permanent fields molded to the ecological conditions of the mountains, wetlands, and river valleys of a particular community's territory. Each community's landscape is different, yet forest cover is consistently substantial. In a sample of 21 communities that mapped their lands and created internal conservation agreements to resist forest or mining concessions between 1996 and 1999, territories ranged from 900 ha to 126 000 ha in size, averaging 12 500 ha, with a median of 4600 ha. Forest cover ranged from 50 to 99 percent, average and median being 70 percent. Primary forest cover averaged 29 percent, with a median of 25 percent.

In the past, millions of hectares were covered by this shifting patchwork, creating a vast, resilient landscape. Patches of disturbed forest transitioned

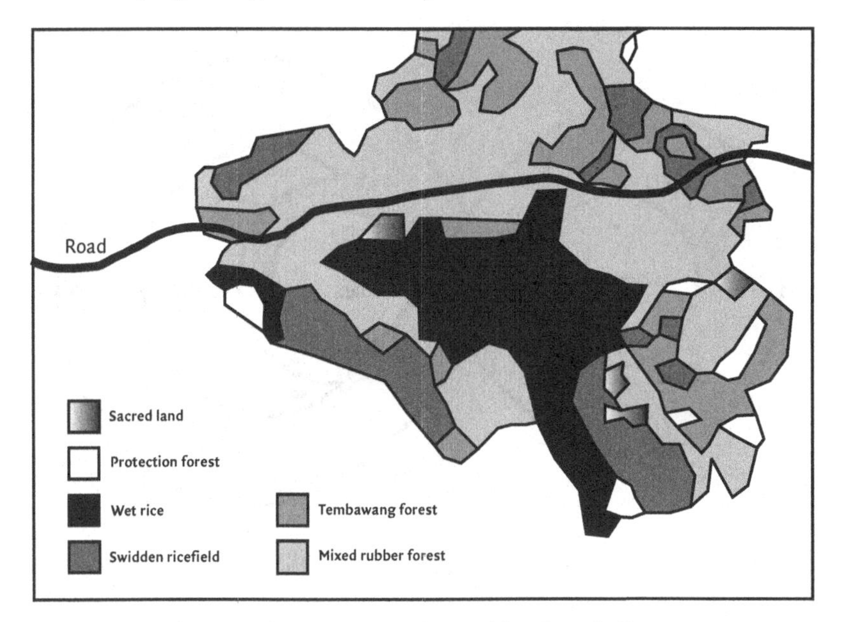

Figure 12.2 Typical Dayak land-use mosaic resulting from indigenous management practices.

into successional forest patches, all the while maintaining ecological services and seed sources so a vast array of native species could recolonize disturbed patches in accordance with natural renewal cycles (as defined by Holling, 1992). Watershed integrity was maintained, and forest and riverine biodiversity flourished.

Although virtually all of Kalimantan has been officially designated as open to exploitation (and illegal concessions are found in nature reserves), some 63 percent is still forested (Potter, 1993) – representing 35 percent of Indonesia's remaining forests. Kalimantan's forest is mostly found on territory claimed by Dayak communities. Larger patches of forest are in inaccessible areas. Communities following indigenous management practices are interspersed with communities that have ceded their lands to oil palm plantations where monoculture has replaced diversity, communities whose forests have been felled by timber or reforestation concessions, and lands taken by migrant non-Dayak communities. Dayak communities' forests sometimes offer isolated patches of refugia habitat in an expanse of monocultured oil palm plantations. Some Dayak community patches are almost identical to those of non-Dayak; others are transitional. Within a given community's patch, the distribution of the smaller patches of land-use types is determined by the historical governance under *adat* and the disruptions suffered in the area. National governance

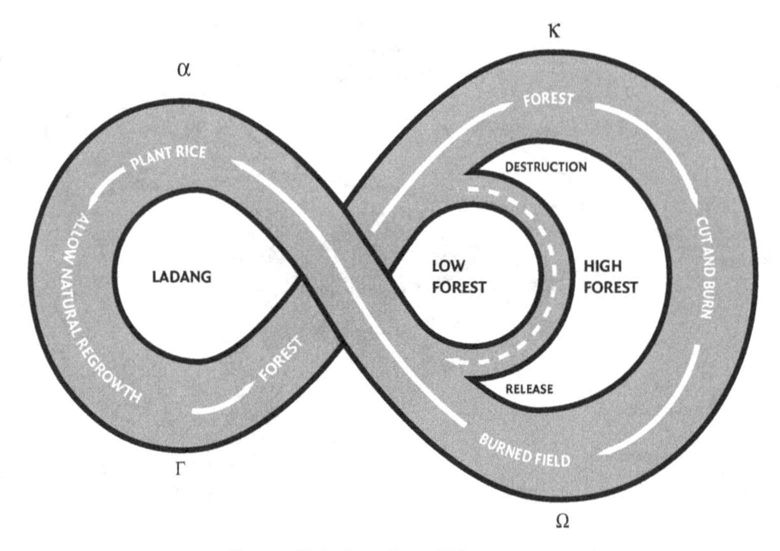

Figure 12.3 Dayak swidden system.

determines the expanding gaps[2] in the landscape where plantations, colonists' farms, and degraded lands replace healthy ecosystems.

The resilience-sustaining practices and 'scripts' of Dayak land-use systems that create this mosaic are similar to those in other forest ecosystems[3] (Alcorn, 1990; Warner, 1991; Messerschmidt, 1993; Alcorn and Toledo, 1998). However, they are richer in diversity – perhaps owing to the lower population density, the historically large market for multiple nontimber forest products, the range of ecological variation available for exploitation in a single community, and the strong indigenous institutions that resisted colonial administration until very recently. Figure 12.3 illustrates the Holling diagram for Dayak swiddens. Detailed descriptions of Dayak ecological knowledge and practices used in resource management have been published (cf., Dove, 1985; Cleary and Eaton, 1992; King, 1993; Padoch and Peters, 1993; Colfer, 1993, 1997). Like other swidden agriculturalists, Dayak use disturbance to create space for food production and use forest succession processes as a production resource.

While Dayak spread risks by depending on a variety of resources through fishing and hunting, forest management and use, and agriculture, community social cohesion also serves to maintain the integrity of the overall system. This is essential to respond to expected but unpredictable events, such as drought, fire, or flood. The widespread use of auguries for such decisions as the selection of a swidden site both supports adherence to an indigenous belief system and throws a randomizing variable into decisions based on recent experience (Dove, 1996a) – thereby enhancing the chances for experimentation in places that would not usually be chosen if the choice were based on existing ecological knowledge. The yield of swidden is believed to depend on the quality of the agreements that the farmer made with the spirits of nature that control the harvest.

Forests and forest products are important, and different forest patches are managed more or less intensively for different products through enriching natural forest (Padoch and Peters, 1993). Dayak use many subsistence products from their forests; for example, a typical Dayak community collects 200 species of medicinal plants (Caniago, 1999). Rituals related to the bounty or scarcity of fruit yields demonstrate the importance of the principles of reciprocity and exchange (Dove and Kammen, 1997). For example, the Dayak view the fruit and nut harvests as indicative of the quality of the relationships among people and the relationship between people and nature. Because many native fruits are mast fruiting species (having a large production of fruit in some years and none in other years), a scarce harvest every few years serves as a reminder of these relationships.

A remarkable number of commercial species have been integrated into Dayak forests over the centuries. Thirty-five native fruits are harvested from Dayak forests for sale in the provincial capital of Pontianak, with an annual market value of over half a million dollars (Arman, 1996). Other native plants produce commercially valuable rattan, resins, and oilseeds (Peters, 1996). Some non-native species have become important as well, such as para rubber, introduced in the early 1900s. Rubber production from the many small Dayak rubber plots has made Indonesia one of the world's leading producers of rubber (Dove, 1996a). Yet Dayak rubber-producing forests are highly species diverse (Penot, 1999), which contributes to the resilience of the Dayak ecosystems.

Dayak have been cautious about maintaining the balance between economic dependence on forest products and subsistence rice production – between maintaining a forest ecosystem and transforming it into a plantation landscape. Rice, which is central to Dayak swidden system and identity, is respected and surrounded by rituals and work activities that bind the community's households together in mutually dependent relationships. To the Dayak, to make swidden is to be human[4] (Dove and Kammen, 1997). Rice has a soul that must be properly cared for and respected through proper swidden management (Djuweng, 1998). These beliefs support the resilience of the overall swidden system, which depends on the proper balance of cleared, successional, and high forest patches. In the 1930s, for example, when rubber prices were fluctuating, one man's dream about rubber that 'ate rice' was widely interpreted across Kalimantan as a symbolic warning to maintain swidden traditions (Dove, 1996b).

12.4 Stresses eroding Dayak solidarity with nature

Stresses create problems at very local scales in local forests, at the larger scale of watersheds, at the regional scale of forests and major rivers of the island, and at the national scale of the Indonesian archipelago. Some are constant, others are spatially and temporally unpredictable. The system and its perturbations, in short, are in dynamic flux, but feedback is constrained by the national level of the social panarchy. The state's 'control response' to rural resistance (Scott, 1990) has been to become increasingly repressive.

While the market produces stresses, the most damaging stresses originate in national policies and laws that do not protect citizens or nature from abuses. The regulations on mass gatherings mean communities no longer have the autonomy to perform swidden rituals without getting a license from the sub-district police office. Indigenous rituals and values are also undermined by an intimidating Indonesian law which says people must choose to follow one of the five approved religions, or be condemned as atheist or communist.

Missionaries actively undermine Dayak solidarity as well. One typical Protestant poster condemns Dayak ritual as the 'road to hell,' extolling church ritual as the 'road to heaven.'

Education, as implemented by the government, is one of the most insidious threats to institutional and ecological resilience in Dayak territory. The curriculum discourages independent, analytical thinking and undermines Dayak cultural values. Courses promote a vision of modernization that supports central government policies, damaging adaptive human resources such as youthful brainpower and collective memory. Television, likewise, has had an astonishing impact. It replaces communication (specifically, sharing of ecologically significant information) that used to occur during the evening. TV programs promote biosphere people's lifestyles and create new 'needs.' For example, one TV spot shows a Dayak farmer's daughter singing in a field, regretting her destiny practicing the swidden agriculture that has brought her family misery. She sings her regrets for causing all the fires[5] and ecological destruction from indigenous agriculture, spreading the government's anti-swidden propaganda.

The destruction of longhouses (traditional housing targeted by GOI since 1960 for being 'unhealthy and communist') has also undermined Dayak solidarity. Today, longhouses exist only in remote areas and in places where the Dayak organized resistance. Without the longhouse, the chief of the longhouse no longer has his full role as head of village government, and that change has disrupted the way that resource allocation decisions are made. The longhouse was central to the functioning of Dayak institutions; its shared space built collective identity and ensured intergenerational knowledge transfer through weaving, dancing, story telling, and conversation.

National policies blame swiddeners for deforestation, while laws legitimize forest concessions, plantations, transmigration, and mining on Dayak lands (Moniaga, 1993; Masiun, 2000). The ecological damage from concessions is extensive; only about 4 percent of logging concessionaires follow ecological guidelines (Potter, 1993). The Agrarian Basic Law No. 5, 1960, gives the state complete control of the land, and states that the GOI will conduct land titling, rather than recognize *adat* tenurial law that regulates the ownership of land. The Basic Forestry Law No. 5, 1967, created forest logging concessions covering *adat* forests. In the early days of the Reformasi opening in 1998, an *adat* forest movement coalition demanded that the state 'revoke the status of state forests by redefining the border between state forests and forests that have been owned and controlled by traditional and local communities,' and 'revoke all regulations and policies regarding the exploitation and violation of community rights to manage natural resources' (KUDETA, 1998: 3). But a new Forest Law No. 41, 1999, fails to significantly change the situation while paying lip service to *adat*, illustrating a growing tendency to introduce limited reform to avoid greater change. Communities are still coerced into accepting the rights asserted by the concessionaires unless they can mobilize unified action to resist the company (Anonymous, 1998).

Adat governance was rejected by Village Governing Law No. 5, 1979, which established a system of village government with a *kepala desa* (headman) under the *camat* that is directly under the control of central government. The *camat* lets it be known who would be acceptable for election and then an election is held to 'elect' that person as headman. The law grouped five to seven villages under the authority of one central 'developing village,' but Dayak villages are scattered up to 10 km from the seat of this village government, so interaction is limited. There is no accountability or transparency. District governments must give their approval of the appointment of any new *adat* chief, whose powers have been restricted because the village group headman and the district head can intervene in his decisions. Many *adat* leaders have kept quiet, or been co-opted to allow logging and palm oil concessions to enter community lands.

In 1997, in response to the revitalization of *adat* by the growing social movement, the GOI Ministry of Internal Affairs produced Regulation No. 3 – ostensibly on 'the empowerment, sustainability and development of customs, people's tradition and *adat* institutions at local government level' – creating a new set of manufactured *adat* institutions, in effect changing the definition of *adat* in national discourse frames and undermining indigenous *adat* autonomy and authority. These puppet '*Adat* Councils' do not include Dayak *adat* chiefs;

nor did GOI consult with the general Dayak population when it unilaterally selected for placement on the "*Adat* Councils" people who are not familiar with *adat* law or traditions – government officers, Golkar politicians, and urban-based businessmen – and who use their positions for their own benefit.

These new councils have urged Dayak to give up their lands for oil palm plantation projects, and made formal statements to high-level government officials condemning *adat* resource management activities, and promising to end swidden practices.

For decades, development projects and transmigration, designed and approved by government planners at the national scale, have struck local areas. Transmigration programs have moved over half a million Javanese and Madurese onto Dayak territories. This program now brings Javanese and Madurese labor for new plantations with the incentive that they will be given 3 ha of plantation to tend and a household plot for nothing. These government policies have led to sporadic violence against settlers, most recently in 2001 when Dayak–Madurese clashes resulted in hundreds of – mostly Madurese – fatalities.

Because these stresses are new, communities generally lack ways to recognize and address them. For example, people are often unaware of the limits of their community's forest territory and, when coerced by concessionaires, sign away rights to vast tracts of forests that actually belong to other communities. This lack of awareness of their territorial borders also contributes to the lack of awareness of ecological feedback or landscape-level problems that they themselves are creating – such as excessive clearance of forest leading to loss of access to useful species that had always been abundant. Upriver communities are unaware of the downriver impacts of their forest-clearing activities. Lack of awareness about pending threats, such as concessions that are invisible until the machinery arrives, means that people literally awake in their beds to the unexpected sound of chainsaws in their forests.

12.5 Social renewal and reorganization: Dayak tools for promoting ecological resilience

Even though Dayak are hampered by lack of adequate knowledge of their rights and the ways in which they can assert their rights – and despite the repressive political environment – a strong Dayak rights movement has grown around environmental concerns (Mayer, 1996), facilitating the communication of political and ecological feedback.

An effective response to the negative ecological feedback caused by bad management at the national scale requires communication between local and larger scales, ideally creating accountability between institutions operating at different scales.

Over the past 20 years, support for the Dayak movement has arisen from a slowly expanding cluster of associations begun by Yayasan Social Karya Pancur Kasih (PK), an education non-governmental organization (NGO) established by a small group of Dayak schoolteachers. PK's founders shared a belief that Dayak self-reliance and solidarity were the key to freedom and empowerment. They slowly expanded their activities, but followed a strategy of supporting the development of many separate associations so that if one unit of the struggle were harassed, other units would not be directly affected. They have been able to reconnect elements of a Dayak social movement that have been in place for centuries, though weakened by the military regime since 1960.

PK-associated sister associations include schools, local credit unions, a bank, a rubber cooperative, a legal assistance group, a cultural support association, a technical mapping unit, and a forest management network. These associations are loosely governed by an umbrella consortium (Consortium for the Empowerment of 'Dayak' Peoples, or KPMD) to ensure that they follow a shared vision. By using the term 'Dayak' in quotation marks, the group committed itself to supporting the struggles of others like the Dayak – in other words, people who are the most oppressed and marginalized from society.

The consortium currently has 17 member organizations with a combined staff of over 240 people. Despite the difficult political environment, the movement has spread from a few villages in West Kalimantan to networks across Borneo and beyond to nurture related movements on other islands and among other cultures in the Indonesian archipelago. Reaching out to build the cross-cultural cohesion that will empower civil society, PK created the NGO Concern Center for Refugees after the Dayak–Madurese conflict in 1997 and, in 2000, the Legal and Human Rights Assistance Network for Reconciliation.

The PK tactics are to consolidate progress in three areas: social solidarity (including people and nature in solidarity), critical thinking, and economic progress. Progress in meeting basic needs, security, and identity renews the Masyarakat (consensus-based community), which in turn generates a new vision to mobilize new associations and progress.

12.6 Critical education

PK – acting on the belief that education and independent thinking are the basis for self-actualization – helps teachers to establish PK schools in interior villages. These are run independently rather than under a PK umbrella. PK also actively negotiated with education agencies of provincial government for the opportunity to develop a Dayak-oriented curriculum for West Kalimantan public schools. It recently completed the development of the curriculum for use by teachers across the province. In addition, PK supports a 'critical education program' for adults to participate in interactive short courses on social analysis, critical thinking, and technical skills useful for building Dayak self-reliance. By integrating elders and other adults into educational programs, PK helps Dayak to maintain species diversity and ecological resilience by preserving and valuing the memory of useful species, ritual protections, sacred areas, and other supportive knowledge.

12.7 Credit unions – access to capital while building self-reliance

PK credit unions meet the demand for small loans, build community solidarity to manage the group's finances, demonstrate the value of accountability mechanisms, and enable people to solve their own problems. PK linked the credit unions (CUs) to the World Council of Credit Unions, which provides technical assistance and advice to CUs in 84 countries. Today, these CUs have over $4 million in assets. CU members create social funds for medical aid, pensions, and other social assistance, and receive insurance benefits. The CU building also provides a shared space – to some degree filling the void created by longhouse destruction. In 1992, 200 Dayak Credit Union members borrowed money from their CU to establish a rural bank, which provides small business loans; it now has over $1 million in assets.

12.8 Cultural revitalization

In response to grassroots interest in cultural and political issues affecting Dayak peoples, PK established the Institut Dayakologi (ID) to promote Dayak identity and dignity through legitimizing respect for Dayak culture and knowledge, with the long-term objective of supporting the Dayak struggle. When PK began its work in the early 1980s, Dayak were afraid to use their own languages in public and ashamed of their identity. In 1992, to counter this trend, ID hosted the Dayak Cultural Exposition, bringing together Dayak subgroups from all over Borneo – the largest Dayak gathering in the twentieth century. They discussed their common problems and agreed to revive the term Dayak for their common identity. Today, ID documents Dayak oral traditions, researches Dayak culture, facilitates the teaching of Dayak languages in schools, publishes books and a monthly Bahasa Indonesia language magazine (Kalimantan Review), and runs a radio program in Dayak languages. The radio program has recently provided a venue for talk shows discussing ideas for policy reforms after Reformasi.

12.9 Legal assistance

After research demonstrated the need for legal assistance, particularly related to land rights cases, two legal aid foundations – LBBT in West Kalimantan and LBBPJ in East Kalimantan – were created. Instead of handling cases, these focus on empowering communities with the knowledge to act themselves. LBBT provides legal training, facilitates communities to share their experiences from confronting logging and plantation companies, and spreads information about how communities have forced payment of compensation by timber companies that broke *adat* rules. They offer advice on how to approach companies, how to face the military, and how to negotiate mitigation after the community is surprised by the concessionaire's arrival. Over the past decade, although there have been hundreds of protests against concessions, only about 10 percent of the protests fully stopped the concessionaire, because most communities did not anticipate the concessionaire's arrival and had not prepared themselves with a resistance plan.

LBBT also works to revitalize *adat* law and courts and their cross-scale links to national-level institutions. This is a challenge in cases in which local leaders have twisted *adat* for their own advantage. LBBT's primary method is facilitation of dialogue on issues of interest to the community – usually leading to a review of *adat* and Indonesian laws so that people will not believe outsiders or local elite who attempt to deceive them.

LBBT collaborates with many NGOs across Indonesia to safeguard human rights and promote agrarian reform for land rights. LBBT also cooperates with NGOs concerned with forest policies to propose alternatives to current laws, to raise public awareness of bad laws, and to facilitate public dialogue with the governor and local 'parliament' (DPRD Tingkat I). A new political opening in 1999, under Local Governing Law No. 22, empowered LBBT to facilitate community involvement in the drafting of district-level policies on natural resources – a major pay-off for the years of community-organizing and awareness-raising work.

12.10 Facilitating exploration of new institutional options

LBBT is considering options to promote the creation of a new Dayak institution that would meet annually to review *adat* laws and the performance of the CU in each community in meetings that bring together 10 representatives from each of 14 different *adat* communities. This is a higher-level collaboration that would expand the scale of Dayak governance and build new cross-scale communication links between local communities and the new higher-level institution – ideally enhancing community members' ability to hold their own local leadership accountable while at the same time enabling the group of communities to address regional-scale concerns.

12.11 Mapping – a tool for social and ecological renewal

PK's community-based mapping unit (PPSDAK) documents Dayak land use based on indigenous knowledge to guarantee conservation and prosperity. During this process (Natalia, 2000), community members produce a large, three-dimensional map and a set of smaller maps including overlays showing land use, animal species, special tree species, rivers, sacred areas, settlement, and topography. Old people teach the youth their land's history, and pass down ecological knowledge and stories containing indigenous wisdom. PK recognizes and nurtures the important backloop function served by this collective memory by documenting it.

The mapping is helping to renew Dayak institutions to (a) maintain resilience in normal times when they alone govern access to the resources; and (b) manage new types of disturbance that happen unexpectedly (e.g., the sudden arrival of a logging concession or the slow increase in local extraction). Maps are also used to calculate productivity and compare the benefits of indigenous farming against the benefits promised by a development project. Working together, people often discover that their territory is smaller than they realized, and that they need to improve their management of it in order to maintain their forests at an adequate level. Some communities expand the size of their protected forests. Others seek assistance from Dayak who have experience with wet rice cultivation so they can increase permanent forest area by reducing the area used for swidden. In some communities, women have proposed new *adat* laws to protect environmental values.

The mapping often concludes with a *musyawarah adat* assembly, which ratifies the map by signatures of the heads of households and creates a consensus agreement to conserve forests and lands for the future. In some communities, PK is asked to compile all the *adat* laws so that everyone in the community can renew their awareness of them. The mapping process empowers the community to confront concessionaires and drive them away, but it also provides a means for adaptively changing local institutions. In one case, 25 communities sharing a watershed worked out a river basin agreement that governs upstream activities impacting water quality downstream.

Local communities often develop new rules as a result of mapping – Box 12.1 illustrates a typical example (Royo, 2000). In some cases, when *adat* rules are broken, PK is called in to bring the violators back into the consensus by

Box 12.1 *Adat* rules from an illustrative community

Some typical *adat* rules, both newly created and renewed, by consensus of community members after mapping their *adat* lands. Violation of rules results in *adat* sanctions and fines.

- The villages will be bound by all customary laws and traditions affecting the *adat* territory.
- All logging concessions, industrial tree plantations, palm oil plantations, mining, and transmigration inside the *adat* territory are to be refused.
- The use of chainsaws by all villagers and outsiders is banned, except when logs are needed by local villagers for housing construction and customary agreement has been reached with the village.
- The burning of forest is banned.
- The felling of *kampung buah* trees is banned whether the tree is owned in common or privately or inherited, except if it is near homes and will gravely endanger the lives of villagers, or for accepted *adat* purposes.
- The felling of *bukit natai*, *gahang-guhak*, and *utung arai* is prohibited.
- The hunting of the following animals is prohibited: *tingang*, *penagung*, *kakh*, *ruik*, *tiung*, *simiaulau*, *kelimpiau*, and *urang hutan*.
- Ironwood trees and other useful wild species will be planted.
- The sale of logs to outsiders is banned.
- *Adat* 'protected forest' cannot be entered or harvested by outsiders. It can only be harvested for family needs by insiders.
- Privately owned forests can only be worked by family or by persons with the family's permission to work there.
- The taking of firewood, rattan, sugar palm, bamboo, and sagu from trees along roads is banned.
- Trees of economic value (e.g., *durian*, *tengkawang*, *nangka*, other non-timber forest products) in Tembawang Forest are owned by individuals and only the owner can harvest.
- Tembawang Forest cannot be converted into fields or other land use.
- Rubber forests cannot be harvested by non-owners or converted into fields or burned.
- Honey trees and the areas where they are found are protected, and the areas cannot be burned or converted into fields.
- Fruit gardens that are traditional burial grounds cannot be sold or violated.
- Fishing with poison is banned.
- Washing or submerging rubber latex upstream of areas used for bathing is banned.
- Plants and plant products cannot be harvested from forest or agricultural fields without the owner's permission.

facilitating a process that enables neighbors to exert social pressure. But if people do not respect *adat* fines and sanctions, it is difficult to enforce their compliance. PK has found that such internal conflicts are harder to address than external conflicts and is currently evaluating approaches to create more sustainable conflict resolution mechanisms.

PK realized the power of maps (Wood, 1992; Alcorn, 2000) in the early 1990s and moved quickly to make mapping a critically important communication tool in its struggle. To date, it has facilitated the mapping of over 160 communities, covering more than 700 000 ha of Dayak territory – protecting over half a million hectares of forest and networking those communities in the process (Natalia, 2000). At the national level, PK has used the maps to take advantage of a political opening provided by Spatial Law No. 24, 1992, which orders provincial authorities to consult with communities when creating land-use plans for the province. West Kalimantan provincial officials have signed the community land-use maps as recognition of their legitimacy as official land-use plans. Also, a US Forest Service survey found that community lands mapped by PK and recognized by local government were protected from the massive fires of 1997–8 (Melnyk, 2001).

PK has also taken advantage of the Ministry of Agrarian Affairs and Land Administration Regulation No. 5, 1999, on *Adat* Land Dispute Handling, which recognizes indigenous rights if the community has evidence of an *adat* territory (a map), *adat* law, and *adat* institutions. As communities realize the advantages of mapping their lands, the demand for assistance has escalated. PK uses apprenticeship methods to train and network people from distant places so they can train others and spread the concept across West Kalimantan and beyond. PK plans to further decentralize the mapping movement by encouraging mapping centers in every district so local communities can handle their own mapping needs.

12.12 Scaling up impacts – from Borneo to Indonesia

To scale up the improvement of local management and to gain national recognition of Dayak forest management rights, PK helped to create the Community-Based Forest System Management Network (SHK). The crisis of the great fires of 1997 opened up new opportunities for PK to negotiate with GOI. PK and SHK joined with other NGOs to use their Geographic Information Systems (GIS) capacity and maps, to work with GOI when technocrats needed accurate information to solve the crisis.

PK associates follow democratic principles among themselves, and the diversity of their team has helped them to adaptively respond to the government's 'control response' so that the political process of the social movement engagement advances. An indigenous social movement is not made up of homogeneous voices; democratic principles adopted by the lead organizations ensure movement solidarity through a dialogue of different community voices. In other words, the key to the movement's success is that PK associations are not attempting to be the decision-making agents driving communities' actions, as is too often the case with NGOs (Pearce, 1993).

For a social movement to be effective, it must become national. This is why PK, though focused in West Kalimantan, has nurtured the NGO movement across Borneo and Indonesia. Institutions founded on various levels with PK involvement include the Dayak Dynamic Network (JDD), Kalimantan Human Rights Network (JAHAMKA), Borneo Indigenous Peoples Networking Program (BIPNP) – which links NGOs in Indonesian and Malaysian Borneo – the Network for the Advocacy of Indigenous Peoples' Rights (JAPHAMA), the Consortium for Supporting Community-based Forest System Management (KPSHK), and the Participatory Mapping Network (JKPP). In 1998, PK initiated a mentoring program to provide support to emerging small, indigenous associations in areas of high biodiversity in Sumatra and Sulawesi.

Dayak have used communication links with national NGOs to build their political strength and draw public attention to their problems shared with other indigenous people across Indonesia (Royo, 2000). PK-associated communities and local government officials created the 'Blue Sky Forum' to raise public awareness about how people destroyed the ecological balance and in turn caused the great forest fires of 1997. From the international level of the panarchy, foreign individuals with technical expertise have brought new technical knowledge, such as GPS and GIS for mapping. Dayak use international NGOs to raise Dayak voices outside Indonesia to protest against foreign investors' actions. The Asian Development Bank (ADB) NGO Working Group, for example, has helped Dayak negotiate with GOI about ADB projects.

The 1997–8 Reformasi social crisis in Indonesia provided a significant opportunity for the Dayak social movement. PK facilitated the emergence of indigenous political organizations, including the West Kalimantan Indigenous Peoples Alliance (AMA Kalbar). AMA Kalbar's members are *adat* chiefs who represent Dayak from all over West Kalimantan. PK and AMA Kalbar worked with other indigenous peoples from all over Indonesia to establish AMAN (the Alliance of Indigenous Peoples of the Archipelago). In April 1999, AMAN met in congress. This first-ever gathering of representatives of indigenous peoples from all of Indonesia included 300 *adat* leaders from 100 indigenous peoples in 22 provinces, who spoke about the problems they face in protecting their land and environment. As a result of this national meeting, Agrarian Ministerial Decree No. 5, 1999, was issued. It outlines ways that an *adat* community can register its land claims with government, legitimizing local government's previously *ad hoc* decisions in favor of *adat* communities. This new policy offers the hope of tenurial rights to the *adat* communities that manage up to 70 percent of the country's lands. PK has joined with other NGOs to work with provincial governments across Indonesia to establish a process for implementing this decree. Regional AMAN subgroups are now active at provincial levels across Indonesia. The tactical advantages and disadvantages of creating a political party for all the indigenous peoples of Indonesia are being debated. One of PK's founders was selected as a parliamentary representative in 2000, sending one Dayak voice to the national level and providing the movement with new insights into national policy making.

As their successes have been more widely appreciated, more donors are interested in supporting their work, but Dayak associations have been wary of becoming dependent on donors. While accepting some support as seed money for new initiatives, they have built in ways for the movement to be supported by Dayak themselves through CUs, better prices for goods, better skills, etc. This has also served to ensure accountability to local people. The Dayak actively avoided the common pitfall of establishing an organization to maintain a social movement, a strategy that often fails because of oligarchization, co-optation by donors, or loss of indigenous support (McAdam, 1982; Fisher, 1997).

In short, across Indonesia, local forces have been joining together to deliver a jolt to powers at the core of national society – with the Reformasi movement's successful overthrow of Suharto in 1998 and subsequent post-Reformasi policy reforms as visible results. This can be expected when political participation through other routes has been denied. Now that the national polity has been broadened to include more voices, it may be able to withstand smaller surprises and stabilize the links that encourage local responsiveness to feedback, which in turn will stimulate appropriate responses at higher levels. However, it is too early to predict whether this opening will be sustained, or whether the resulting turbulence and unrest will be met with a military crackdown or national disintegration over the next 5 years.

The Dayak social movement has not yet fully realized its goals, but it is successfully interacting with the 'control agent' in a continuing ballet toward them. The Dayak movement continues to build a broad consensus to counter internal changes and external threats peacefully and thoughtfully. While setbacks can be expected to occur, the current trend is positive. More forests are being protected by communities despite the challenges accompanying decentralization. More people and villages are defending their territory from various projects and holding their leaders accountable for decisions that run counter to community interests. There are more and more demands from communities to withdraw laws that do not respect indigenous rights – and some bad laws have been removed after years of advocacy. Finally, NGO facilitation is welcomed by various parties, including government agencies and members of parliament, which gives us hope for future prospects.

12.13 Lessons for social renewal movements working to revive ecological resilience

Our case study offers some valuable lessons for those who are interested in supporting movements to recouple society to ecological feedback. Globally, local, national, and international institutions are largely insensitive to the negative ecological feedback evident at local levels. In the face of urgent need, there are few studies of the tactics and strategies used by effective civil associations toward the goal of recoupling. Many NGOs and other civil associations function instead as self-serving interest groups that are ineffective in promoting change (Pearce, 1993; Sinaga, 1995; Fisher, 1997; Edwards, 1999), and few achieve what Dayak associations have achieved.

The Dayak case bears out McAdam's social movement principles, derived from the study of the American Civil Rights Movement. According to McAdam (1982), four sets of factors are crucial to the emergence and development of a social movement: (1) strength of organization; (2) collective assessment of prospects for success; (3) political opportunities at any given time; and (4) responses by groups in power.

First, indigenous organizations are essential because they provide the members, the numbers that give a movement its strength, as well as solidary incentives, a communication network, and leaders. PK has created organizations that meet Dayak needs, thus providing solidary incentives. Because people belong to PK organizations that meet their other needs, it is not necessary to create incentives to be part of the social movement. If members do not support the movement, they could lose the benefits that come through the PK-associated organizations. Indigenous social networks are critical resources for a social movement, because networks serve for communication, as a source of leaders, and as a group from which strategy and action plans emerge. Established organizations form a communication infrastructure, 'the strength and breadth of which largely determine the pattern, speed and extent of movement expansion' (McAdam, 1982: 46). Communication networks beyond local borders enable the movement to become more than local spontaneous protests. PK has consciously interwoven a fabric of multiple local organizations that meet different needs, and thereby fashioned a means for consensus building across diverse communities. This is key to the strength of the movement.

McAdam's second factor refers to the process by which groups create meaning as they react to events and attribute their problems to a system rather than to isolated individual actions (McAdam, 1982). PK's educational processes promote cognitive liberation and recognition of shared injustice. As people think together about their problems, they discover and construct a shared injustice frame (Klandermans, 1992) – and then they develop a discourse for change. By having a new, shared frame for discourse, people are more likely to 'chang[e] the prevailing economic, political, moral, cultural, and social dispositions of society which support environmental degradation' (Wapner, 1995: 311).

Many analysts have observed that the creation of collective identity is one of most central tasks facing a movement (Friedman and McAdam, 1992; Mueller, 1992). Indigenous groups have an advantage, not because they are inherently in harmony with nature, but because they already have a culturally based discourse that supports links to ecological feedback and a collective cultural identity on which to build a movement. The public dialogue and consensus processes supported by the PK associations provide mechanisms for collective identities to be transformed from seeing themselves as weak to seeing themselves as resourceful shapers of their own destinies.

Finally, Dayak's close identification with their geography is a strong asset to the movement. Ecological resilience support movements gain from the symbolic value of a terrain of resistance which is both an actual site of contestation and 'an interwoven web of specific symbolic meanings, communicative processes, political discourse, religious idioms, cultural practices, social networks, economic relations, physical settings, envisioned desires and hopes' (Routledge, 1996: 516). The terrain of resistance is at once metaphoric and literal.

Despite the need for a global ecological resilience support movement to change the pressures that flow down from the global level of the social panarchy, despite the potential for indigenous peoples' movements to form the basis for a global movement, and despite the growth of small, regional, indigenous networks and nascent, global networks, the world's indigenous peoples cannot focus their energies on nurturing a global social movement from the grassroots. They are struggling to survive. However, increased international support for indigenous peoples' survival and social movements in many countries – based on the principles used by PK to mobilize many local communities' support for one movement within one nation-state (Alcorn *et al.*, 2000) – could nurture allies for a global social movement to recouple Earth's societies to ecological feedback across scales.

Acknowledgements

We thank all those who have provided comments and inspiration for this chapter, especially Fikret Berkes, Carl Folke, Johan Colding, and Judee Mayer. We have produced this text from our various perspectives as Dayak social movement actors and supporters – perspectives that have often been lacking in the discourse on social movements, as noted by Routledge (1996). John Bamba is the Executive Director of Institut Dayakologi; Stefanus Masiun is Director of LBBT; and Ita Natalia was, until recently, Director of Pancur Kasih's PPSDAK mapping unit. Antoinette Royo and Janis Alcorn supported the Dayak social movement through their work at the World Wildlife Fund's Biodiversity Support Program. We thank Aziz Gokdemir for editing our manuscript to consolidate our points under a single writing style. The research for this chapter was partially supported by the PeFoR and KEMALA projects of the Biodiversity Support Program – a USAID-funded consortium of WWF-US, World Resources Institute, and The Nature Conservancy. The opinions expressed herein are those of the authors and do not necessarily reflect the opinions of the United States Agency for International Development, WWF, WRI or TNC.

Notes

1. ABRI was split into TNI and National Police in 1999, as a reform after Suharto was forced from power.
2. These expanding gaps are creating new agricultural problems. In 1999, Dayak farmers faced their first locust plague, evidence that the plantation clearing has crossed a threshold where forest patches are no longer large enough to protect agricultural fields from pest outbreaks.
3. Unlike Mexico (Alcorn and Toledo, 1998), Indonesia offers no tenurial shell to protect indigenous systems. Readers are encouraged to compare the Indonesian and Mexican cases on their own, as this is beyond the scope of our chapter.
4. This belief, common among other swiddeners (cf., Alcorn and Toledo, 1998), supports the integrity of values, rules, and practices sustaining ecologically sound swidden.
5. Despite this propaganda, investigations revealed that the massive fires of 1997–8 were primarily caused by concessionaires clearing land for oil palm plantations.

References

Alcorn, J.B. 1990. Indigenous agroforestry systems in the Latin American tropics. In *Agroecology and Small Farm Development*, pp. 203–20, ed. M.A. Altieri and S.B. Hecht. Boca Raton, FL: CRC Press.

Alcorn, J.B. 2000. *Borders, Rules and Governance: Mapping to Catalyze Policy and Management Change*. Gatekeeper Series No. 91. London: IIED.

Alcorn, J.B., Bamba, J., Masiun, S., Natalia, I., and Royo, A.G. 2000. Lessons about tactics and strategies: recommendations for supporting social movements to recouple society to ecological feedback. In *Indigenous Social Movements and*

Ecological Resilience: Lessons from the Dayak of Indonesia, pp. 87–96, ed. J.B. Alcorn and A.G. Royo. Washington DC: Biodiversity Support Program.

Alcorn, J.B. and Toledo, V.M. 1998. Resilient resource management in Mexico's forest ecosystems: the contribution of property rights. In *Linking Social and Ecological Systems. Management Practices and Social Mechanisms for Building Resilience*, pp. 216–49, ed. F. Berkes and C. Folke. Cambridge: Cambridge University Press.

Anonymous 1998. Plantation projects and logging concessions: impacts and resistance. *Kalimantan Review* English Language No. 1: 22–8.

Appell, G.N. 1992. The history of research on traditional land tenure and tree ownership in Borneo. Working Paper No. 6. Phillips, ME: Social Transformation and Adaptation Research Institute.

Arman, S. 1996. Diversity and trade of market fruits in West Kalimantan. In *Borneo in Transition*, pp. 256–65, ed. C. Padoch and N.L. Peluso. New York: Oxford University Press.

Bamba, J. 2000. Land, rivers, and forests: Dayak solidarity and ecological resilience. In *Indigenous Social Movements and Ecological Resilience: Lessons from the Dayak of Indonesia*, pp. 35–59, ed. J.B. Alcorn and A.G. Royo. Washington DC: Biodiversity Support Program.

Barber, C.V. 1997. *Case Study of Indonesia*. Project on Environmental Scarcities, State Capacity, and Civil Violence. Cambridge, MA: American Academy of Arts and Sciences and the University of Toronto.

Berkes, F. 1999. *Sacred Ecology: Traditional Ecological Knowledge and Resource Management*. Philadelphia: Taylor & Francis.

Berkes, F. and Folke C., eds. 1998. *Linking Social and Ecological Systems. Management Practices and Social Mechanisms for Building Resilience*. Cambridge: Cambridge University Press.

Brookfield, H., Potter, L., and Byron, Y. 1995. *In Place of the Forest: Environmental and Socio-economic Transformation in Borneo and the Eastern Malay Peninsula*. Tokyo: United Nations University Press.

Caniago, I. 1999. The diversity of medicinal plants in secondary forest post-upland farming in West Kalimantan. In *Management of Secondary and Logged-Over Forests in Indonesia*, pp. 13–20, ed. P. Sist, C. Sabogal, and Y. Byron. Bogor, Indonesia: Center for International Forestry Research (CIFOR).

Cleary, M. and Eaton, P. 1992. *Borneo: Change and Development*. Oxford: Oxford University Press.

Cleary, M. and Eaton, P. 1996. *Tradition and Reform: Land Tenure and Rural Development in South-East Asia*. Oxford: Oxford University Press.

Colfer, C.J.P. 1993. *Shifting Cultivators of Indonesia: Marauders or Managers of the Forest? Rice Production and Forest Use Among the Uma' Jalan of East Kalimantan*. Rome: FAO.

Colfer, C.J.P. 1997. *Beyond Slash and Burn: Building on Indigenous Management of Borneo's Tropical Rain Forests*. Bronx: New York Botanical Garden.

Dasmann, R.F. 1991. The importance of cultural and biological diversity. In *Biodiversity: Culture, Conservation, and Ecodevelopment*, pp. 7–15, ed. M.L. Oldfield and J.B. Alcorn. Boulder, CO: Westview Press.

Djuweng, S. 1998. The Dayak: children of the soil. *Kalimantan Review*, English Language Version No. 1: 6–8.

Dove, M.R. 1985. *Swidden Agriculture in Indonesia*. New York: Mouton Publishers.

Dove, M.R. 1996a. Process versus product in Bornean augery: a traditional knowledge system's solution to the problem of knowing. In *Redefining Nature: Ecology, Culture, and Domestication*, pp. 557–96, ed. R. Ellen and K. Fukui. Oxford: Berg.

Dove, M.R. 1996b. Rice-eating rubber and people-eating governments – peasant versus state critiques of rubber development in Colonial Borneo. *Ethnohistory* 43(1): 33–63.

Dove, M.R. and Kammen, D.M. 1997. The epistemology of sustainable resource use – managing forest products, swiddens, and high-yielding variety crops. *Human Organization* 56(1): 91–101.

Edwards, M. 1999. NGO performance – what breeds success? New evidence from South Asia. *World Development* 27: 361–74.

Fisher, W.F. 1997. Doing good? The politics and antipolitics of NGO practices. *Annual Review of Anthropology* 26: 439–64.

Folke, C. and Berkes, F. 1998. *Understanding Dynamics of Ecosystem–Institution Linkages for Building Resilience.* Beijer Discussion Paper Series No. 122. Stockholm: Beijer International Institute of Ecological Economics.

Folke, C., Pritchard, L. Jr, Berkes, F., Colding, J. and Svedin, U. 1997. *The Problem of Fit Between Ecosystems and Institutions.* Beijer Discussion Paper Series No. 108. Stockholm: Beijer International Institute of Ecological Economics.

Friedman, D. and McAdam, D. 1992. Collective identity and activism: networks, choices and the life of a social movement. In *Frontiers in Social Movement Theories*, pp. 156–73, ed. A.D. Morris and C.M. Mueller. New Haven: Yale University Press.

Gamson, W.A. 1975. *The Strategy of Social Protest.* Homewood, IL: The Dorsey Press.

Gunther, F. and Folke, C. 1993. Characteristics of nested living systems. *Journal of Biological Systems* 1: 257–74.

Holling, C.S. 1992. Cross-scale morphology, geometry and dynamics of ecosystems. *Ecological Monographs* 62: 447–502.

Holling, C.S. and Sanderson, S. 1996. Dynamics of (dis)harmony in ecological and social systems. In *Rights to Nature*, pp. 57–86, ed. S.S. Hanna, C. Folke, and K-G. Mäler. Washington DC: Island Press.

International Labour Organization (ILO) 1996. *Indigenous and Tribal Peoples: a Guide to ILO Convention No. 169.* Geneva: ILO.

King, V.T. 1993. *The Peoples of Borneo.* Cambridge: Blackwell Publishers.

Klandermans, B. 1992. The social construction of protest and multiorganizational fields. In *Frontiers in Social Movement Theories*, pp. 77–103, ed. A.D. Morris and C.M. Mueller. New Haven: Yale University Press.

KUDETA (Coalition for the Democratization of Natural Resources) 1998. Return Natural Resources to the People! Position statement released to the media 11 June 1998, Jakarta.

Mayer, J. 1996. Environmental organizing in Indonesia: the search for a newer order. In *Global Civil Society and Global Environmental Governance: the Politics of Nature from Place to Planet*, pp. 169–216, ed. R. Lipschutz and J. Mayer. Albany: State University of New York Press.

Masiun, S. 2000. National frameworks affecting Adat governance in Indonesia, and Dayak NGO responses. In *Indigenous Social Movements and Ecological Resilience: Lessons from the Dayak of Indonesia*, pp. 17–33, ed. J.B. Alcorn and A.G. Royo. Washington DC: Biodiversity Support Program.

McAdam, D. 1982. *Political Process and the Development of Black Insurgency 1930–1970.* Chicago: University of Chicago Press.

Melnyk, M. 2001. Resource rights: A condition for effective community-based fire management. In *International Workshop on Community Based Fire Management*, p. 25, ed. D. Ganz, P. Moore and B. Shields. RECOFTC Training and Workshop Report Series 2001/1. Bangkok: RECOFTC.

Messerschmidt, D.A., ed. 1993. *Common Forest Resource Management: Annotated Bibliography of Asia, Africa, and Latin America*. Rome: Food and Agricultural Organization of the United Nations.

Moniaga, S. 1993. Toward community-based forestry and recognition of *adat* property rights in the Outer Islands of Indonesia. In *Legal Frameworks for Forest Management in Asia*, pp. 131–50, ed. J. Fox. Honolulu: East-West Center.

Mueller, C.M. 1992. Building social movement theory. In *Frontiers in Social Movement Theories*, pp. 3–25, ed. A.D. Morris and C.M. Mueller. New Haven: Yale University Press.

Natalia, I. 2000. Protecting and regaining Dayak lands through community. In *Indigenous Social Movements and Ecological Resilience: Lessons from the Dayak of Indonesia*, pp. 61–71, ed. J.B. Alcorn and A.G. Royo. Washington DC: Biodiversity Support Program.

Ngo, T.H.G.M. 1996. A new perspective on property rights: examples from the Kayan of Kalimantan. In *Borneo in Transition*, pp. 137–49, ed. C. Padoch and N.L. Peluso. New York: Oxford University Press.

Padoch, C. and Peluso, N. 1996. Borneo people and forests in transition: an introduction. In *Borneo in Transition*, pp. 1–9, ed. C. Padoch and N.L. Peluso. New York: Oxford University Press.

Padoch, C. and Peters, D. 1993. Managed forest gardens in West Kalimantan, Indonesia. In *Perspectives on Biodiversity: Case Studies of Genetic Resource Conservation and Development*, pp. 167–76, ed. C.S. Potter, J.I. Cohen and D. Janczewski. Washington DC: American Association for the Advancement of Science.

Pearce, J. 1993. NGOs and social change: agents or facilitators? *Development in Practice* 3: 222–7.

Peluso, N. 1996. Fruit trees and family trees in an anthropogenic forest – ethics of access, property zones, and environmental change in Indonesia. *Comparative Studies in Society and History* 38(3): 510–48.

Peluso, N.L. and Padoch, C. 1996. Changing resource rights in managed forests of West Kalimantan. In *Borneo in Transition*, pp. 121–36, ed. C. Padoch and N.L. Peluso. New York: Oxford University Press.

Penot, E. 1999. Prospects for conservation of biodiversity within productive rubber forests in Indonesia. In *Management of Secondary and Logged-Over Forests in Indonesia*, pp. 21–32, ed. P. Sist, C. Sabogal, and Y. Byron. Bogor, Indonesia: Center for International Forestry Research (CIFOR).

Peters, C. 1996. Illipe nuts (*Shorea* spp.) in West Kalimantan: use, ecology, and management potential of an important forest resource. In *Borneo in Transition*, pp. 230–44, ed. C. Padoch and N.L. Peluso. New York: Oxford University Press.

Potter, L. 1993. The onslaught on the forests in South-East Asia. In *South East Asia's Environmental Future*, pp. 103–28, ed. H. Brookefield and L. Potter. New York: United Nations University Press.

Routledge, P. 1996. Critical geopolitics and terrains of resistance. *Political Geography* 15: 509–31.

Royo, A.G. 2000. The power of networking: building force to navigate cross-scale turbulence where solo efforts fail. In *Indigenous Social Movements and Ecological Resilience: Lessons from the Dayak of Indonesia*, pp. 73–85, ed. J.B. Alcorn and A.G. Royo. Washington DC: Biodiversity Support Program.

Scott, J.C. 1990. *Domination and the Arts of Resistance: Hidden Transcripts*. New Haven: Yale University Press.

Sinaga, K. 1995. *NGOs in Indonesia*. Saarbrucken: Verlag Entwicklungspolit.

Wapner, P. 1995. In defense of banner hangers: the dark green politics of Greenpeace. In *Ecological Resistance Movements*, pp. 300–14, ed. B.R. Taylor. Albany: State University of New York Press.
Warner, K. 1991. *Shifting Cultivators: Local Technical Knowledge and Natural Resource Management in the Humid Tropics*. Rome: FAO.
Wood, D. 1992. *The Power of Maps*. New York: Guilford Press.

13

Policy transformations in the US forest sector, 1970–2000: implications for sustainable use and resilience

RONALD L. TROSPER

13.1 Introduction

State bureaucracies managing forests are known to develop problems because of command-and-control mentalities and political links to legislatures and industry (Baskerville, 1995; Holling and Meffe, 1996). Command-and-control thinking leads to simple models of ecosystem behavior, such as the sustained yield of a single ecosystem product like wood fiber. Such thinking, by suppressing complications, can lead to unpleasant surprises. Economic dependence on a short list of products, linked via politicians in powerful legislative positions who are supported by economic interests, can reinforce the thinking in a bureaucracy managing a natural resource. When the side-effects of the dominant approach create a crisis, either natural or political, will the crisis lead to changes in the social and political configurations that caused the problem? Gunderson, Holling, and Light (1995) emphasize the importance of crisis as a *necessary* condition for change. A crisis is not *sufficient*, however. This chapter explores conditions preventing and allowing change by examining the forest sector in the USA from 1970 to 2000.

Between 1970 and 2000, two similar scenarios occurred in the management of national forests in the USA. Each time, action began when a local problem of ecosystem imbalance induced litigation that led a court to stop Forest Service timber harvesting. After the first halt, Congress passed the National Forest Management Act, but the Forest Service continued high levels of wood fiber production. The impact of these cuts on an endangered species led to the second halt. In the second scenario, the President intervened and assisted the Forest Service to create the Northwest Forest Plan. Subsequently, the Forest Service revised its planning regulations (US Department of Agriculture, 2000). Harvests have not returned to the previous high levels.

Can the contrast between the two scenarios help in understanding the socio-cultural conditions under which a linked social and ecological system can move toward greater resilience? Stated differently: when will response to a crisis lead to a configuration of agents and a system of ideas that promote activities which provide resilience? Theorists studying the policy process have suggested that changes occurring outside of a policy sector are necessary but not sufficient for major policy change (Sabatier and Jenkins-Smith, 1999: 147–8); what additional is required? Does the magnitude of the external perturbation matter, or are other factors internal to the sector also important? Although a case study cannot fully answer these questions, the contrast of the two scenarios offers an opportunity to search for possible factors.

In the first case, no significant change occurred because the dominant elites in the forest sector were able to continue to emphasize 'yield,' not 'sustained' in the internally contradictory idea of 'sustained yield.' High production of timber inevitably raises questions about sustaining that yield and about the condition of the forest in general. If the yield of one resource is driven too high, the yields of other uses fall and the future yield of the dominant resource is also threatened. The 1897 Organic Act of the Forest Service, the Sustained Yield Forest Management Act of 1944, and the 1960 Multiple Use and Sustained Yield Act all provide the contradiction. In 1976, the National Forest Management Act continued the contradiction by both protecting forested lands and giving rules for planning timber harvest.

In the presence of a contradiction such as sustained yield, powerful elites must continue to provide lip service to the subordinate idea. While the harvest continued, many interests outside of the forest sector developed ideas that emphasized the idea of 'sustain,' which became 'ecological sustainability.' A successful challenge to the Forest Service's 'yield' policy occurred under the Endangered Species Act, allowing the formerly suppressed idea to rise to the top and displace the previously dominant idea. 'Sustained' became a concept independent of the idea of yield. A national polarization resulted, with outright opposition between commodity interests and environmentalists. Both desired completely to dominate policy.

In the midst of the political conflict, a Committee of Scientists drawn from the forest sector provided a re-interpretation of the provisions of the National Forest Management Act and other environmental legislation. The committee proposed that complementarities exist among the interests of all groups, particularly when forest condition is seriously degraded. It proposed redefining the planning units at regional and local levels in ecosystem terms. Its proposal also emphasized communication among parties, which would change the Forest Service from an arbiter among competing interests to a facilitator of local collaborative efforts. Scientists would review plans and monitoring strategies. Whether these changes create a system of continued conflict or the realization of complementarities in forest management remains undetermined.

If the changes create a realization of complementarities, they may also improve the resilience of the linked social and ecological system by improving feedback from ecosystem condition to the human cultural system and by changing the characteristics of forest biomass through collaborative planning. Different configurations of ideas create different conditions for the treatment of information about ecosystem condition. When a constraining contradiction such as the idea of sustained yield dominates, information that would reveal the contradiction and upset policy tends to be denied. Concepts that minimize the contradiction are supported; these concepts place barriers in the processing of objective information. In the case of forest management, the timber-dominated paradigm of sustained yield limits information to data on timber types, volume, age structure, and so forth.

When contradictory conceptions are in conflict, and are not necessarily related as in the case of sustained yield, each party promotes the data that support its ideas. The parties which triumph will, after their victory, have an easier time of suppressing the contradictory data and ideas, because there is no necessary relationship to their own ideas. During the time of contest among the ideas, new data will be developed; once the contest is settled, problems may arise, because the adherents of the dominant ideas will support only data collection favorable to that approach.

If the Committee of Scientists' emphasis on complementarities is accepted, useful monitoring should increase as a consequence of a new configuration of ideas about forests. If the contending parties agree that ecosystem condition matters for all interests, the result of accurate feedback is mutually beneficial. This chapter tells the story of these shifts in the structure of political conflict in the USA and proposes some possible lessons about navigating nature's dynamics in an industrial nation's forest sector. If correct, this chapter's analysis suggests that even in an industrial nation such as the USA, responses to damaged ecosystems can lead to increased social–ecological resilience.

13.2 The contradiction of sustained yield

The internally contradictory nature of the 'sustained yield' concept drives the analysis of this chapter. To explain the point, Figure 13.1 shows a usual depiction of the idea of sustained yield. The vertical axis is the annual consumption of a renewable resource such as timber or fish. The horizontal axis is the stock of the resource. In the case of timber, often the horizontal axis is the rotation age of trees of a particular species. Because older trees are larger, higher

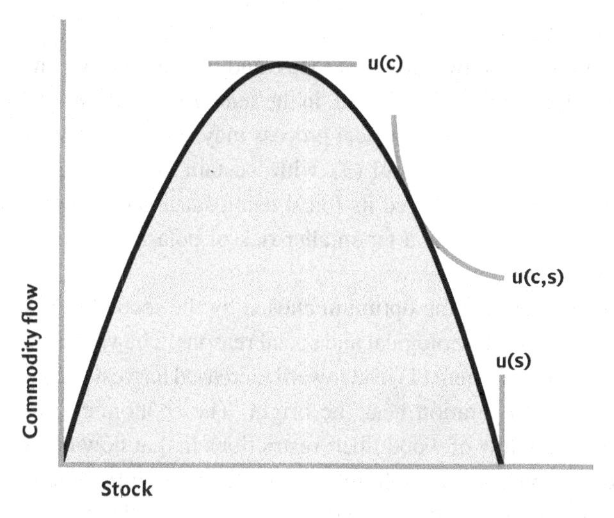

Figure 13.1 Three optima: commodity, commodity and stock, and stock as valued goals.

ages also mean higher biomass. The yield graph expresses annual growth as a function of the stock of the resource. If consumption occurs on the yield line, stock neither increases nor decreases. If consumption is above the yield line, the stock decreases. If consumption is below the yield line, the stock increases. The highest point on the yield line is usually described as 'maximum sustained yield.'

The contradiction can be described by comparing three utility functions. Let c_t represent consumption of the resource in year t; let s_t represent the stock in year t. Three alternative ways to understand society's valuation of the annual consumption and the stock are as follows:

	Utility expression	Short title	Example
(1)	$u(c_t)$	'Commodity viewpoint'	Timber plantation
(2)	$u(c_t, s_t)$	'Balanced viewpoint'	Menominee Tribe
(3)	$u(s_t)$	'Preservation viewpoint'	Wilderness reserve

The first expression says society values only consumption; this is the commodity viewpoint. The last expression says society values only the stock; this may be called the preservation viewpoint. The middle expression says society values both annual consumption and the stock of the resource. Valuing the stock in this situation is a shorthand way of stating that society receives noncommodity values from the standing forest biomass – such as aesthetic value, preservation of biodiversity, or ecosystem services (Heal, 1998: 14–20). If people in general understand the term 'sustained yield' in the sense of (2) but a profession such as forestry uses it in the sense of (1), then the same term is being used to describe two different ideas, and it is internally contradictory.

The peril of using 'sustained yield' in the sense of (1) – with 'yield' dominating 'sustained' – is that the political process may lead to an inversion of terms to the expression in the sense of (3), with 'sustain' dominating. The Menominee Tribe, which has managed its forest using sustained yield in the sense of (2) for over a century, runs a far smaller risk of polarization between (1) and (3) occurring.

One can characterize the optimum chosen by the second utility function as more resilient for both ecological and social reasons. The values that lead to the optimum for utility system (1) lead toward increased harvest, simplified species composition, and operation near the origin. The economic system becomes dependent on the flow of wood fiber; restrictions in that flow or a change in its value lead to economic crisis. Some would argue that forest monocultures are also vulnerable ecologically. Operation of a forest at full biomass, as specified by utility function (3), creates an ecosystem at risk of 'natural' catastrophe – insect or fire events that create a 'release' in the adaptive cycle.

In the Menominee approach (Pecore, 1992; Huff and Pecore, 1995), forest biomass and tree ages are held below their maximum but well above the stock levels consistent with maximum sustained yield.[1] Forest managers determine what the mill will harvest. No one species is favored, resulting in a diverse species mix. The Menominee forest has survived numerous crises, from windstorms to loss of chestnuts and elms. The federal government terminated reservation status, which was reversed in less than two decades, partly because the newly created corporation began to manage the forest exclusively for wood fiber. In recent decades, the tribe has moved away from uneven-age harvest systems toward more even-age systems in order to avoid reducing the pine component in their inventory. The Menominee have managed their forest on a sustained basis for more than 150 years, which may be the North American record for a publicly owned forest (Floyd, Vonhof, and Seyfang, 2001).

Federal legislation in the USA which established the Forest Service in 1897, and the subsequent acts in 1944 and 1960, used the term 'sustained yield' in the sense of (2); but the Forest Service interpreted it in the sense of (1) after 1950. With the National Forest Management Act of 1976, Congress attempted to define the idea in the sense of (2), but the Forest Service was able to continue to use the idea in the sense of (1) into the 1990s. In the 1990s, application of the Endangered Species Act seemed to imply that the proper idea was (3), the preservation of old growth in order to preserve species such as the Northern Spotted Owl (*Strix occidentalis caurina*). A fight ensued between the advocates of $u(c_t)$ and the advocates of $u(s_t)$. The Committee of Scientists proposed that management of the national forests should follow the middle idea.

13.3 Cultural change

Figure 13.1 is an economist's depiction of three ideas which have been in contention in the forest sector of the USA throughout the twentieth century. The picture is extremely simplified, typical of the type of analysis that labels economics the 'dismal science.' To add to this criticism, economists often use a diagram such as Figure 13.1 to illustrate alternative 'steady states;' society, through an unexplained process, selects one of the three alternatives, based upon society's preferences as illustrated by one of three utility functions.[2] How is the utility function selected? How might it change over time?

This case study provides an opportunity to study such change because different outcomes resulted from two plays of the same type of opening event in similar scenarios in the management of national forests in the USA. Using a metaphor of a stage play, the act begins with a stage, lit by dominant ideas. The actors interact during the particular act, limited by the props on stage and the lighting provided. During the act, they can change props and affect the lighting, as well as state their lines. At the end of the act they have provided a show as well as rearranged the props and the lighting for the subsequent act. Of particular interest is the set of ideas that are lit at the start of each act, and the supply of alternative lights stored just off stage. This approach, conceptually separating the stage from the actors but allowing the actors to rearrange the stage, draws upon the 'realist social theory' explained by Margaret Archer in two recent books (Archer, 1995, 1996). Although agents act and cause results, they are limited both by the characteristics of emergent cultural properties (the lights) and by emergent social and ecological properties (the props).

13.4 The US forest sector, 1950–70

The situation that led up to the decade of the 1970s is a prelude to the main two acts. The idea of 'sustained yield' of 'multiple uses' of the forest operated as a constraining contradiction during the 1950s, 1960s, and 1970s. The main organizations in the forest sector all agreed with the dominant interpretation of the idea of sustained yield, as defined in the Sustained Yield Forest Management Act of 1944, and the 1960 Multiple Use and Sustained Yield Act. The contradictions between a commodity orientation and intrinsic valuation of the forest are clear in two key texts from the Multiple Use Sustained Yield Act of 1960:

(a) 'Multiple use' means: the management of all the various renewable surface resources of the national forests so that they are utilized in the combination that will best meet the needs of the American people; making the most judicious use of the land for some or all of these resources or related services over areas large enough to provide sufficient latitude for periodic adjustments in use to conform to changing needs and conditions; that some land will be used for less than all of the resources; and harmonious and coordinated management of the various resources, each with the other, without impairment of the productivity of the land, with consideration being given to the relative values of the various resources, and not necessarily the combination of uses that will give the greatest dollar return or the greatest unit output.

(b) 'Sustained yield of the several products and services' means the achievement and maintenance in perpetuity of a high-level annual or regular periodic output of the various renewable resources of the national forests without impairment of the productivity of the land.

(16USC § 531)

Despite the balanced view in the above language, the dominant interpretation became the production of timber for the expanding American economy, with its emphasis on expanding homeownership among all who could afford homes (Hirt, 1994). Given the contradiction, the adherents of the post-Second World War dominant interpretation had to keep the related subordinate idea under control. If they failed to do this, the subordinate idea – preservation of the productivity of the land – would threaten change. In the 1940s, Aldo Leopold had advocated preserving the integrity of ecosystems (Leopold, 1988). In his view, a forest is a system of interrelated parts, and excessive utilization of one part would threaten the whole.

Containing the idea of ecosystems required an alternative forest definition. In the Forest Service and in forestry schools, the alternative defined any forest as a collection of stands. The primary decision in forest management is selection of the schedule of harvests among the stands. Cross-stand events are excluded. With the exact location of the stands unknown, impacts on other forest uses – streams, views, wildlife – cannot be identified. This conception of the forest as a collection of stands means that *any* collection of stands can be a forest. For instance, in its third edition, the widely used textbook by Davis and Johnson identified a forest with a management unit and defined a management unit as follows:

Management Unit. A geographically contiguous parcel of land containing one or more stand types and usually defined by watershed, ownership, or administrative boundaries for purposes of locating and implementing prescriptions. A management unit is usually larger than a stand and typically contains many stand types and individual stands.

(Davis and Johnson, 1987: 29)

This is from a revised textbook originally published in 1954. All of Part 4 of the book defines a forest as a collection of stands (Davis and Johnson, 1987: 475–715). Although watershed is mentioned as one way to define a management unit, neither it nor any other ecological concept is *necessary*; administrative boundaries can suffice. The light available for illuminating forests was limited to this concept of forests, which successfully subordinates ecological ideas.

On the structural side – the props for the stage – there was a contradiction as well. As mills were built for the liquidation of old growth, mill capacity usually exceeded the long-term prospect for harvest of timber from the national forests. The investment in capital and labor skills and other components of the private economies next to forests were not configured with long-term sustainability. Rather, the local economies were oriented to the use of mature timber. The Forest Service became enamored of the idea of 'intensive forestry,' the idea that with intensive management and manipulation of the forest, high cuts consistent with the actual cuts could be sustained. National political and economic leaders demanded increases in the cut, which the assumptions of intensive management could justify (Hirt, 1994).

At the national level, the following configuration of agents (an 'iron triangle') created lack of change in the cultural system and social structure (Wilkinson, 1992: 169–71). Timber companies received wood fiber. Politicians at both the national and local levels received support. At the national level, they provided a steady flow of timber for home construction. At the local level, the sale of timber provided 'in lieu of' tax payments to county governments. Forestry schools modeled forests as producers of timber, and the graduates of the schools staffed the Forest Service. Trained to supervise timber harvest and timber management, foresters were happy to provide the materials and financial flows desired by timber companies and national politicians. The United States Forest Service (USFS) prospered, due to political support and a flow of budgets for the support of timber harvest. But increases in cut intensified the contradiction between maximizing $u(c_t)$ and maximizing $u(c_t,s_t)$. Forest biomass, while remaining large, shifted to young age-classes, moved toward the origin in Figure 13.1, and thus both the forests and the economies dependent on them were moving out of the resilient region. The forests were less able to supply biomass in the quantity and quality required by installed lumber mill capacity.

13.5 Containment strategy tested: the 1970s

The apparently stable cultural and social system in the forest sector received a jolt from the federal courts and from Congress in the early 1970s. Outside of the forest sector, substantial change occurred with the passage of the National Environmental Protection Act, the Endangered Species Act, and other environmental acts such as the Clean Air and Clean Water Acts.

In West Virginia, on the Monongahela Forest, high timber harvests had caused impacts on other forest uses. The Isaac Walton League sued. The courts interpreted the 1897 Organic Act to prohibit the cutting of trees not biologically mature and not individually marked. Although the decision did not prohibit clearcutting, its restrictions effectively would have stopped clearcutting if applied nationwide (Wilkinson, 1992: 143).

In another part of the country, visual concerns on the face of the Bitterroot Mountains, plus additional concerns about wildlife and water by local hunters and local farmers led to the establishment of a committee of professors from the University of Montana. Chaired by the dean of the forestry school, the resulting Bolle Report concluded that 'multiple use management does not exist as the governing principle on the Bitterroot National Forest' (Bolle *et al.*, 1970) and defined 'timber mining' as harvest above the levels justified by economic analysis of regeneration costs. Senator Metcalf from Montana used the Bolle Report to initiate legislative action; that action became mandatory when timber harvest was halted by the Monongahela decision.

The 1976 passage of the National Forest Management Act (Public Law 94–588) repealed the language in the Organic Act which had halted timber harvest. The new law restated the constraining contradiction between timber harvest and ecosystem sustainability. In addition to referring to the 1960 Multiple Use and Sustained Yield Act for definitions, the National Forest Management Act provided statements such as the following:

> The Forest Service . . . has both a responsibility and an opportunity to be a leader in assuring that the Nation maintains a natural resource conservation posture that will meet the requirements of our people in perpetuity.
>
> *(16 U.S.C. § 1600(6))*

> It is the policy of the Congress that all forested lands in the National Forest System shall be maintained in appropriate forest cover with species of trees, degree of stocking, rate of growth, and conditions of stand designed to secure the maximum benefits of multiple use sustained yield management in accordance with land management plans.
>
> *(16 U.S.C. § 1601(d)(1))*

Various statements in section 6 (16 U.S.C. § 1604) of the act, on planning, repeat similar ambiguities. Ostensibly, the National Forest Management Act required a change in the operation of the Forest Service. Multiple uses of the forest were to be considered in a system of planning that would involve the public in determining the priorities of the planning process. The new National Environmental Protection Act, with its requirements that public hearings be held on impact statements, would seem also to require a change in Forest Service behavior. But no change in behavior occurred, because both acts left considerable discretion to the Forest Service. As long as proper procedures were followed, the contradictions in the law could justify continued high harvest.

New planning regulations were to be written with advice from a Committee of Scientists. The committee appointed for this purpose was drawn from existing schools of forestry. Ideas prevalent in schools of forestry at the time were not suitable for the change in direction that at least some of the senators wanted. Planning was an exercise in rational deduction from a set of premises, best done by planners. A number of tools existed to ensure that the idea of 'sustained' could be contained. Foremost among these tools was the linear programming model, FORPLAN, which exactly embodied the definition of a forest as a sum of stands, location not identified. Utilization of the FORPLAN model, with its inability to keep track of stand locations, meant that nontimber concerns would be hard to handle in the fundamental database for forest planning. President Reagan's appointed officials increased Allowable Sale Quantities above those recommended by forestry staff based on FORPLAN.

In addition, Forest Service discretion under both the National Forest Management Act and the National Environmental Policy Act led to many side-bars on public participation. The alternatives to be presented to the public were determined by the agency, which provided ways to contain both the idea that timber supply would be unsustainable and that other uses of the land be considered. For instance, in the Pacific Northwest, all alternatives presented to the public embodied 'minimum management requirements' that were not subject to debate. Ostensibly, these requirements would meet the needs of nontimber species. In fact, they allowed continued high timber harvest.

Despite isolated examples of resistance within the forest sector, policy continued to emphasize high timber harvests. Harvests from the national forests were 6 billion board feet in 1960 and rose to 12 billion board feet in 1970. Harvests fell somewhat in the early 1980s, and rose above 12 billion board feet in the late 1980s (Committee of Scientists, 1999: 7).

Meanwhile, forest ecosystems were continuing to degrade as a consequence of the massive harvests, road networks, and policies of the Forest Service under the continuing dominance of the timber-oriented ideas. The stage was set for a loud, clear ecosystem signal to be sent to the social system. How could that happen? Local signals through the court system had been ruled out by the discretion granted to agencies in the planning system under the National Forest Management Act and the National Environmental Policy Act. Signals through the legislative or executive branches were not possible because of the iron triangle configuration of members of Congress, leaders of the USFS, and the timber industry.

13.6 The Endangered Species Act and the ecosystem approach

The 1991–3 District Court enforcement of the Endangered Species Act represents the start of a transformation of the fundamental contradiction. From 1976 to the early 1990s, the Forest Service and Republican administrations had been able to contain the idea of 'sustain' as a modifier of the idea of 'yield.' But developments outside of the forest sector, as well as the consequences of proposed continued harvest of timber in the Pacific Northwest, created a crisis that could not be contained. Extinction of a species is an ecosystem signal that cannot be explained away with complicated rationalizations. The National Forest Management Act allowed the Forest Service to ignore local ecosystem signals. The Endangered Species Act provided a potential 'national' ecosystem signal.

The forest plans of Region 6 of the Forest Service were 'dead on arrival,' for failure to comply with protection of the Northern Spotted Owl. Just as with the start of the first act of this drama, the courts stepped in and brought northwest timber harvest to a full stop. District Court Judge Dwyer ruled that the forest plans paid inadequate attention to the habitat of the Northern Spotted Owl. The court's database was scientific, drawn in great part from the work of the research branch of the Forest Service.

Repeated attempts by the Forest Service and a Republican administration to comply with Judge Dwyer's requirements failed, and may have intensified the gridlock. After taking office in 1993, President Clinton organized a Forest Summit in order to solve the deadlock and lift the harvest moratorium. During a day televised nationally, top administration officials heard directly from regional interests. Members of the northwest congressional delegation and the Forest Service listened in the audience (Yaffee, 1994). The President framed the issue in broader terms than just the survival of the Northern Spotted Owl; problems with salmon harvests and other issues could be raised. This created an opening for those speaking to agree on the idea of ecosystem management, although one observer noted that environmentalists heard 'ecosystem' and business interests heard 'management.' The President ended the meeting by providing five guiding principles that emphasized development of a plan that provided predictable timber supplies without harming the long-term health of the environment (Tuchmann *et al.*, 1996: 28–30).

The resulting Northwest Forest Plan implemented ideas of ecosystem management which were very new to the forest sector. An attempt was made to reach a solution that combined a lower level of commodity harvest as well as some stock preservation. Advocates of either extreme position were not pleased. Deeper issues of ecosystem structure and function are added to issues of species viability in the scientific understanding that underpins the Northwest Forest Plan (Tuchmann *et al.*, 1996).

In terms of utility functions, the Northwest Forest Plan proposed to evaluate policy in terms of both commodity production and changes to the forest. Timber growing stock as a measure of s_t is replaced with ecosystem health (Tuchmann *et al.*, 1996: 76–7). Rather than the cut driving management, other goals given priority by the Northwest Forest Plan determine the cut. The proof of this assertion is that more timber comes from lands being treated for ecosystem reasons than from lands reserved only for timber harvest under the Northwest Forest Plan. Restoration of ecosystem integrity now runs the show, and the court approved of the new plan, validated by science.

Why, after 20 years of successful resistance to a change in the dominant scientific theory of forest management, did change occur? Part of the answer is that the national strength of the Endangered Species Act could not be nullified, even though Congressional action delayed it by 2 years (Tuchmann *et al.*, 1996: 22). Substantial public support existed for enforcement of the act. This is only part of the answer, however, because the Northwest Forest Plan was temporarily modifed by the salvage rider in 1995. For 1 year, the rider allowed the harvest of old growth without enforcement of the constraints of the Northwest Forest Plan (Tuchmann *et al.*, 1996: 107). However, due to lack of public support, the advocates of timber harvest were not able to perpetuate the salvage rider. The President removed one side of the triangle by appointing successively two individuals as Chief of the Forest Service who were not from the dominant timber paradigm. These officials supported ecosystem management, which had become scientifically credible since the passage of the National Forest Management Act.

13.7 Polarization emerges

With the application of the Endangered Species Act, the protection offered by the Multiple Use Sustained Yield concept simply disappeared. Total polarization developed at the national level. The procedural republic (Kemmis, 1990) encourages a fight between those who want plantation forestry and those who want pristine nature. In a majority-rule system such as the USA, a coalition that can prevail in the legislature and executive branch can take over all policy. This encourages those on the extremes, those who favor $u(c_t)$ and $u(s_t)$.

This polarization resulted from the dynamics of the tactics used to keep the idea of timber production dominant over the related subordinate idea of protection of the land and production of other commodities. Having been subordinate for so long, the idea of protecting the land, present in the law, found adherents who want to repeat the strategy used by the timber interests: place preservation firmly on top, with complete subordination for the idea of timber yield.

The temptation existed for the idea of preservation to dominate; but the law still contained a statement represented by $u(c_t,s_t)$. Ecosystem management of a forest does require timber harvest – but not an even flow or a sustained timber harvest. Vegetation treatments are driven by the goals of ecosystem management (which include other uses and preservation of future options). Consequently, those desiring high timber harvests, once out of control, are in a position of competing with other uses.

Politically, no middle position appeared feasible in the debate, even if middle positions exist logically. Polarization tends to obscure any complementarity or potential for compromise. Roger Sedjo (1996) portrayed the choices as between custodial or commodity management of the forests, leaving out an option that combined the two. This polarization remained, as shown by examination of the websites of the Sierra Club (2000) and the American Forest and Paper Products Association (2000b).

The nature of the distortions in each position changed. Whereas when they were connected, the desire was to emphasize elements of consistency and to suppress inconvenient ideas and data; when the ideas are in competition, the desire is to emphasize elements of difference, of truth on one side and falsity on the other. As a result of competition, both sides begin to improve their analysis. This is a positive-feedback loop, making the competition even stronger as each side resists elimination. This is good for the receipt of signals, until one side dominates.

At the beginning of this second act, actors on the stage picked up the light called the 'Endangered Species Act,' pointed it at the data on the Northern Spotted Owl, and convinced another actor, a federal judge, to tell the Forest Service to cease harvest. Previous agents had put this light on stage, and the district judge used it. The result was that major interests lined up on either side of the stage, polarized over whether or not the Endangered Species Act should be enforced.

As the change occurred in the Pacific Northwest, the defenders of timber interests were able to deploy some containment strategies, as shown by the fate of the idea of protecting 'forest health.' Among ecologists, ecosystem health has a strong following; many scientists regard it as a useful way to think about ecosystem integrity (Costanza, Norton, and Haskell, 1992). But in the political sphere, the idea of protecting forest health can also mean cut the dead trees: if trees will die from disease or fire, then cut them to prevent their death. The salvage rider is a good example of this rhetoric. When timber interests emphasized forest health and fire danger, leaders of environmental organizations heard an excuse for timber harvest.

13.8 The committee of scientists recognizes complementarities

While the salvage rider was a hot issue, the Seventh American Forest Congress convened in February, 1996. The Congress provided 'Vision Elements' and 'Principles' that contained very strong support for sustainability. One can argue that the previously subordinate idea had by the mid-1990s risen to the top in the cultural system of the forest sector (Langbein, 1996). To adherents of the old idea of timber yield, this was a rapid and revolutionary event, which complemented the shock administered by Judge Dwyer in the Pacific Northwest. The American Forest and Paper Association came to recognize the profound nature of the shift (Wallinger, 1995).

Proponents of ecosystem approaches to management used the National Forest Management Act's authorization to consult scientists to appoint a second Committee of Scientists. While the charge to the committee was neutrally stated, the changes in public values regarding forest management, as represented by the results of the Seventh Forest Congress, opened the door for the new committee to provide a re-interpretation of the provisions of the National Forest Management Act and the other environmental legislation as applied to forest planning.

With national polarization at a high level, which side would the new Committee of Scientists select? The committee took a new track, by proposing to unify ecological, social, and economic sustainability. The committee pointed to fundamental complementarities, asserting that protection and promotion of ecosystem integrity are needed for all of the multiple uses of the forest. Using language in the National Forest Management Act, the committee connected the idea of ecological sustainability to another idea prevalent in American politics: the idea of local control.

To some observers, complementarity is not an empirical truth: they see the choice as either using the forest for man's purposes or excluding commodity use altogether. This can be illustrated with the figure used to compare different sustainability goals, as in Figure 13.2. In that figure, a point below the annual yield curve, a low level of ecosystem integrity such as A, might be used as an analogy to the situation in the 1990s, when there were limited harvest possibilities because of the political stalemate.

From a point like A, three directions are possible. One is to harvest the current increment of fiber, by moving to a point like H. Another is to leave the forest alone, and allow all of the increment to add to the timber stock, ultimately moving toward a point like E. A third possibility is to increase both the stock and the level of timber harvest, moving to a point somewhere in the area marked

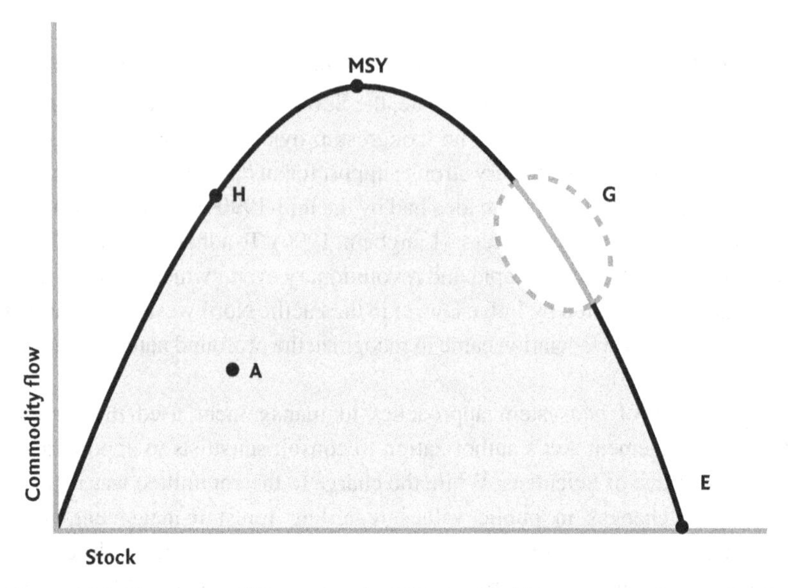

Figure 13.2 Comparing different sustainability goals: different targets from point A.

G. The Committee of Scientists emphasized the importance of the possibility of a move to a point like G. A study by the Society of American Foresters reached a similar conclusion (Floyd, 1999).

The graph, however, does not capture adequately the latest ideas about ecological systems. These ideas emphasize the complexities of ecosystems, which are dynamic, hard to predict, and relatively poorly understood when management aims at achieving a balance of uses (Kay, 1991; Costanza *et al.*, 1992; Holling and Meffe, 1996). Such ecosystems need to be considered at multiple scales both in space and in time. For instance, after a century of controlling fires, many western forests in the USA have a high risk of catastrophic crown fires which would reduce the growing stock to zero, and completely reset forest conditions. Such fires are a disaster for both those who wish to save timber volumes and those who wish to hold the forest in a condition suitable to particular species that are in danger of extinction. Figure 13.2 is too deterministic to capture this aspect of ecosystems. If the horizontal axis measures 'ecosystem integrity' rather than 'stock,' increases in integrity can move the system further from the origin, and maximization of commodity flows move the system toward the origin. But a large biomass of trees ready to burn could be portrayed as a large stock.

The public discussion of forest policy is perhaps too simplistic to capture such complexities in forest ecosystem management. Rather, simple alternatives such as those in Figure 13.2 capture the nature of the public discussion. The committee addressed this potential simplicity by emphasizing desired future conditions as a focus of planning. Attention to desired future conditions would assist in helping the public and the Forest Service reach consensus about desired management activities. Desired future condition could represent a point like H, or one like E, or any point in between. If the public values the condition of the forest as well as the flow of commodities from the forest, then a planning process that places attention on both is likely to succeed, probably identifying a point in the area marked G in Figure 13.2.

A profound difference exists in the language used to describe moving to the multiple sustained yield point at the top of the curve or to a point in the range identified by area G. Authors Regier and Baskerville (1986), for instance, use the term 'redevelopment' to describe public actions aimed at addressing low productivity of forests in New Brunswick and fisheries in the Great Lakes. In both cases the starting point is one like A, which resulted from a period of 'development.' Their article continues to use the language of development while describing a move toward greater ecosystem productivity. As the new millenium approached, the hidden assumptions of 'development' came under scrutiny (Escobar, 1995; Rist, 1997). Gilbert Rist, for instance, defines development as a process that requires the destruction of the natural environment. In these terms, 'redevelopment' cannot be consistent with what is now called ecosystem restoration. Policy aimed at the multiple sustained yield point would be called 'redevelopment;' policy aimed at a point in the G range would be called 'restoration;' the ecosystem is restored for its noncommodity values as well as for its commodity values.

The Committee of Scientists used the language of restoration and also proposed that a successful planning effort would produce 'stewardship capacity' among the participants in the planning process. Stewardship capacity is the existence of both the ability and the willingness to care for the condition of forests. The ability to care for forests results both from knowledge and from the maintenance of needed capital, such as harvesting equipment. The willingness results both from the desires of those involved and from the existence of social networks that promote the trust needed for public cooperative action.

13.9 Forest certification

With governmental processes apparently deadlocked, the development of two different certification efforts illustrates an alternative response that can be compared to the proposal of the Committee of Scientists. Both of the certification processes recognize the existence of complementarities in the values provided by forested lands. They differ in emphasis, however. The principles proposed by the Forest Stewardship Council (2000) lean toward the quality of the stock – or the integrity of the ecosystems – whereas the principles of the American Forest and Paper Association (2000a) tend to emphasize commodity production. Both, however, could be broadly construed to be statements of the form $u(c_t,s_t)$ – both include attention to commodities and to the quality of the stock. Either could be aimed at results in the circle labeled G in Figure 13.2.

An emphasis on monitoring is another similarity between the certification processes and the proposal of the Committee of Scientists. Certification involves third parties that audit or evaluate the management results of forest owners. The Forest Stewardship Council emphasized third-party audits from the start. The American Forest and Paper Association had self-monitoring in their initial certification process, but by the fourth year of its operation they had started to include optional audits by third parties.

The Committee of Scientists proposed scientific review panels at each level of planning within the Forest Service system. Planning involves a tension between the interests and values of people living near forests and the interests and desires of the nation, which owns national forests and grasslands. How is this tension to be addressed? One solution is to have local people dominate in the formation of plans, but to have the national interest represented by external review of those plans. External review also addresses the distrust generated by the Forest Service in its past devotion to timber production.

The ultimate outcome for forest management on the national forests and grasslands is unknown as the twenty-first century opens. The iron triangle which supported the $u(c_t)$ interpretation of the laws has weakened considerably, but could be re-established, depending on electoral outcomes. On the other hand, other electoral outcomes could support environmental concerns allied with the idea of supporting the $u(s_t)$ interpretation. A third possibility is that collaborative efforts as envisioned by the Committee of Scientists, which are consistent with forest certification, could dominate.

The fate of monitoring plans depends upon which of these views wins. In the fight between two positions, neither side is really interested in the facts of forest behavior. Data about the forest are evidence to be used in the battle of position. In the collaborative and certification positions, the forests' responses to treatments matter for everyone. Uncertainty about forest ecosystem processes is not caught up in a debate about who is right or wrong; rather, uncertainty obscures the choice of policy which will benefit everyone.

13.10 Lessons and questions regarding resilience

The static diagrams in Figures 13.1 and 13.2 provide one way to define and address resilience. The dynamics of the four-phase adaptive cycle provide another approach to resilience (Holling, 1986, 1995; Gunderson *et al.*, 1995). In the long phase of the adaptive cycle, as a natural ecosystem or a linked human–nature ecosystem exploits resources and becomes more complex in terms of accumulated capital and increased connections, crisis potential develops. If a crisis occurs, connections are broken and capital is released from existing structures to become available for renewal and reorganization. When humans are involved, might it be that the dominant ideas and coalitions that governed the long phase, the move from exploitation to conservation, will affect choices and possibilities during the subsequent renewal and reorganization phase? In the first scenario just described, the ideas and coalitions existing prior to the crisis re-established old patterns. In the second scenario, new ideas and a broken iron triangle make new patterns possible. The shackles of the old ideas and coalitions remain, however, in affecting the way new choices are conceived. The logical characteristics of reasons given to justify actions during the previous growth and crisis phases condition the processes of public discussion during the ensuing renewal phase.

The issue of interest for navigating nature's dynamics is the ability of negative feedback from ecosystems to cause responses in the socio-cultural system that move humankind's use of the ecosystem away from destabilizing behavior and toward actions that restore system resilience. In the USA, negative feedback was delivered twice in the same manner: a cessation of harvest ordered by a court because of the ecological consequences. But the response was different in the two cases, and the second response seems to be better than the first one was: why?

In the first case, there was a particular kind of orderliness in the cultural system of the forest sector: that created by the logic of a constraining contradiction. In spite of efforts to make national policy clearly favor commodity interests, the political appeal of 'sustained yield' and 'multiple use' prevented elimination of the contradictory statements. Thus, the dominant interest groups in the forest sector had to accept the rhetoric of sustained yield, even while emphasizing 'yield' (Hirt, 1994: 182–92). The people running the Forest Service and those in the forestry schools and industry were in general agreement about how to think about forests. This agreement in approach supplemented the iron triangle coalition: forestry committees in Congress supported the Forest Service, the Forest Service provided timber, and industry supported the members of Congress.

Disorderliness in the social sector was in the area of local groups concerned about the ecological impact of the cut. But these people were not controlling much wealth, coercive power, or knowledge. They were uneasy about policy consequences but had insufficient power to change the result. The strong agents on both the cultural and structural sides agreed, supported each other, and contained the situation, using the leverage provided by Forest Service discretion under the National Forest Management Act and the procedures of the National Environmental Protection Act. The result was removal of the legal leverage which the local actors had managed to use to stop harvests. FORPLAN dominated technical analysis, and was vulnerable to administrative increases in Allowable Sale Quantity amounts. The pattern of ignoring legal requirements to consider forest conditions continued, as did high harvests.

In the second case, disorderliness existed in the dominant cultural ideas. Ecosystem ideas had supplanted much of the old single-resource analysis. Two decades of scientific research had uncovered information that had been systematically excluded from management decisions but that was now available should crisis develop (Behan, 1990). Leopold's views were held by many more people and had become socially salient. National environmental groups had grown in membership and wealth. Change in the cultural sector had moved emphasis to *sustain* and away from *yield*.

When Judge Dwyer halted timber harvest in the Pacific Northwest in 1991, the iron triangle was faced with a crisis much like that of the early 1970s. But several conditions had changed. First, the Endangered Species Act had real teeth, and considerable support among the public. Second, the tactics of manipulation of public participation in forest planning had made the resulting plans extremely unpopular. Third, a national election brought to office a Democrat, Bill Clinton, who was not as supportive of continued timber harvest as his predecessors had been. The President convened a meeting of major agents in the Pacific Northwest, placing national-level representatives of the iron triangle in the audience rather than behind the microphones. The resulting guidelines led to a new Northwest Forest Plan.

Reaction to the Northwest Forest Plan reveals a consequence of decades of containment of the contradiction called 'multiple use sustained yield.' These responses revealed extreme polarization. Neither commodity interests nor environmentalists were pleased. This polarization continues to characterize debates in Washington, DC, about forest policy (Cooper, 1999). Any proposal to allow some timber harvest, even if justified in terms of restoring ecosystem integrity, can be characterized by environmental interests as a return to the old policy of excessive commodity harvest. Any proposal that is justified by restoring ecosystem integrity can be attacked by commodity interests as out of compliance with the law and a new mission for the Forest Service that excludes commodity outputs (Sedjo, 1999; Geisinger, 1999). Both sides attack those interested in a balanced approach. Neither side is really particularly interested in the facts about forest condition; both are seeking total victory for their viewpoint. If one side wins, no significant monitoring of forest condition is likely to occur.

Perhaps national deadlock, as provided by the federal elections of November 2000, will allow local collaborative efforts to continue.

The interesting and potentially beneficial result for the receipt of negative feedback from ecosystems is the emergence of a desire to move away from polarization toward collaboration. Such collaboration involves developing new concepts of the relationship between forests and human communities (Carey, 1999). Ideas have changed in the forest sector, as shown by the fourth edition of the *Forest Management* textbook (Davis *et al.*, 2001) with its subtitle: *To Sustain Ecological, Economic, and Social Values*.[3] The new textbook provides an ecological definition of forest, replacing the previous narrow definition used in earlier editions.

In terms of the adaptive cycle, this story suggests that one should examine the cultural dynamics created by the policies of the cultural and structural elites during periods of exploitation, conservation, and release in order to understand the dynamics of political and social decision-making during the following period of renewal and reorganization. Maintenance of an internally contradictory policy signals that elites can stifle negative feedback for a considerable period of time. When that time expires, the ensuing polarization may itself inhibit development of the ability to respond to ecosystem signals, as contending parties seek victory for their viewpoint. Victory by one side or the other will create another period of weak receipt of signals. When ecosystems are sufficiently degraded that people can see the potential for win–win solutions, the emergence of collaboration among the groups offers some hope.

This chapter opens by posing a question: under what conditions can a crisis create a social response that improves the resilience of a linked social–ecological system? Although a single case study cannot answer such a complex question, repetition of two similar scenarios provides ideas for further examination. In the first scenario, nationally dominant agents used the idea of sustained yield to appear to respond to the crisis without changing policy actions. In the second scenario, ecological ideas such as ecosystem management and political ideas such as local collaboration were sufficiently developed to challenge the previously dominant idea of sustained yield, which was vulnerable because of its own internal contradiction.

The new ecosystem ideas enhance the influence of local agents while contributing to deadlock at the national level. Local communities in Montana and Ohio had acted to create the first crisis; their efforts were subsequently neutralized at the national level. In the second crisis, national industrial and environmental interests canceled each other out, leaving room for some local people to pursue policies which were appropriate to their ecosystem conditions (Wondalleck and Yaffee, 2000).

The new planning regulations for national forests and grasslands recognize and legitimize these local efforts. In terms of Figure 13.2, the regulations enable planning to focus on desired conditions, an emphasis on the condition of the forest stock (US Department of Agriculture, 2000, §219.12). They also enable planners to examine the impact of forest uses on local economies. The new regulations emphasize three concepts that are new to forest planning: social and ecological sustainability, local collaboration in forest decision-making, and increased monitoring and scientific review.

What was the relationship between the development of these new ideas and the dynamics operating in the forest sector in the USA? Did these ideas originate inside of or outside of the forest sector? If the origin was inside, then this may be important for understanding the adaptive cycle as it operates at one level. If the origin was outside (from a higher level), then this particular story provides an example to support the hypothesis by Sabatier and Jenkins-Smith (1999) that coalitions in a sector require outside perturbations for change to occur. When conditions inside the forest sector allowed it to resist the perturbation, no change occurred. In the second scenario, ideas both inside and outside of the forest sector had seriously weakened the dominant interpretation of sustained yield. Combined with more evident changes in forest condition – species threatened with extinction – the change in ideas created great change in the forest sector. Comparison to other cases should show if this is a general pattern or an historical accident.

Acknowledgements

The author was a member of the Committee of Scientists described in this chapter, and thanks are due to the US Department of Agriculture for support during that participation. Thanks are also due to the Pew Fellows Program in Conservation and the Environment and Northern Arizona University for support during a sabbatical year. The views in the chapter are those of the author.

Notes

1. The Menominee Tribe, whose forest management was among the first in the USA to be certified under Forest Stewardship Council criteria, has been criticized by commodity-oriented observers for using long rotation ages, and by preservation-oriented observers for not leaving old enough trees. Menominee forest managers dislike the fact that 'sustained yield' has come to be interpreted narrowly as $u(c_t)$. In the Menominee cultural system, disconnected from that of the USA in general, $u(c_t, s_t)$ has been dominant for over a century.
2. Most economists, in analyzing Figure 13.1, would include the impact of discounting utility into the future and would seek the optimum using control theory. The optimum point for $u(c_t)$ would move to the left, as would the point for $u(c_t, s_t)$.

For an explanation of the analysis for $u(c_t)$, see Bowes and Krutilla (1989). For an explanation of $u(c_t, s_t)$, see Heal (1998: 46–56). Figure 13.1 uses the green golden rule point for the dynamic solution to the maximization problem for $u(c_t, s_t)$. The green golden rule applies when the far future receives some weight in the social welfare function; see Chichilnisky (1997) and Heal (1998). Colin Clark (1990: 4–10) pointed out, using a model based on $u(c_t)$, that a high enough interest rate would justify harvesting a slowly reproducing resource – such as whales – to extinction even under private ownership. This result is not possible when the evaluation of the optimum is governed by $u(c_t, s_t)$, an approach Clark does not consider.

3. The chair of the Committee of Scientists, K. Norman Johnson, is a coauthor of the textbook.

References

American Forest and Paper Association 2000a. SFI Program. Web page [accessed 6 March 2000]. Available at http://www.afandpa.org/forestry/forestry.html

American Forest and Paper Association 2000b. Welcome to AF&PA's 'Washington Watch.' Web page [accessed 6 March 2000]. Available at http://www.afandpa.org/legislation/legislation.html

Archer, M.S. 1995. *Realist Social Theory: the Morphogenetic Approach*. Cambridge: Cambridge University Press.

Archer, M.S. 1996. *Culture and Agency: the Place of Culture in Social Theory*. Cambridge: Cambridge University Press.

Baskerville, G.L. 1995. The forestry problem: adaptive lurches of renewal. In *Barriers and Bridges to the Renewal of Ecosystems and Institutions*, pp. 37–102, ed. L.H. Gunderson, C.S. Holling, and S.S. Light. New York: Columbia University Press.

Behan, R.W. 1990. Multiresource forest management: a paradigmatic challenge to professional forestry. *Journal of Forestry* 88(4): 12–18.

Bolle, A.W., Behan, R.W., Browder, G., *et al.* 1970. *Report on the Bitterroot National Forest*. Senate Document 91–115. Washington DC: US Government Printing Office.

Bowes, M.D. and Krutilla, J.V. 1989. *Multiple-Use Management: the Economics of Public Forestlands*. Washington DC: Resources for the Future.

Carey, H. 1999. The guiding star of ecological and rural sustainability. *Journal of Forestry* 97(5): 42–3.

Chichilnisky, G. 1997. What is sustainable development? *Land Economics* 73: 467–91.

Clark, C.W. 1990. *Mathematical Bioeconomics: the Optimal Management of Renewable Resources*, 2nd edn. New York: John Wiley & Sons.

Committee of Scientists 1999. *Sustaining the People's Lands: Recommendations for Stewardship of the National Forests and Grasslands into the Next Century*. Washington DC: United States Forest Service.

Cooper, A.W. 1999. The *Second* Committee of Scientists: moving forward while looking backward. *Journal of Forestry* 97(5): 24–5.

Costanza, R., Norton, B.G., and Haskell, B.D., eds. 1992. *Ecosystem Health: New Goals for Environmental Management*. Washington DC: Island Press.

Davis, L.S. and Johnson, K.N. 1987. *Forest Management*, 3rd edn. New York: McGraw Hill.

Davis, L.S., Johnson, K.N., Bettinger, P.S., and Howard, T.E. 2001. *Forest Management: To Sustain Ecological, Economic, and Social Values*, 4th edn. New York: McGraw-Hill.

Escobar, A. 1995. *Encountering Development: the Making and Unmaking of the Third World*. Princeton, NJ: Princeton University Press.

Floyd, D.W., ed. 1999. *Forest of Discord: Options for Governing Our National Forests and Federal Public Lands*. Bethesda, MD: Society of American Foresters.

Floyd, D.W., Vonhof, S.L., and Seyfang, H.E. 2001. Forest sustainability: a discussion guide for professional resource managers. *Journal of Forestry* 99(2): 8–28.

Forest Stewardship Council 2000. Certification. Web page [accessed 19 March 2002] available at http://fscus.org/html/certification/index.html

Geisinger, J. 1999. Nonscience from the Committee of Scientists. *Journal of Forestry* 97: 24–5.

Gunderson, L.H., Holling, C.S., and Light, S.S. 1995. Barriers broken and bridges built: a synthesis. In *Barriers and Bridges to the Renewal of Ecosystems and Institutions*, pp. 489–532, ed. L.H. Gunderson, C.S. Holling, and S.S. Light. New York: Columbia University Press.

Heal, G. 1998. *Valuing the Future: Economic Theory and Sustainability*. New York: Columbia University Press.

Hirt, P.W. 1994. *A Conspiracy of Optimism: Management of the National Forests since World War II*. Lincoln, NB: University of Nebraska Press.

Holling, C.S. 1986. Resilience of ecosystems; local surprise and global change. In *Sustainable Development of the Biosphere*, pp. 292–317, ed. W.C. Clark and R.E. Munn. Cambridge: Cambridge University Press.

Holling, C.S. 1995. What Barriers? What Bridges? In *Barriers and Bridges to the Renewal of Ecosystems and Institutions*, pp. 3–34, ed. L.H. Gunderson, C.S. Holling, and S.S. Light. New York: Columbia University Press.

Holling, C.S. and Meffe, G.K. 1996. Command and control and the pathology of natural resource management. *Conservation Biology* 10: 328–37.

Huff, P.R. and Pecore, M. 1995. Case Study: Menominee Tribal Enterprises. Web page. Available at http://www.menominee.com/sdi/articles/csstdy.htm

Kay, J.J. 1991. A nonequilibrium thermodynamic framework for discussing ecosystem integrity. *Environmental Management* 15: 483–95.

Kemmis, D. 1990. *Community and the Politics of Place*. Norman, OK: University of Oklahoma Press.

Langbein, W.D. 1996. *Final Report, Seventh American Forest Congress*. Office of the Seventh American Forest Congress, New Haven, 205 Prospect St, New Haven, CT 06511.

Leopold, A. (1988). *A Sand County Almanac, with Essays on Conservation from Round River*. New York: Ballantine Books.

Pecore, M. 1992. Menominee sustained yield management: a successful land ethic in practice. *Journal of Forestry* 90(7): 12–16.

Regier, H.A. and Baskerville G.L. 1986. Sustainable redevelopment of regional ecosystems degraded by exploitive development. In *Sustainable Development of the Biosphere*, pp. 75–101, ed. W.C. Clark and R.E. Munn. Cambridge: Cambridge University Press.

Rist, G. 1997. *The History of Development: from Western Origins to Global Faith*. London: Zed Books.

Sabatier, P.A. and Jenkins-Smith, H.C. 1999. The advocacy coalition framework: an assessment. In *Theories of the Policy Process*, pp. 117–66, ed. P.A. Sabatier. Boulder, CO: Westview Press.

Sedjo, R.A. 1996. Toward an operational approach to public forest management. *Journal of Forestry* 94(8): 24–7.

Sedjo, R.A. 1999. Mission impossible. *Journal of Forestry* 97(5): 13–15.

Sierra Club 2000. Stop logging our National Forests! Web page [accessed 6 March 2000]. Available at http://www.sierraclub.org/forests/

Tuchmann, E.T., Connaughton, K.P., Freedman, L.E., and Moriwaki, C.B. 1996. *The Northwest Forest Plan: a Report to President and Congress.* Washington DC: US Department of Agriculture, Office of Forestry and Economic Assistance.

US Department of Agriculture, Forest Service 2000. 36 CFR Parts 217 and 219: national forest system land resource management planning: final rule. *Federal Register* 65 (218): 67513–81.

Wallinger, S. 1995. A commitment to the future: AF&PA's Sustainable Forestry Initiative. *Journal of Forestry* 93(1): 16–19.

Wilkinson, C.F. 1992. *Crossing The Next Meridian: Land, Water, and the Future of the West.* Washington DC: Island Press.

Wondalleck, J.M. and Yaffee, S.L. 2000. *Making Collaboration Work: Lessons from Innovation in Natural Resource Management.* Washington DC: Island Press.

Yaffee, S.L. 1994. *The Wisdom of the Spotted Owl.* Washington DC: Island Press.

14

Synthesis: building resilience and adaptive capacity in social–ecological systems

CARL FOLKE, JOHAN COLDING, AND FIKRET BERKES

14.1 Introduction

A weekly magazine on business development issued an analysis of Madonna, the pop star, and raised the question 'How come Madonna has been at the very top in pop music for more than 20 years, in a sector characterized by so much rapid change?' A few decades ago, successful companies developed their brand around stability and security. To stay in business this is no longer sufficient, according to the magazine. You must add change, renewal, and variation as well. However, change, renewal, and variation by themselves will seldom lead to success and survival. To be effective, a context of experience, history, remembrance, and trust, to act within, is required. Changing, renewing, and diversifying within such a foundation of stability and maintaining high quality have been the recipe for success and survival of Madonna, and for rock stars such as Neil Young and U2. It requires an active adaptation to change, not only responding to change, but also creating and shaping it. In the same spirit, Sven-Göran Eriksson, coach of several soccer teams in Europe, claimed that it is the wrong strategy not to change a winning team. A winning team will always need a certain amount, but not too much, of renewal to be sustained as a winning team. Sustaining a winning team requires a context for renewal, or 'framed creativity,' borrowing from the language of the advertiser.

These metaphors get to the very core of resilience, the concept in focus of this volume in relation to the dynamics of complex and coupled social–ecological systems. Resilience, the capacity to lead a continued existence by incorporating change (Holling, 1986), stresses the importance of assuming change and explaining stability, instead of assuming stability and explaining change (van der Leeuw, 2000). The latter perspective dominated twentieth-century resource and environmental science and management (Gunderson, Holling, and Light, 1995; Berkes and Folke, 1998), and has been successful in producing stability and security of resource flows in the short term. However, doing so has simplified ecosystems and reduced functional diversity (Holling and Meffe, 1996), and eroded resilience (Gunderson and Pritchard, 2002). Strategies for controlling environmental variability and natural disturbance become key in such systems, because fluctuations impose problems on meeting predicted production goals. Thus, managers seek to command and control these processes in an attempt to stabilize ecosystem outputs (Carpenter and Gunderson, 2001). Short-term success of increasing yield in homogenized environments contributes to creating mental models of human development that are superior and largely independent of nature's services. According to this thinking, nature can indeed be conquered, controlled, and ruled. Short-term success makes navigating nature's dynamics a nonissue and, as a consequence, knowledge, incentives, and institutions for monitoring and responding to environmental feedback erode (Holling, Berkes, and Folke, 1998). The belief system of humanity as superior to and independent of nature is reinforced.

The separation of social systems and natural systems is more of a recent mental artifact than an observation of the real world (Nelson and Serafin, 1992; see also Chapter 3). The Earth has for long been transformed by human action (Turner, Clark, and Kates, 1990). Throughout history, humanity has shaped nature and nature has shaped the development of human society (Tainter, 1988; Grimm *et al.*, 2000). In that sense, there are neither natural or pristine systems, nor social systems without nature. Instead, humanity and nature have been co-evolving in a dynamic fashion (Norgaard, 1994; Berkes and Folke, 1998). In the present era of the human-dominated biosphere, co-evolution now takes place also at the planetary level and at a much more rapid and unpredictable pace than previously in human history.

Facing complex co-evolving systems for sustainability requires the ability to cope with, adapt to, and shape change without losing options for future adaptability. The irony is that the mental model of optimal management of systems assumed to be stable and predictable has in many respects reduced options and removed the capacity of life-support systems to buffer change (Jackson *et al.*, 2001). The insurance for dealing with the unexpected has been driven down by suppressing disturbance and reducing the diversity of the environment.

Actions toward sustainability will require understanding and appreciation of the dynamics of complex life-support ecosystems – a new level of ecological literacy – and not just among scientists, but also the general public at large. A fundamental challenge in this context is to build knowledge and incentives into institutions and organizations for managing the capacity of local, regional, and global ecosystems to sustain societal development (Ostrom *et al.*, 1999;

Scoones, 1999) in the context of uncertainty, surprise, and vulnerability (Kates and Clark, 1996; Gunderson and Holling, 2002).

The focus of the volume is the study of the adaptability of social–ecological systems to meet change and novel challenges in navigating ecosystem dynamics without compromising long-term sustainability. Throughout this volume, we argue that resilience is a key property of sustainability; that loss of resilience leads to reduced capacity to deal with change. Ecological resilience has been defined as the magnitude of disturbance that can be experienced before a system moves into a different state and different set of controls (Holling, 1973). Social resilience has been defined as the ability of human communities to withstand external shocks to their social infrastructure, such as environmental variability or social, economic, and political upheaval (Adger, 2000). Systems may be ecologically resilient but socially undesirable, or they may be socially resilient but degrade their environment. Here, we are concerned with the combined systems of humans and nature, with emphasis on social–ecological resilience. We are concerned with management that secures the capacity of ecosystems to sustain societal development and progress with essential ecosystem services. Successful ecosystem management for social–ecological sustainability requires institutional capacity to respond to environmental feedback, to learn and store understanding, and be prepared and adaptive to allow for change. The challenge is to anticipate change and shape it for sustainability in a manner that does not lead to loss of future options. It involves enhancing the capacity for self-organization.

In this chapter we explore the above hypotheses and present some tentative conclusions on the dynamics of linked social–ecological systems for resource and ecosystem management, drawing on the chapters of the volume. We are interested in periods of change and how people and nature relate to and organize around change. Are there elements that sustain social–ecological systems in a world that is constantly changing? In that sense, we are focusing on periods of change caused by disturbance, surprise, or crisis, followed by periods of renewal and reorganization (referred to in Chapter 1 as 'the back-loop'). Such periods of change are the most neglected and the least understood in conventional resource management.

In discussing resilience building for adaptive capacity, the synthesis identifies and expands on four critical factors highlighted in many of the chapters, factors that interact across temporal and spatial scales and that seem to be required for dealing with nature's dynamics in social–ecological systems:

- learning to live with change and uncertainty;
- nurturing diversity for reorganization and renewal;

Table 14.1 *Building resilience and adaptive capacity in social–ecological systems*

Theme	Components
Learning to live with change and uncertainty	
	Evoking disturbance
	Learning from crises
	Expecting the unexpected
Nurturing diversity for reorganization and renewal	
	Nurturing ecological memory
	Sustaining social memory
	Enhancing social–ecological memory
Combining different types of knowledge for learning	
	Combining experiential and experimental knowledge
	Expanding from knowledge of structure to knowledge of function
	Building process knowledge into institutions
	Fostering complementarity of different knowledge systems
Creating opportunity for self-organization	
	Recognizing the interplay between diversity and disturbance
	Dealing with cross-scale dynamics
	Matching scales of ecosystems and governance
	Accounting for external drivers

- combining different types of knowledge for learning; and
- creating opportunity for self-organization toward social–ecological sustainability.

The way that we address these factors in relation to building resilience for adaptive capacity is presented in Table 14.1. The first part of the chapter emphasizes the necessity of accepting change and living with uncertainty and surprise, and provides examples of strategies of social–ecological management for taking advantage of change and crisis and turning it into opportunity for development. The second part illuminates the importance of nurturing diversity for resilience, recognizing that diversity is more than insurance to uncertainty and surprise. It also provides the bundle of components, and their history, that makes development and innovation following disturbance and crisis possible, components that are embedded in the social–ecological memory. The third part addresses the significance of peoples' knowledge, experience and understanding about the dynamics of complex ecosystems, their inclusion in management institutions, and their complementarity to conventional management. The fourth part brings these issues together in the context of self-organization, scale, governance, and external drivers, stressing the significance of the dynamic interplay between diversity and disturbance. Both diversity and disturbance are parts of sustainable development and resilience and their interaction needs to be explicitly accounted for in an increasingly globalized and human-dominated biosphere.

14.2 Learning to live with change and uncertainty

Change and crisis are parts of the dynamic development of complex co-evolving social–ecological systems (see Chapter 2). It is impossible to lock a system in a steady state for eternity, or to manage it for stability and security in a command-and-control fashion. Policies aimed at removing change and variation will instead cause an accumulation of such disturbance and a more widespread crisis. For example, suppression of forest fires locally will cause an accumulation of fuel on the forest floor as well as of tree biomass. When a fire event finally occurs, it will be hot and intensive, burning deeper into the soil and affecting seed viability, microorganisms, organic content, and nutrients. An ecosystem that can withstand a small, low-intensity fire may be severely affected by a large, hot fire that can change soil conditions and water-holding capacity, and destroy old, seed-bearing trees important for reorganization.

Similarly, suppression of institutional or organizational change can create gridlock in the capacity to adapt to change, and may lead to erosion of both natural and social capital (Gunderson *et al.*, 1995). In Chapter 13, Trosper illustrates the suppression of change in the US forest sector. The response of the iron triangle coalition to external disturbance was to try to avoid it by tightening the internal structure. This strategy, successful in the short term, led to a major management crisis later on. The polarization of US forest policy between commodity interest groups and environmental interest groups removes attention from monitoring and managing forest condition and responding to ecosystem dynamics. To avoid social traps (Costanza, 1987) that erode resilience, there have to be knowledge, practices, and social mechanisms that recognize that disturbance, surprise, and crisis are part of development and progress (see Chapters 2, 5, 6, and 8). These mechanisms include conflict resolution, negotiation, participation, and other mechanisms for collaboration (Ostrom, 1990; Röling and Wagemakers, 1998) with rules aimed at maintaining the process of learning and adaptation in situations facing uncertainty and external change (see Chapters 4, 7, 9, 11, and 12). It also requires a social network with trust and respect (Chapter 10) and social nestedness for ecosystem management operating at multiple scales (Chapter 12 and 13; Folke *et al.*, 1998; Gibson, Ostrom, and Ahn, 2000; Cash and Moser, 2000; Gunderson and Holling, 2002).

14.2.1 Evoking disturbance

Many local communities have recognized the importance of disturbance for securing ecosystem services and have developed management practices that mimic disturbance regimes in nature. There are practices that evoke release in ecosystems by creating small-scale pulse disturbance (Berkes and Folke,

2002). By imitating fine-scale natural perturbations, these practices speed up local renewal cycles in the larger ecosystem and help avoid the accumulation of disturbance – the revolt connection – that moves across scales and further up in the panarchy (Gunderson and Holling, 2002), the nested set of adaptive renewal cycles introduced in the introductory chapter.

Examples include burning of pastures for pest control and pulses of herbivore grazing in a rotational manner, as practiced by some African agro-pastoralists (Chapter 6; Niamir-Fuller, 1998), that contribute to the capacity of the land to function under a wide range of climatic conditions. Fire management has been practiced widely by traditional societies in Australia and North America to open up clearings (meadows and swales), corridors (trails, traplines, ridges, grass fringes of streams and lakes), and windfall areas (Chapter 7; Lewis and Ferguson, 1988). It was not used on a large scale, but rather in a patchy distribution and on targeted species and habitats (Fowler, 2000). Fire management is also practiced in contemporary forest and protected area management.

Patch clearing through swidden–fallow management and associated agroforestry systems among Amazon area tribes in cycles of up to 30–40 years provided a diversity of resources and ecosystem services over the long term (Denevan *et al.*, 1984; Posey, 1985; Irvine, 1989). Small-scale patch management and shifting cultivation in agriculture are a common measure, as illustrated in Chapters 6 and 7. The short periods of additional openings of the channel to the Ibiraquera Lagoon by local fishers is another example of resource flows being secured and large-scale crisis at a later stage being avoided (Chapter 11).

Such management practices, embedded in institutions and often initiated by rituals, are frequently guided by stewards and have a cultural and an ethical dimension (Chapters 10 and 12; Berkes, 1999). For example, Native Americans of the Great Basin of Western North America viewed fire as part of an ethic of caring for the land and its resources, and fire was considered to be a beneficial and cleansing force (Fowler, 2000). The annual ritual of 2 days of intensive crayfishing (pulse fishing) in August in Sweden builds social capital in the community and supports the development of institutions for the management of a common pool resource in a watershed context (Chapter 8; Olsson and Folke, 2001).

Building social–ecological resilience also requires evoking change in social structures. In a study of the US Fish and Wildlife Service, Jeffrey (2000) highlights the need for organizational change as a component of ecosystem management, and puts forward the role of leadership in actively initiating change within organizations. In Chapter 9, Blann, Light, and Musumeci stress the key role of social networks of 'practitioners' in which support, trust, and sharing of lessons learned can facilitate processes of change at multiple scales.

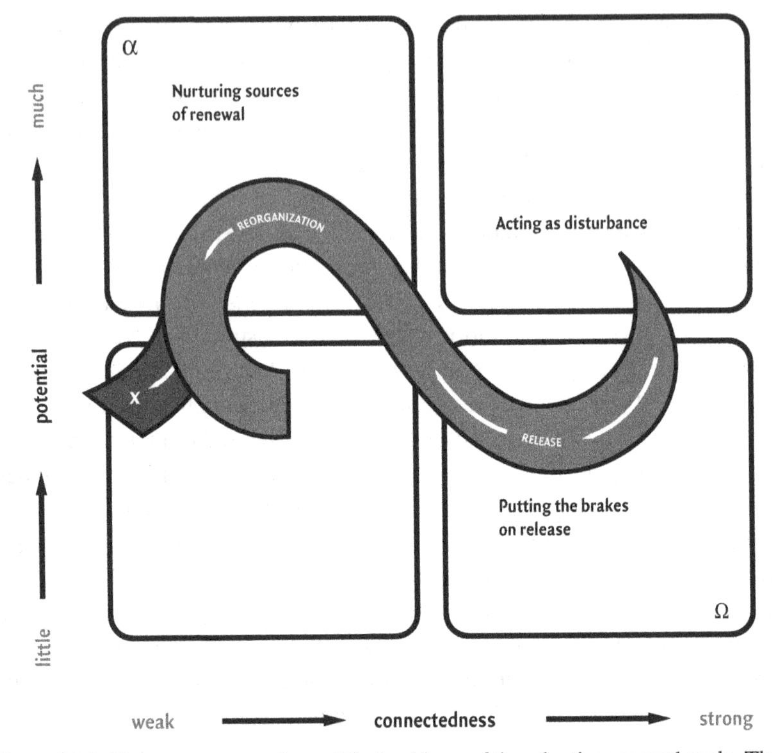

Figure 14.1 Management practices of the backloop of the adaptive renewal cycle. There are management practices of local communities that 'act like small-scale disturbance' to trigger renewal, 'put the brakes' on release to cope with uncertainty and surprise, and 'nurture diversity' and ecological memory to secure sources of reorganization. Modified from Berkes and Folke (2002).

Management that actively behaves like disturbance is one of a sequence of practices, ecological and social, that generates resilience. It appreciates the role of disturbance in development and includes monitoring and ecological understanding of ecosystem condition and dynamics embedded in social institutions. Evoking disturbance is the first sequence of management practices that relate to 'the backloop phases' of the heuristic adaptive renewal cycle model (see the left part of Fig. 14.1). We assume that such practices among resource users have evolved as a response to crisis.

14.2.2 Learning from crisis

For our purposes, a crisis may be broadly defined as a large perturbation; it may be human induced (e.g., a resource collapse) or natural (e.g., the effects of a

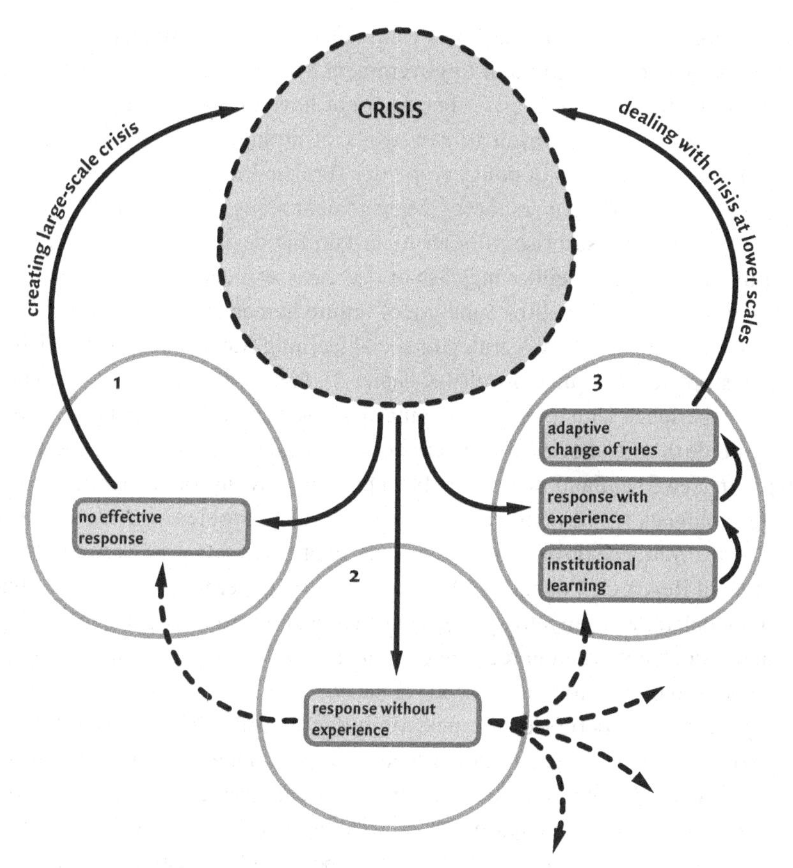

Figure 14.2 Three generic responses to resource and environmental crisis. Modified from Berkes and Folke (2002).

hurricane). In Chapter 2, Gunderson states that crises are a special type of surprise. A surprise (a qualitative disagreement between ecosystem behavior and *a priori* expectations) becomes a crisis when it reveals an unambiguous failure of management actions and policy. Three generic responses are possible when a crisis occurs (Fig. 14.2). 'No effective response' is one possible management reaction. This aims at getting rid of the disturbance by blocking out change. Institutions and organizations that have a lot of inertia or vested political or economic interests characterize the agents of this reaction (Gunderson *et al.*, 1995). Such a response, often based on a presumption of a system near equilibrium, tends to create the conditions for a larger-scale and widespread crisis later on. It creates the revolt connection; trying to preserve the *status quo* often leads to organizational and political brittleness, as well as to ecological brittleness.

A second possibility is 'response without experience,' a frequently seen reaction in which the institution (a government agency or a local resource-user institution) responds to a crisis but does not have previously tested policies at its disposal. It may result in new types of arrangements of management institutions and a series of policy responses (broken lines in Fig. 14.2), including that of 'no effective response.' Management recognizes the need to relate to uncertainty and surprise in order to sustain the desirable stability domain, but the response may either increase or decrease resilience. At the same time, resilience is needed to allow a margin of failure in management. Alternatively, it can lead to institutional learning or social learning (Lee, 1993). If the crisis is a true surprise or genuine novelty (Chapter 2), the institution will have no previous experience with it; or the crisis may have been predictable but may be of a magnitude that had never been experienced. An example is the cod resource collapse in Newfoundland, which had been predicted by inshore fishers and some field biologists (Finlayson and McCay, 1998). The problem was exacerbated by 'an overreliance on the science and culture of quantitative stock assessment' (National Research Council, 1998: 35) by central government agency population modelers, who (in retrospect) misused or misjudged their data, and precipitated a stock collapse unprecedented in its magnitude in the North Atlantic.

Would the Newfoundland cod collapse help the management agency 'respond with experience' the next time that a similar crisis looms? There is no clear answer to the question, because 'responding with experience,' as we formulate it in Figure 14.2, depends on institutional learning based on previous crises and social–ecological memory. Evoking disturbance practices such as those discussed in the previous section may be a result of institutional learning stored in the memory of local communities (Chapters 5–8) and may help avoid unwanted qualitative shifts in stability domains of resource systems (Chapter 2; Carpenter, Ludwig, and Brock, 1999; Scheffer *et al.*, 2001).

The chapter on policy transformation in the US forest sector (Chapter 13) illustrates the inertia and tension of adaptive learning for sustainable forest management at regional and national scales of governance, but the findings of the interviews with resource managers leading diverse groups of constituents through resource management crisis and change in Minnesota move beyond the tension between conserving the old structures and creating new approaches by responding to environmental feedback (Chapter 9). Creating social and institutional space or platforms for dialogue and innovation (Röling and Wagemakers, 1998) following crises is key to stimulating learning and to resolving social uncertainties (Chapter 2). It opens the way for double-loop learning, drawing on social–ecological memory and visioning, expanding the temporal frame of reference, and reorganizing conceptual models and paradigms based on a revised understanding of the preconditions generating the crisis (Chapter 9). Hence, institutions seem to emerge as a response to crisis and are reshaped by crisis.

14.2.3 Expecting the unexpected

There are numerous local management practices and associated institutions that avoid large-scale crises by relating to uncertainty and surprise in order to survive their effects. Such practices 'put the brakes on release' (Berkes and Folke, 2002). Instead of trying to get rid of disturbance, the existence of uncertainty and surprise and their unpredictable nature are an accepted part of development, and management actions evolve to cope with their effects by spreading risks through diversification of both resource use patterns and alternative activities (Chapters 3 and 7). Such responses seem to contribute to social–ecological resilience by aiming at protecting a desirable stability domain in the face of change. Several of these local resource and ecosystem management strategies resemble risk spreading and insurance building within society, similar to portfolio management in financial markets (Costanza *et al.*, 2000), and deal with the three types of surprises outlined by Gunderson in Chapter 2.

Strategies designed to improve survival when faced with resource uncertainty and surprise include practices that manage biodiversity through redundancy at several levels, from the species to the landscape level. There are many traditional agricultural groups that conserve low production crop varieties as insurance to climate and pest events that impact high-yield crop varieties (Chapter 6). Such ecological strategies include investment in emergency food crops (Turner, Boelscher Ignace, and Ignace, 2000), use of polyculture as a bet-hedging strategy to reduce vulnerability to tropical cyclones, and planting disturbance-tolerant crop varieties (Chapter 7). Multiple-disturbance-resistant species among the char-dwellers in Bangladesh serve as emergency foods (Chapter 7).

Conserving patches in the landscape to serve as emergency resource supply in the face of change is a common practice. The establishment of range reserves within the annual grazing areas among the African agro-pastoralists is one example (Chapter 6). These reserves provide an emergency supply of forage, which functions to maintain resilience in the face of disturbance of both the ecosystem and the social systems. They serve as 'savings banks' when drought challenges the processes and functions of the dryland ecosystem (Niamir-Fuller, 1998). In a similar fashion, Swedish forest commons served and still serve as a 'savings account' for many local farmers (Chapter 5). These practices deal with ecological uncertainty and surprise and accept disturbance and crisis as a part of reality in managing complex ecosystems and living with them. They reduce the impacts and improve survival during the duration of a disturbance (Fig. 14.2). We believe that these practices are the result of a long-term trial-and-error process of social–ecological response and adaptation to environmental unpredictability.

In traditional societies and other communities, there is an array of social mechanisms, such as systems of flexible user rights and land ownership, in combination with coping mechanisms, such as reciprocal gift giving, that help members of these communities survive periods of resource and ecological crises. As suggested by Low *et al.* in Chapter 4, diversity and redundancy of institutions and their overlapping functions may play a central role in absorbing disturbance and in spreading risks.

Building social–ecological resilience in the face of uncertainty and surprise is about promoting the capacity to expect the unexpected and absorb it (Kates and Clark, 1996). Accumulating experience through institutional and social learning is important in this context (Lee, 1993; Olsson and Folke, 2001; Berkes and Folke, 2002). If experience embedded in institutions provides a context for the modification of management policy and rules, the institution can act adaptively to deal with the surprise. It can navigate nature's dynamics and do so by diversification and redundancy rather than simplification (Chapter 4). Furthermore, surprise and crisis create space for reorganization, and for renewal and novelty (Chapter 2), and provide opportunities for new ways of social and institutional self-organization for resilience (Chapter 9).

However, a dynamic balance is needed between experience and memory, on the one hand, and the amount of renewal and novelty, on the other. Creativity needs to be framed. The frame is the history, the accumulated experience and memory of the systems, and it involves interactions across spatial and temporal scales (Holling, 2001; Gunderson and Holling, 2002). The dynamic balance between memory and novelty is dealt with in the next section.

14.3 Nurturing diversity for reorganization and renewal

We have emphasized the common response of managing for diversity when faced with disturbance, uncertainty, surprise, and crisis. Diversity as part of resilience provides complex social–ecological systems with the ability to persist in the face of change. Managing complex systems implies spreading risks and creating buffers and not putting all eggs in one basket. In addition to the insurance aspect dealt with in the previous section, diversity also plays an important role in the reorganization and renewal process following disturbance. It is in this context that the memory – ecological and social – becomes significant, because it provides a framework of accumulated experience for coping with change. It provides the frame for creativity and adaptive capacity. We start with ecological memory and management practices that nurture this memory. These are followed by an identification of functional aspects of social memory of importance for sustaining adaptive capacity and of how the diversity of social and ecological memory relates to social–ecological resilience.

14.3.1 Nurturing ecological memory

Species perform key functions in ecosystems. However, it is not the number of species *per se* that sustains an ecosystem in a certain stable state or stability domain, but rather the existence of species groupings, or functional groups (e.g., predators, herbivores, pollinators, decomposers, water-flow modifiers, nutrient transporters), with different, sometimes overlapping, characteristics or multiple, nonidentical, in-use copies (Chapter 4). Species that may seem redundant and unnecessary for ecosystem functioning during certain stages of ecosystem development may become of critical importance for regenerating and reorganizing the system after disturbance and disruption (Holling *et al.*, 1995). Others may not. Seemingly redundant species connect habitats through their overlapping functions within and across scales (Peterson, Allen, and Holling, 1998). Overlapping functional diversity increases the variety of possible alternative reorganization patterns and pathways following disturbance and disruption and contributes to ecosystem resilience.

The ecological memory is an important component of overlapping functional diversity. Ecological memory is the composition and distribution of organisms and their interactions in space and time, and includes the life-history experience with environmental fluctuations (Nyström and Folke, 2001). Ecological memory consists of at least three basic and interacting assemblages and their overlapping functional diversity (Fig. 14.3a). The first is *biological legacies* or species and patterns that persist within an area affected by disturbance, like a tree surviving a fire or a seed that requires fire to germinate (Franklin and MacMahon, 2000). The second is *mobile links*, i.e., species of functional groups that migrate between areas (Lundberg and Moberg, in review). These links include species that passively spread from one area to another, like larvae through currents or seeds through wind, or actively move between areas, like fish, birds, bats, and mammals, and that contribute to reorganization of the area hit by disturbance. The third is the *support areas* for the functional links, i.e., a diversity of habitats in the landscape of which the disturbed area is a part.

Each of these assemblages consists of several functional groups, interacting with overlapping functions as a dynamic ecological community and linked with hydrological, biogeochemical, and other abiotic flows. The vulnerability

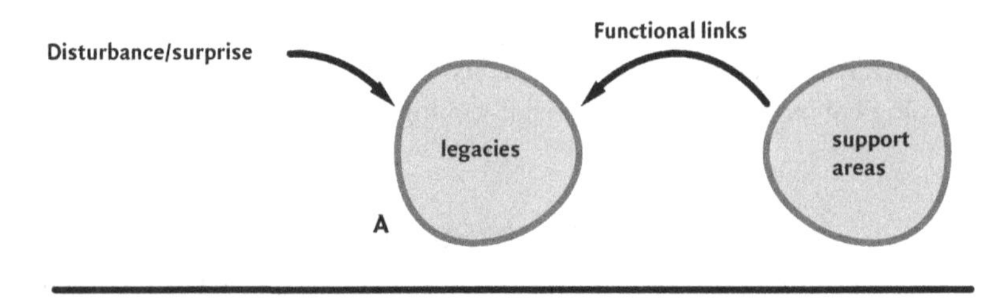

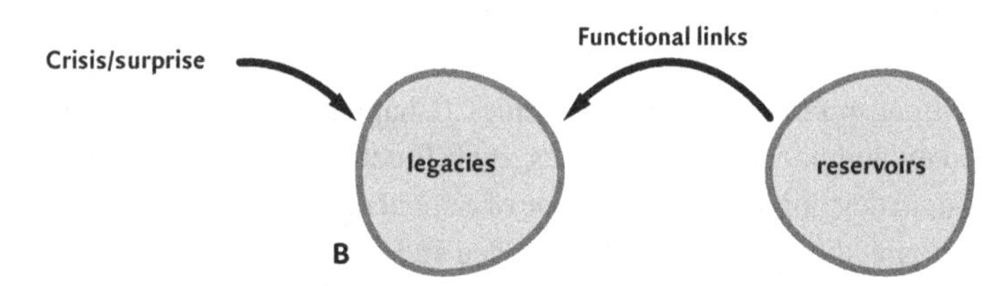

Figure 14.3 Components of memory for adaptive capacity in ecological and social systems. The ecological memory (A) consists of a diversity of species within and between functional groups (e.g., photosynthesizers, pollinators, seed dispersers, grazers, predators, decomposers, nutrient transporters, water flow modifiers). These groups exist as biological legacies in the area subject to disturbance, and as mobile species in functional groups that link the disturbed area to other areas. Support areas provide habitat for the functional groups of mobile links. The social memory (B) consists of a diversity of individuals, institutions, organizations, and other actors with different and overlapping roles within and between critical groups (e.g., knowledge carriers and retainers, interpreters and sense makers, stewards and leaders, networkers and facilitators, visionaries and inspirers, innovators and experimenters, entrepreneurs and implementers, followers and reinforcers; see Box 14.1). The groups exist as legacies on the arena subject to crisis, and as functional links that enter from outside to take advantage of the space created by the crisis. They draw on reservoirs of practices, knowledge, institutions, cultural values, and worldviews. The combination of such sources of ecological and social resilience, and their overlapping functions and redundancy, provides potential for reorganization, novelty, and adaptive capacity in the face of disturbance and crisis. Functional social–ecological diversity is a key ingredient of adaptive capacity and a key characteristic of resilience.

of functional groups seems to be related to redundancy of species within each group and suggests that ecosystems have lower resilience to disturbance when functional groups have low diversity (Nyström, Folke, and Moberg, 2000). The presence and dynamics of this network of interacting species with overlapping functions are the result of previous experience and accumulated information in the life history of species to environmental change (Bengtsson *et al.*, in press). The memory connects a system's present to its past and to its neighbors. The rates of reorganization and the level of similarity between the old and the new ecosystem are a function of the extent and availability of ecological memory in the landscape or the seascape. This does not imply tightly connected systems, but systems with redundant and loosely connected subsystems, which allow persistence in the face of change (Chapter 4).

As illustrated in several chapters of the volume, there are many ecological practices in both traditional and contemporary societies supporting ecological memory and biological diversity, thereby contributing to the regeneration and renewal of disturbed ecosystems (see also Fig. 14.2). Obvious ones include the protection of species, including keystone species, as well as a ban on harvesting during certain stages in their life cycle, and the setting aside of reserves, protected areas, and other socially fenced areas (e.g., sacred groves, spirit sanctuaries, and buffer zones).

To be effective, these habitats need a social context. Contemporary society tends to create reserves for the conservation of biodiversity *per se*. It is only recently that such reserves are being implemented to enhance production in surrounding areas (Allison, Lubchenco, and Carr, 1998), and there are still many problems with the monitoring, enforcement, and sanctioning of reserves and protected areas. Protected areas common in indigenous cultures include sacred groves and other forms of taboo systems (Ramakrishnan, Saxena, and Chandrashekara, 1998; Colding and Folke, 2001). There are communities that provide temporal and spatial refugia to a number of ecologically viable species involved in the generation of ecosystem services. These species may hide, forage, and reproduce in the vicinity of the local ecosystems belonging to these communities. Protection of species, especially keystone ones, is important for communities prone to large-scale and frequent natural disturbances (Chapter 7). Smaller terrestrial and aquatic ecosystems, such as forest patches, coastal reefs, and river stretches, may receive protection from habitat taboos that conserve ecological services on which the local community depends (Chernella, 1987; Gadgil, 1987; Ruddle, 1988; Ramakrishnan *et al.*, 1998).

Protection of species and habitats helps nurture ecological memory through diversity and supports the reorganization phase of ecosystem development. Other measures include management practices during ecological succession that make use of and support biological diversity in the production of food, timber, and other ecological services (Chapter 6). They play a significant role in building resilient landscapes for adaptive capacity (Gadgil, Berkes, and Folke, 1993; Berkes, Folke, and Gadgil, 1995). The biological diversity such practices support helps sustain redundancy within and between functional groups of species and provides ecological memory for reorganization in a patch or an area following disturbance (Fig. 14.3A).

Such management in the context of landscape dynamics, with patchy mosaic of ecosystems in different development stages, is in stark contrast to land-use transformations aiming at optimal production and control of commodities over vast areas (Bengtsson *et al.*, in press). In homogenized and intensively managed landscapes, the ecological memory is degraded, more distant and reduced, leading to much longer time periods for reorganization. One may speculate that simplified landscapes, with low levels of ecological memory and homogeneous spatial patterns, will be more prone to opportunistic invasive organisms and may more easily shift between stability domains. Such systems will probably be subject to higher variability and lower predictability concerning their capacity to sustain a flow of ecosystem services. Homogenization leads to ecosystems that are more susceptible to disturbance and consequent regime shifts (Scheffer *et al.*, 2001; Gunderson and Pritchard, 2002). Due to reduced ecological memory, simplified landscapes will probably require management interventions to secure ecosystem services and avoid shifts into undesirable stability domains. As human actions continue to remove or degrade ecological memory that provides redundancy, there comes a time when the buffering capacity is lost, and management is confronted with a surprise, often leading to a crisis (Chapter 4). As stated by Gunderson in Chapter 2, ecological memory and its functional diversity maintain the capacity for renewal in a dynamic environment by providing a buffer that protects the system from the failure of management actions that are based upon incomplete understanding, and thereby allow managers to learn and actively adapt their resource management policies.

14.3.2 Sustaining social memory

Through institutional and social learning (Lee, 1993), resource users develop a collective memory of experiences with resource and ecosystem management. This memory provides a context for social responses to ecosystem change, increases the likelihood of flexible and adaptive responses, and seems to be particularly important during periods of crisis, renewal, and reorganization. It draws on experience, but allows for novelty and innovation within the framework of accumulated experience. The institutional memory is an accumulation of a diversity of experiences concerning management practices and rules-in-use at the level of institutions, as opposed to that of individual resource user. Institutional memory provides the foundation for the modification of rules, and typically refers to decadal time scales, as opposed to months or a year.

Institutional memory is a subset of social memory that bridges the deepest values and symbolic truths of a society and the social or ecological environment on which members of that society have to act. Social memory is the arena in which captured experience with change and successful adaptations, embedded in a deeper level of values, is actualized through community debate and decision-making processes into appropriate strategies for dealing with ongoing change (McIntosh, 2000). It is a part of the cultural capital of human society (Berkes and Folke, 1992). Social memory embeds long-term historical and cultural observations (McIntosh, Tainter, and McIntosh, 2000), of which cultural diversity (Gadgil, 1987) and a diversity of worldviews (Chapter 10) linked to cultural evolution (Boyd and Richerson, 1985) may play an essential role in nurturing resilience for adaptive capacity.

In Figure 14.3B we initiate the identification of key aspects of social memory, in parallel to ecological memory (Fig. 14.3A). The social memory consists of a diversity of individuals, institutions, organizations, and other players with different but overlapping roles within and between critical functional groups. Some suggested 'functional groups' of social memory are listed and exemplified in Box 14.1.

When there is a crisis, space is created for renewal, reorganization, and novelty. The crises may be external markets and tourism pressure (Chapter 11), floods and flood management or changes in property rights (Chapter 7), acidification (Chapter 8), resource failures like the caribou crisis (Chapter 10), rigid, conventional paradigms of resource management (Chapter 9), new legislation (Chapter 13), or governmental policies that do not take into account local contexts or traditional societies (Chapter 12).

Groups such as the ones identified in Box 14.1 play a part (a) among the survivors on the arena subject to crisis, (b) among the functional links that enter from outside to take advantage of the space created by the crisis, and (c) among the reservoirs of practices, knowledge, institutions, cultural values, and worldviews supporting the functional links (Fig. 14.3B). In real life, some of these functions are combined. For example, among the Cree of James Bay, elders combine the functions of knowledge carriers, sense makers, and stewards (Berkes, 1998). We hypothesize that the combination of functional groups related to social memory, their diversity, overlapping functions, and their redundancy provides resilience for reorganization and novelty, and thereby enhances adaptive capacity in the face of disturbance and crisis.

But their combination may also cause barriers, collision, and erosion of memory, as may be the case when different cultural value systems, worldviews, and discrepancies in conceptualization are brought together and interact (Chapters 6 and 10), or when the cultural dynamics created by the policies of those in power during earlier periods may inhibit the development of the ability to respond to disturbance and surprise through building resilience (Chapter 13; Gunderson

Box 14.1 Functional groups for social memory

A number of distinct, but often overlapping, roles exist regarding social memory in social–ecological systems. Note that we are not trying to characterize all social roles in society, but merely the functional roles by which resource users develop and retain a collective memory of resource and ecosystem management. Based on the chapters of the volume, we identify eight such roles: knowledge carriers and retainers; interpreters and sense makers; networkers and facilitators; stewards and leaders; visionaries and inspirers; innovators and experimenters; entrepreneurs and implementers; followers and reinforcers.

Knowledge carriers and retainers of memory include the wise men of the traditional society in Tanzania, who have regular meetings regarding such things as the status of the land, issues on drought, diseases, and land fertility (Chapter 6). Other examples include the institution in Samoa that sustains the practical knowledge of polyculture as a resilience strategy in the face of large unpredictable disturbances (Chapter 7), and the sense-of-place project of cultural revival and remembrance in the Sonora watershed (Chapter 3).

Interpreters and sense makers include the biology teacher in the local community in Sweden who continuously processes ecological information into practical knowledge and makes it accessible for decision-making (Chapter 8), and the fishers who monitor pulses of fish migrations in the Brazilian lagoon fishery (Chapter 11).

Networkers and facilitators include the resource managers of Minnesota leading diverse groups of constituents through crises and change (Chapter 9). Other examples include the networkers and organizers in Peoples' Biodiversity Registers in India (Chapter 8), organizers of forest commoners in Sweden (Chapter 5), the second Committee of Scientists in the US forestry sector in the attempt to facilitate the ecosystem approach in forestry (Chapter 13), and Dayak educational program organizers who bring together elders and adults to help preserve the memory of useful species, ritual protections, and sacred areas (Chapter 12).

Stewards and leaders include those in the tiny Inuit community in eastern Hudson Bay who organized a major aboriginal peoples' assessment of regional environmental change, and the two stewards who triggered the evolution of the local Swedish fishing association for watershed management (Chapter 8).

Visionaries and inspirers who initiate and create incentives for renewal and reorganization following crisis in resource management are illustrated in the Minnesota example (Chapter 9), and in the aboriginal forest management case in British Columbia (Pinkerton, 1998).

Innovators and experimenters, such as those in the caribou co-management case, bring together social memories of different cultures for mutual learning and novelty, a process that requires respect and trust building (Chapter 10).

Entrepreneurs and implementers often take innovations and apply them, as in the indigenous social network in Indonesia that developed and implemented a new social structure to respond to crisis from external drivers (Chapter 12).

Followers and reinforcers are those people and groups, the willing participants, who make a project work, such as those in the local fishing association in Sweden initiated by two key stewards, and the participants of the Peoples' Biodiversity Registers (Chapter 8).

et al., 1995). In this sense, the underlying worldview of resource management imposes a grid on social memory for managing ecosystem dynamics.

14.3.3 Enhancing social–ecological memory

Sustaining ecosystem capacity that fosters ecological memory requires continual social and institutional adjustments to environmental dynamics and ecological feedback. The experience, of the role of disturbance, uncertainty and surprise, and the need to nurture biodiversity and conserve ecological memory for maintaining adaptive capacity, must be stored in the social memory of resource users and managers and be expressed in management practices that build resilience. For example, elders with extensive ecological knowledge (Berkes, 1999) and other similar 'stewards of wildlife habitats' (Nabhan, 1997) are carriers of social memory of resource and ecosystem dynamics, with observations that often include an understanding of long-term and large-scale changes (Chapter 3; Berkes and Folke, 2002). Rituals in both traditional and contemporary society play a role in transforming social memory into practical resource and ecosystem management (Chapter 8; Alcorn and Toledo, 1998). Local monitoring can provide a more effective response to signals of change in ecosystem dynamics than does monitoring by centralized agencies (Chapter 5). Key individuals who act as visionary leaders create incentives for cross-scale institutional and organizational collaboration and contribute to sustaining and building institutional and social memory for ecosystem management (Chapter 9; Pinkerton, 1998; Olsson and Folke, 2001).

Enhancing social–ecological memory also involves sharing experiences of crisis at wider levels of society. This is the case in West and East Kalimantan, where the organization operating at broader scales provides legal training and facilitates local communities to share experiences from confronting logging and plantation companies. It also involves nurturing the opportunity for self-organization in the institutional or political space created by crisis. For example, the great fires of Indonesia in 1997 opened up new opportunities for negotiation among non-governmental organizations (NGOs) and the political elite (Chapter 12).

Institutional and social memory for resilience should not be uniform and static, but diverse and dynamic. We propose that systems with uniform and static memory, with limited carriers of memory, or few structures for storing and developing memory, are more vulnerable to change and surprise. They have lower adaptive capacity. In this sense, redundancy among individuals and institutional and social redundancy may serve functions similar to the overlapping functions of ecological redundancy for reorganization and renewal after disturbance (Peterson *et al.*, 1998), recognizing that redundancy implies similar but not identical functional roles. In Chapter 4, Low *et al.* suggest that policies and institutions that recognize, and respond to, the inherent redundancy of ecosystems are much less likely to be surprised by cumulative erosive actions or loss of ecological resilience. They propose that institutions organized in ways that parallel the structure of ecosystems are more likely to receive accurate and timely information about the state and dynamics of the systems and be able to respond in constructive ways. The existence of seemingly redundant local institutional variations may imply that institutional responses can be more potent and rapid than otherwise, and that local variations may be able to meet unforeseen contingencies.

In a metaphorical sense, such characteristics of social memory may be compared with the proposed role of the diversity of species with and between functional groups in ecosystem reorganization (Chapter 4; Walker, 1992; Peterson *et al.*, 1998; Nyström *et al.*, 2000). For example, the system may do well with only one steward under periods of slow and stable development, but when faced with change, other stewards may be needed to buffer surprise and crisis, renew and innovate, and reorganize. Similarly, seemingly redundant rules may be waiting in the wings to secure the ability of ecosystem management to respond to uncertainty and surprise (Chapter 4). Without diversity and redundancy in functional groups, the resource management system may be fragile and vulnerable to disturbance and crisis, as illustrated by the bifurcation policies in crayfish management in Sweden, balancing on the edge of shifting into another stability domain of management (Chapter 8). Such sliding may lock the system in a certain development trajectory, with an irreversible loss of adaptive capacity.

There has to be a potential for combining and recombining social memory, adding and filtering external influences (Chapters 6, 11, and 12), as well as transferring knowledge and experience in space and time (Chapter 8) to enhance overlapping functions and redundancy for resilience. Based on the discussion of social–ecological memory in resilience (Fig. 14.3), we propose that human actions and innovations framed by a dynamic, diverse, and evolving social memory in tune with ecosystem dynamics have the potential to foster adaptive capacity in social–ecological systems. We also argue that social memory needs to encompass ecological knowledge and understanding, including knowledge of local resource users. In the introductory chapter, we stated that such ecological knowledge and understanding are a key link between ecosystem dynamics and their successful management, and that this knowledge could fruitfully be combined with other knowledge systems, the subject of the next section.

14.4 Combining different types of knowledge for learning

Sustainable use of the capacity of ecosystems to generate services is unlikely without improved understanding of ecosystem dynamics. It has been argued that all forms of relevant information should be mustered to increase knowledge and understanding for improved management of complex ecosystems, including different systems of knowledge and their combination (Berkes and Folke, 1998). Here we address the relationship between experiential and experimental knowledge, the need to expand knowledge from structure to function of nature, the incorporation of knowledge of ecological processes and dynamics into institutions, and the increased potential for learning and building social–ecological resilience by making use of and combining different knowledge systems.

14.4.1 Experiential and experimental ecological knowledge

Only a fraction of the dynamics of ecosystems is likely to have been the subject of careful observations within the framework of formal science. A large proportion would be part of the experience of the people living, observing, and using the systems in a variety of contexts (Chapters 5–8 and 11). Monitoring change is key to increasing the ability to respond to change and shape institutions and management practices. Resource users' local ecological knowledge incorporates knowledge derived from historical observations of 'natural experiments' and the dynamics of these systems. It is practical and place based. In Chapter 8, Gadgil *et al.* argue that such 'experiential' knowledge of many local communities and indigenous societies may play an important role in the understanding of the behavior of ecological systems, particularly in situations of crisis and reorganization. Such practical working knowledge may be a valuable complement to scientific 'experimental' knowledge in addressing the dynamics of complex adaptive ecosystems and their management (Johannes, 1998; Levin, 1999). In this sense, practice informs theory as much as theory informs practice. The strength of conventional experimental science is in the collection of synchronic (simultaneously observed) data, whereas the strength of many experiential management systems is in diachronic information (long time series of local observations).

Local knowledge systems are not static, but are in continuous development. There are knowledge systems with conceptualizations of the world that are quite different from science-based resource management knowledge (Chapter 10; Berkes 1999). However, actually held ecological knowledge of local communities facing a globalized world increasingly tends to become a combination of locally generated experiential knowledge and outside knowledge, of which scientific knowledge plays an important role (Olsson and Folke, 2001). Furthermore, local knowledge systems in themselves will not be sufficient for maintaining sustainable resource use, because the world is increasingly human dominated and interconnected across spatial and temporal scales, both ecologically and socially (Chapter 12). The important point to make is that an adaptive learning process for managing ecosystems for social–ecological resilience should not dilute, homogenize, or diminish the diversity of experiential knowledge systems for ecosystem management, with lessons for how to respond to change and how to nurture diversity.

14.4.2 Expanding from the knowledge of structure to the knowledge of function

The bulk of ethnobiological knowledge analyzed by scientists concerns taxonomic knowledge or knowledge about species or particular resources. In this volume, we have been more concerned with ecological knowledge of ecosystem processes and functions. Several chapters analyze the capacity of ecosystems to sustain human well-being, including the role of functional biodiversity underlying the generation of particular resources or ecosystem services, as well as the role of society in managing this capacity. By analyzing management practices, we have been able to focus on what is sometimes referred to as 'tacit' ecological knowledge and understanding (Chapter 9; Polanyi, 1958) and how it contributes to maintaining the stability of human communities in the face of change. We have also illustrated that local knowledge of ecosystem dynamics is not static, but in continuous development, drawing on social–ecological memory.

14.4.3 Building process knowledge into institutions

Management practices of local resource users and communities do not exist in a vacuum but are framed by a social context. Hence, they tend to be embedded in institutions and other forms of social mechanisms (Chapters 5–8 and 11). Those that work in synergy with ecosystem dynamics are thought to be supported by worldviews and cultural values that do not decouple people from their dependence on natural systems (Chapters 3, 9, and 12).

The importance of coupled social–ecological management of ecosystem dynamics can be exemplified from a study on frontier colonist farmers in the Brazilian Amazon (Muchagata and Brown, 2000). People moving from one area to another rapidly gained detailed knowledge of particular resources, but their knowledge of the processes and functions of the underlying ecosystem was very patchy. Humans seem to acquire fairly rapidly the knowledge of biological structures and the taxonomy of resources, but securing the flow of resources requires the capacity of the ecosystem to sustain its processes and functions, the knowledge of which takes a longer time to develop. Securing the capacity requires dwelling for long periods of time in the ecosystem and a sense of place perspective (Chapter 3). Such knowledge acquisition is an ongoing, dynamic learning process; perhaps most importantly, it seems to require social networks and an institutional framework to be effective. Thus, it is not effective to separate ecological studies aimed at management from the institutional framework within which management takes place. Understanding ecosystem processes and managing them is a progression of social–ecological co-evolution, and it requires learning and accumulation of ecological knowledge and understanding in the social memory. In this sense, a collective learning process that builds knowledge and experience with ecosystem change evolves as part of the institutional and social memory, and it embeds practices that nurture ecological memory. We propose that, over the longer term, environmental wisdom evolves as part of such a process.

If this is true, knowledge generation in itself will not be sufficient for building adaptive capacity in social–ecological systems to meet the challenge of navigating nature's dynamics. Learning how to sustain social–ecological systems in a world of continuous change needs an institutional and social context within which to develop and act (Chapters 2, 9, and 10). The knowledge system itself becomes part of the processes of institutional and social learning and memory to deal with ecosystem dynamics.

14.4.4 Fostering complementarity of different knowledge systems

Local ecological knowledge systems of resource users and local communities can complement conventional resource management in at least three ways. Here we discuss them in relation to the adaptive renewal cycle presented in the introductory chapter:

- qualitative monitoring and management during the exploitation and conservation phases of the adaptive renewal cycle;
- building resilience during the release and reorganization phases; and
- providing long time series of local observation and institutional memory for understanding ecosystem change.

The exploitation and conservation phases are the two stages that have been the main concern of conventional resource management science, with emphasis on the collection of quantitative data. Many locally developed management systems, however, collect and use qualitative information, as the examples in Chapters 6–9 indicate. The two approaches, and the kind of information they collect, are complementary in that they may be used to add to the strengths of one another.

The release phase of the adaptive renewal cycle seems to be largely ignored by conventional resource management, and the reorganization phase is largely understudied. By contrast, many experiential-knowledge-based systems seem to accord a great deal of emphasis to these backloop phases, judging by the rich variety of practices that exists in a variety of cultures and geographic areas (Folke *et al.*, 1998). Many societies interpret and respond to feedback from complex adaptive ecosystems, and their practices are not only locally adaptive, but generalizable over wide regions (Chapter 8; Berkes, 1999). These practices provide insights for contemporary resource management, and complement existing approaches (Berkes, Colding, and Folke, 2000).

Conventional science does not have a great time depth of environmental data. For example, instrumental data on climate change in the Canadian Western Artic only date back to the mid-1950s. Because assessing climate change requires baseline information and a longer time series of observations than is available, a study was initiated in 1998 to tap the knowledge of Inuvialuit hunters and fishers. The results of the project showed the feasibility of using traditional knowledge to complement scientific data in five areas. These are the use of traditional knowledge (i) as local-scale expertise; (ii) as a source of climate history and baseline data; (iii) in formulating research questions and hypotheses; (iv) as insights into impacts and adaptation in Arctic communities; and (v) for long-term, community-based monitoring (Riedlinger and Berkes, 2001).

Thus, scientific understandings of complex adaptive systems and their change could be enriched by insights from local management systems. Several chapters of the volume support this conclusion. Conceptually, one may think of a sequence of knowledge systems for ecosystem management. In some, knowledge for ecosystem management can be based on either traditional ecological knowledge (e.g., Chapters 6, 7, and 12; Berkes, 1999) or scientific/governmental knowledge (e.g., Chapter 13; Gunderson *et al.*, 1995). Some aim at bringing insights from traditional ecological knowledge into scientific knowledge and *vice versa* (e.g., Chapter 8). In others it may be hard to separate the content of practical or local ecological knowledge from the influence of scientific knowledge (e.g., Chapter 11; Olsson and Folke, 2001). Furthermore, there may be deliberate attempts to co-manage and combine different knowledge systems. For example, in Chapter 9, Kendrick discusses the role of co-managing knowledge among interest groups with different worldviews, characterizing them as mutual learning systems. Blann *et al.*, in Chapter 10, emphasize the importance of bringing diverse interest groups together in temporary learning systems, or platforms for learning, in the management of complex dynamic ecosystems. Adaptive management (Holling, 1978) is about creating platforms and involving user groups and interest groups for knowledge sharing about complex ecosystem management and for relating to uncertainty and surprise. Adaptive co-management explicitly recognizes the necessity of combining adaptive management with institutions across scales. There is appreciation throughout the volume of the tight coupling between knowledge of ecosystem dynamics and of institutions in social–ecological systems. Such coupling is an important characteristic of self-organized complex systems (Levin, 1999); process knowledge is integral to institutions in an ongoing feedback and learning process.

14.5 Creating opportunity for self-organization toward social–ecological sustainability

Resilience may be considered a precondition for adaptive capacity, the capacity to respond to and shape change. Adaptive capacity includes learning and changing of resource management rules as experience accumulates and social–ecological memory develops. Adaptation in self-organized systems is related to rules-of-thumb or simplified rules (Holland, 1995). These rules govern how the system adapts in response to past and present conditions. In the self-organized process, patterns of behavior and framework emerge that help facilitate future learning processes (Levin, 1999). However, responding to and shaping change can take different pathways. For example, adaptation may concentrate on reducing the impacts of change, or it may take advantage of new opportunities created by change. Environmentalists and the nature conservation movement have tended to focus on reducing the impacts of change on nature, and preserving nature as it is, whereas those promoting progress see conservation as a constraint, and tend to focus on the new opportunities that change creates for economic development. A conflict is created, the positions are locked, and there is a battle between perspectives and worldviews. The interplay between sustaining and developing is not recognized. In this section we discuss the significance of the interplay between diversity and disturbance for social–ecological resilience and adaptive capacity, as well as cross-scale issues in coupled social–ecological systems, and the relation to external drivers of change that may support or contest resilience.

14.5.1 Recognizing the interplay between diversity and disturbance

Resilience buffers change, thereby contributing to sustaining the system in a desirable stability domain. Memory is a key component of resilience. It supplies the experience of previous self-organizations and the ingredients for new self-organization embedded in a historical and evolutionary context. Change, surprise, and crisis, here referred to as disturbance, open up space for change. The creation of space allows for opportunity and renewal by recombining memories and by bringing in novelty. Diversity of memories provides the potential for alternative ways to maintain functioning when faced with change. Diversity provides insurance to cope with change. It also provides the potential for reorganizing following change. Hence, there is a dynamic symbiosis between diversity and disturbance that is part of resilience and key to sustainable development. Disturbance can trigger positive change when there is memory, but, with lack of memory for resilience, a similar disturbance may cause severe consequences. Human actions that alter and accumulate disturbance and reduce memory seem to generate vulnerable social–ecological systems through loss of resilience – vulnerable in the sense of balancing on the edge of undesirable stability domains.

Hence, there is a dynamic interplay between reducing the impacts of change and at the same time taking advantage of the opportunities created by change. Systems where change is not allowed will almost certainly generate surprise and crisis. Systems that allow too much change and novelty will suffer loss of memory. The interplay between change, capacity to respond to and shape change that allows for renewal is key in self-organization. The dynamic process requires social–ecological memory with functional diversity (including redundancy) for turning disturbance into options for renewal and novelty. In this sense, memory frames creativity. Holling (2001) coined the concept panarchy (a heuristic model of a hierarchy of nested adaptive renewal cycles, see Chapter 1) to capture the scale dimension of the interplay between diversity and disturbance. Each level in the panarchy operates at its own pace, protected from above by slower, larger levels with spatial and temporal memory, but revitalized from below by faster, smaller cycles of innovation. The whole panarchy therefore both creates and conserves, combining learning and innovation with continuity. Systems can develop, and at the same time be sustained (Gunderson and Holling, 2002), by operating in a context of framed creativity.

The dynamic interplay between disturbance and diversity is stressed in Figure 14.4. Based on the findings of the volume, we argue that the interplay of disturbance and functional diversity, as a part of memory, is a prerequisite for building resilience for adaptive capacity in linked social–ecological systems.

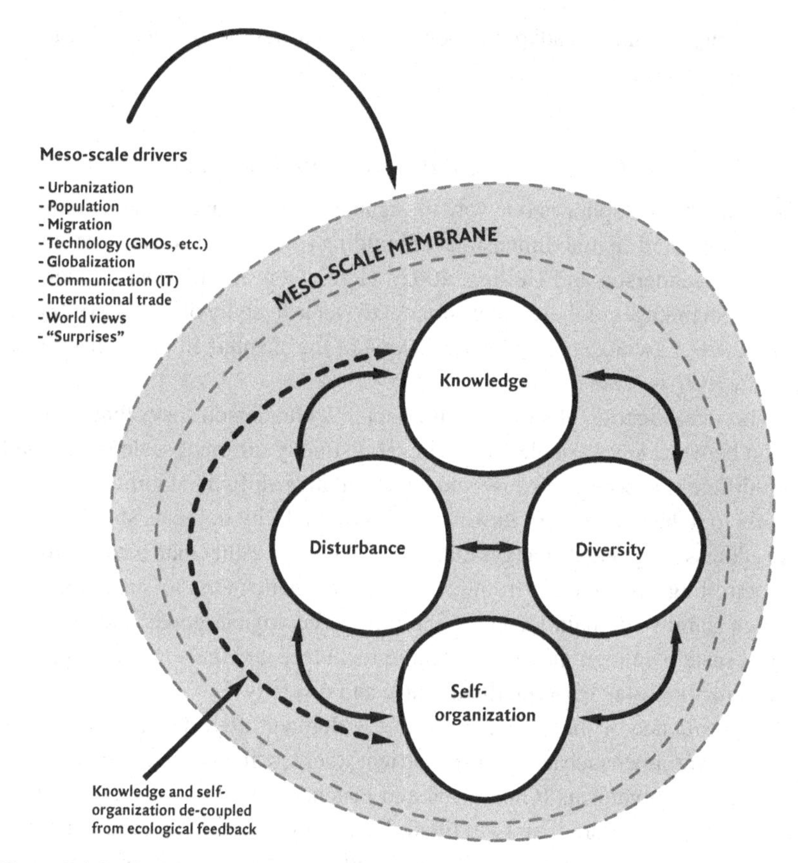

Figure 14.4 The interplay between disturbance and diversity and their relationship to knowledge systems and self-organization as a prerequisite in the building of resilience for adaptive capacity in social–ecological systems. The figure also addresses cross-scale dynamics and the matching to meso-scale drivers of human domination.

Creating platforms for conflict resolution and participation by various interest groups for knowledge creation will not be sufficient for sustainability. It requires the context of the interplay between diversity and disturbance in resilience. Ecological knowledge and understanding of this interplay are a necessity as well.

Learning to live with change and uncertainty and nurturing diversity for reorganization and renewal are an ongoing dynamic process of sustaining development. But they require a fundamental shift in thinking and perspective from assuming that the world is in a steady state and can be preserved as it is, by focusing on preventing and controlling change, to a recognition of change being the rule rather than the exception, and thereby concentrating on managing the capacity in complex adaptive social–ecological systems to live with change and shape change.

14.5.2 Dealing with cross-scale dynamics

The panarchy metaphor is one way to organize thinking and understanding of the temporal and spatial dimensions of building resilience in social–ecological systems (Gunderson and Holling, 2002). Here, we deal with institutional and ecosystem linkages of these dimensions and over a diversity of scales (McIntosh *et al.*, 2000; Costanza *et al.*, 2001). Several of the chapters in this volume deal with such cross-scale dynamics.

In his classification of surprises in Chapter 2, Gunderson states that an interaction between key variables operating at distinctly different scale ranges and with different speeds can cause sudden qualitative shifts in stability domains. Efforts to reduce the risk of unwanted shifts in stability domains should therefore address the gradual changes that affect resilience, rather than merely aiming at controlling disturbance. In other words, the challenge is to sustain a large enough stability domain to secure the flow of ecological goods and services, and to sustain enough memory to secure resilience and the adaptive capacity-enhancing interplay between disturbance and diversity.

The resilience of the ecological stability domain often depends on slowly changing variables such as land use, nutrient stocks, soil properties, and biomass of long-lived organisms (Gunderson and Pritchard, 2002). These factors may be predicted, monitored, and modified. By contrast, stochastic events that can trigger threshold effects and shifts in stability domains (such as hurricanes, droughts, or floods) are usually difficult to predict or control. Therefore, building and maintaining resilience of desired ecosystem states may be the most pragmatic and effective way to manage ecosystems in the face of increasingly human-driven environmental disturbances across scales from local to global levels.

Human adaptability can be disadvantageous in this context. Loss of resilience is often masked by support from socio-economic infrastructures at other scales that make it possible to maintain a business-as-usual strategy in a situation of crisis. A common response to deal with surprises is trying to remove or ignore them. Such responses impede social–ecological learning. Examples include capital markets that provide loans and financial insurance to fishermen and farmers for periods of resource crisis, thereby removing incentives for responding to environmental feedback and avoiding the building of an ecological knowledge system into local institutions that may tighten social–ecological feedback loops (Chapter 7). There are local communities aware of the pitfalls of relying on outside inputs and incentives. In Chapter 12, Alcorn *et al.* describe how the Dayak have been cautious about maintaining a balance between economic dependence on forest products and products used for subsistence. Dayak associations have become wary of dependence on external donors. While accepting some support, they have built in ways for the movement to be supported by the Dayak themselves, for example through credit unions. Such strategies buffer against economic surprises and create local incentives for developing strategies to cope with change and learn.

In Chapter 4, Low *et al.* argue that diversity and redundancy among local-level resource management systems enhance performance as long as there are overlapping units of government that can resolve conflicts, aggregate knowledge across scale, and insure that when problems occur in smaller units, a larger unit can temporarily step in. This seems to be the case in the lagoon fishery in Brazil (Chapter 11), where repeated crises combined with externally driven change led to the involvement of higher-level institutions. In the crayfish watershed management system in Sweden (Chapter 8), management self-organized ecologically from individual crayfish to the whole watershed and institutionally from a few individuals to a nested set of organizations, facilitated by rules and incentives at the national level.

In Chapter 4, Low *et al.* conclude that the presence of larger overlapping jurisdictions is an important complement to the work of parallel, smaller-scale units. Ostrom (1998) uses the term polycentric management for management that involves a diversity of local as well as higher level of governance, and aims at finding the right balance between decentralized and centralized control. Polycentric management allows decisions to be made at different levels in society and increases the possibilities to create feedback loops at various scales. In polycentric systems the skills, networks of human relationships, and mutual trust that are incrementally developed in one setting represent social memory that could be transformed for reorganization in another setting (Chapters 8, 9, and 12).

Another such cross-scale transformation in social–ecological systems concerns the evolution of ecological knowledge systems in settings external to the dominating US forestry management paradigm. In Chapter 13, Trosper describes how the ecosystem approach, developing over decades in science and policy (Dale *et al.*, 2000), was available and activated for implementation in a situation of crisis and restructuring in the US forestry sector. The case illustrates that the process of generating and accumulating knowledge of complex ecosystem management over temporal and spatial scales in one setting can be made accessible in another setting. In this case, the transfer of knowledge from one panarchy to another required functional groups of leaders and networkers for the activation and creation of social memory in a new setting.

14.5.3 Matching scales of ecosystems and governance

Several chapters in the volume stress the capability of local users and their organizations to constantly reshape and adapt their institutions to ecosystem dynamics (Chapters 5–8, 10, and 12). In Chapter 5, Carlsson concludes from his study of a boreal forest area in Sweden that well-organized groups of local common pool resource users, closely connected to the resource system, are in a better position to adapt to and shape ecosystem change and dynamics than remote levels of governance. Such systems involve social mechanisms to spread risks over time. A diversity of different types of property rights provides opportunities for local users to gain the benefits of each property-rights system. As illustrated throughout this volume, insights are available from local management systems, including those with cross-scale interactions, on how to relate to the dynamic interplay between diversity and disturbance, an interplay that seems critical for building resilience.

The learning process is of central importance for social–ecological capacity to build resilience. It is important that learning processes include operational monitoring and evaluation mechanisms in order to generate and refine ecological knowledge and understanding into management institutions. This is the focus of adaptive co-management in which institutional arrangements and ecological knowledge are tested in an ongoing trial-and-error process. Adaptive co-management draws on social–ecological memory and is informed by both practice and theory. It relies on the participation of a diverse set of stakeholders operating at different scales (Chapters 8–10).

Adaptive co-management requires management practices that are different from conventional ones (Chapter 4; Gunderson and Holling, 2002), practices with the ability to tighten environmental feedback loops and build resilience to allow for disturbance and change (Berkes and Folke, 1998). Adaptive co-management benefits from combining the ecological knowledge of local resource users, scientists, and other interest groups, often with different conceptualizations of the issues and even different worldviews and belief systems, in mutual learning systems (Chapter 10). They should not start from scratch, i.e., trying to be adaptive without experience (see Fig. 14.2), but draw on and nurture ecological and social memory of the area and region to shape and turn surprise and crisis into renewal and opportunity. This we refer to as the process of framed creativity. Adaptive co-management designs should make use of information technology and create novel platforms for learning how to build resilience under uncertainty in landscapes increasingly transformed by human actions. Such design may be important for reducing ecological illiteracy in urbanized regions and reconnecting people whose livelihoods are no longer closely related to the land. Incentives from urban areas need to be created to support local areas that practice resiliency management of the resource base that sustains urban lifestyles, so that people attentive to the land can continue to nurture social–ecological systems (Chapter 3).

14.5.4 Accounting for external drivers

Even if a resource management system is dynamic in its response to ecosystem change and surprise and builds social–ecological resilience, it may be fragile and vulnerable to external social and economic drivers (Dasgupta and Mäler, 1991; Baland and Platteau, 1996). This is exemplified by Seixas and Berkes in Chapter 11, where the local system is overwhelmed by shocks derived from the larger scale, in this case related to tourism. In contrast, in Chapter 12, Alcorn *et al.* describe how social renewal as a response to crisis can counteract the erosion of social–ecological resilience. In this case, the response by the Dayak of Borneo was to develop a cluster of dynamic associations between local and national levels to influence external social and economic drivers. The response involved diversification across scale, with many loosely connected associations with overlapping functions derived from a bottom-up perspective.

In Figure 14.4 we referred to such institutional designs as meso-scale membranes. They may be generated from the bottom-up or constructed from above, or be a combination of the two. They may function to filter out major external socio-economic drivers entering from larger scales, such as urbanization, side effects of international trade, new technologies, surprises, disruptive world views, climate change, and human health impacts.

Many aspects of globalization tighten intersystem linkages, hierarchies, and interdependencies between local resource users and the wider society. These effects may operate through the market, political control, and social networks, and tend to distance resource users from their dependence on dynamic and complex adaptive life-support ecosystems, disconnect the production from consumption, and disconnect the production of knowledge from its application (Folke *et al.*, 1998). There is a risk that the tightening of processes of globalization weakens societal feedback loops to ecosystem dynamics essential for sustaining and building adaptive capacity and for securing the flow of critical ecosystem services for societal development. Therefore, local solutions in an increasingly globalized world cannot work by themselves. New levels of governance are required. Cash and Moser (2000) propose that governance for linking global and local scales should: utilize boundary organizations, like the one described in Chapter 12; utilize scale-dependent comparative advantages, like polycentric management systems; and employ adaptive assessment and management strategies, like the adaptive co-management approach. Such governance should focus on sustaining ecosystem stability domains that generate essential support to society in the face of change.

The paradox is that the processes of economic diversification, liberalization, and globalization ultimately depend on nature's subsidies, on diversity and resilience of ecosystems, but tend to create increasingly fragile ecosystems, as witnessed in modern food, fiber, and timber production systems. Modern belief systems and associated institutional frameworks often seem to create simplified ecosystems with impoverished diversity, low resilience, and reduced capacity to adapt to environmental change. They create their own vulnerability.

The issue of interest for navigating nature's dynamics, according to Trosper (Chapter 13), is the ability of surprise and crisis to precipitate responses in sociocultural systems that move away from destabilizing behavior and polarization, toward collaboration and actions that restore resilience. Given the pace of global change, human welfare is utterly dependent on forward-looking, adaptive, and informed management decisions for building social–ecological resilience for adaptive capacity and sustainability.

14.6 Concluding remarks

Previously in human history there were major events and local changes in the ability of the ecosystem to support social systems, but resilience was so high that in many respects nature could be seen as fairly stable. Change was buffered by resilience. Our view is that the situation is different today due to the human dominance of ecosystems. Widespread human alteration of ecological interactions and biogeochemical processes, from local to global levels, result in modified ecological resilience, increased likelihood of surprises, and unpredictable and enhanced variability in essential resource flows. The situation requires a shift to a view of the world as consisting of complex systems. This implies a shift in management toward those social institutions and organizations that can deal with nature's dynamics in a fashion that builds not only ecological or social but also social–ecological resilience. Otherwise, the development and well-being of human societies will become increasingly vulnerable to environmental change.

The chapters of the volume all deal with this issue and address management practices and characteristics of social mechanisms that build social–ecological resilience in complex systems. Building resilience includes: 'learning to live with change and uncertainty;' 'nurturing diversity for reorganization and renewal;' 'combining different types of knowledge for learning;' and 'creating opportunity for self-organization' (see Table 14.1). These four overall categories – or, as we propose, *principles for building resilience* – interact and are in many ways interdependent. For example, learning for self-organization will not be sufficient by itself and will not lead to social–ecological resilience. It will require the dynamic interplay between diversity and disturbance, along with recognition of cross-scale dependencies. We suggest that recognition of the four principles and their interactions is a prerequisite for directing society toward sustainability.

References

Adger, W.N. 2000. Social and ecological resilience: are they related? *Progress in Human Geography* 24: 347–64.

Alcorn, J.B. and Toledo, V.M. 1998. Resilient resource management in Mexico's forest ecosystems: the contribution of property rights. In *Linking Social and Ecological Systems. Management Practices and Social Mechanisms for Building Resilience*, pp. 216–49, ed. F. Berkes and C. Folke. Cambridge: Cambridge University Press.

Allison, G.W., Lubchenco, J., and Carr, M.H. 1998. Marine reserves are necessary but not sufficient for marine conservation. *Ecological Applications* 8: 79–92.

Baland, J.M. and Platteau, J.P. 1996. *Halting Degradation of Natural Resources: is there a Role for Rural Communities?* Oxford: FAO of the United Nations and Oxford University Press.

Bengtsson, J., Angelstam, P., Elmqvist, T., *et al.* Reserves, resilience, and dynamic landscapes. *Ambio*, in press.

Berkes, F. 1998. Indigenous knowledge and resource management systems in the Canadian subarctic. In *Linking Social and Ecological Systems. Management Practices and Social Mechanisms for Building Resilience*, pp. 98–128, ed. F. Berkes and C. Folke. Cambridge: Cambridge University Press.

Berkes, F. 1999. *Sacred Ecology: Traditional Ecological Knowledge and Management Systems*. Philadelphia and London: Taylor & Francis.

Berkes, F., Colding, J., and Folke, C. 2000. Rediscovery of traditional ecological knowledge as adaptive management. *Ecological Applications* 10: 1251–62.

Berkes, F. and Folke, C. 1992. A systems perspective on the interrelations between natural, human-made and cultural capital. *Ecological Economics* 5: 1–8.

Berkes, F. and Folke, C., eds. 1998. *Linking Social and Ecological Systems. Management Practices and Social Mechanisms for Building Resilience*. Cambridge: Cambridge University Press.

Berkes, F. and Folke, C. 2002. Back to the future: ecosystem dynamics and local knowledge. In *Panarchy: Understanding Transformations in Systems of Humans and Nature*, pp. 121–46, ed. L.H. Gunderson and C.S. Holling. Washington DC: Island Press.

Berkes, F., Folke, C., and Gadgil, M. 1995. Traditional ecological knowledge, biodiversity, resilience and sustainability. In *Biodiversity Conservation*,

pp. 269–87, ed. C. Perrings, K.-G. Mäler, C. Folke, C.S. Holling, and B.-O. Jansson. Dordrecht: Kluwer Academic Publishers.
Boyd, R. and Richerson, P.J. 1985. *Culture and the Evolutionary Process*. Chicago: Chicago University Press.
Carpenter, S.R. and Gunderson, L.H. 2001. Coping with collapse: ecological and social dynamics in ecosystem management. *BioScience* 51: 451–7.
Carpenter, S.R., Ludwig, D., and Brock, W.A. 1999. Management of eutrophication for lakes subject to potentially irreversible change. *Ecological Applications* 9: 751–71.
Cash, D.W. and Moser, S.C. 2000. Linking global and local scales: designing dynamic assessment and management processes. *Global Environmental Change* 10: 109–20.
Chernella, J. 1987. Endangered ideologies: Tukano fishing taboos. *Cultural Survival* 11: 50–2.
Colding, J. and Folke, C. 2001. Social taboos: 'invisible' systems of local resource management and biological conservation. *Ecological Applications* 11: 584–600.
Costanza, R. 1987. Social traps and environmental policy. *BioScience* 37: 407–12.
Costanza, R., Daly, H., Folke, C., *et al.* 2000. Managing our environmental portfolio. *BioScience* 50: 149–55.
Costanza, R., Low, B.S., Ostrom, E., and Wilson, J., eds. 2001. *Institutions, Ecosystems, and Sustainability*. Boca Raton, FL: Lewis Publishers.
Dale, V.H., Brown, S., Haeuber, R.A., *et al.* 2000. Ecological principles and guidelines for managing the use of land. *Ecological Applications* 10: 639–70.
Dasgupta, P. and Mäler, K.-G. 1991. The environment and emergent development issues. In *Proceedings of the World Bank Annual Conference on Development Economics 1990*, pp. 101–31. Washington DC: World Bank.
Denevan, W.M., Treacy, J.M., Alcorn, J.B., Padoch, C., Denslow, J., and Paitan, S.F. 1984. Indigenous agroforestry in the Peruvian Amazon: Bora Indian management of swidden fallows. *Interciencia* 9: 346–57.
Finlayson, A.C. and McCay, B.J. 1998. Crossing the threshold of ecosystem resilience: the commercial extinction of northern cod. In *Linking Social and Ecological Systems. Management Practices and Social Mechanisms for Building Resilience*, pp. 311–37, ed. F. Berkes and C. Folke. Cambridge: Cambridge University Press.
Folke, C., Pritchard, L., Berkes, F., Colding, J., and Svedin, U. 1998. The problem of fit between ecosystems and institutions. IHDP Working Paper No. 2. Bonn: International Human Dimensions Program on Global Environmental Change.
Fowler, C.S. 2000. 'We live by them': native knowledge of biodiversity in the Great Basin of Western North America. In *Biodiversity and Native America*, pp. 99–132, ed. P.E. Minnis and W.J. Elisens. Norman, OK: University of Oklahoma Press.
Franklin, J.F. and MacMahon, J.A. 2000. Enhanced: messages from a mountain. *Science* 288: 1183–90.
Gadgil, M. 1987. Diversity: cultural and biological. *Trends in Ecology and Evolution* 12: 369–73.
Gadgil, M., Berkes, F., and Folke, C. 1993. Indigenous knowledge for biodiversity conservation. *Ambio* 22: 151–6.
Gibson, C.C., Ostrom, E., and Ahn, T.K. 2000. The concept of scale and the human dimensions of global change: a survey. *Ecological Economics* 32: 217–39.
Grimm, N., Grove, M., Pickett, S.T.A., and Redman, C.L. 2000. Integrated approaches to long-term studies of urban ecological systems. *BioScience* 50: 571–84.
Gunderson, L.H. and Holling, C.S., eds. 2002. *Panarchy: Understanding Transformations in Systems of Humans and Nature*. Washington DC: Island Press.

Gunderson, L.H., Holling, C.S., and Light, S.S., eds. 1995. *Barriers and Bridges to the Renewal of Ecosystems and Institutions*. New York: Columbia University Press.
Gunderson, L.H. and Pritchard, L., eds. 2002. *Resilience and the Behavior of Large-Scale Ecosystems*. Washington DC: Island Press.
Holland, J.H. 1995. *Hidden Order: How Adaptation Builds Complexity*. Reading, MA: Addison-Wesley.
Holling, C.S. 1973. Resilience and stability of ecological systems. *Annual Review in Ecology and Systematics* 4: 1–23.
Holling, C.S., ed. 1978. *Adaptive Environmental Assessment and Management*. London: Wiley.
Holling, C.S. 1986. The resilience of terrestrial ecosystems: local surprise and global change. In *Sustainable Development of the Biosphere*, pp. 292–317, ed. W.C. Clark and R.E. Munn. Cambridge, UK: International Institute for Applied Systems Analysis (IIASA), Cambridge University Press.
Holling, C.S. 2001. Understanding the complexity of economic, ecological, and social systems. *Ecosystems* 4(5): 390–405.
Holling, C.S., Berkes, F., and Folke, C. 1998. Science, sustainability and resource management. In *Linking Social and Ecological Systems. Management Practices and Social Mechanisms for Building Resilience*, pp. 342–62, ed. F. Berkes and C. Folke. Cambridge: Cambridge University Press.
Holling, C.S. and Meffe, G.K. 1996. Command and control and the pathology of natural resource management. *Conservation Biology* 10: 328–37.
Holling, C.S., Schindler, D.W., Walker, B.W., and Roughgarden, J. 1995. Biodiversity in the functioning of ecosystems: an ecological synthesis. In *Biodiversity Loss: Economic and Ecological Issues*, pp. 44–83, ed. C.A. Perrings, K.-G. Mäler, C. Folke, C.S. Holling, and B.-O. Jansson. Cambridge: Cambridge University Press.
Irvine, D. 1989. Succession management and resource distribution in an Amazonian rain forest. In *Resource Management in Amazonia: Indigenous and Folk Strategies*, pp. 223–37, ed. D.A. Posey and W.L. Balee. Bronx, NY: New York Botanical Garden.
Jackson, J.B.C., Kirby, M.X., Berher, W.H., *et al.* 2001. Historical overfishing and the recent collapse of coastal ecosystems. *Science* 293: 629–38.
Jeffrey, D.K. 2000. Organizational change as a component of ecosystem management. *Society and Natural Resources* 13: 537–48.
Johannes, R.E. 1998. The case of data-less marine resource management: examples from tropical nearshore finfisheries. *Trends in Ecology and Evolution* 13: 243–6.
Kates, R.W. and Clark, W.C. 1996. Expecting the unexpected. *Environment* March: 6–11, 28–34.
Lee, K. 1993. *Compass and the Gyroscope*. Washington DC: Island Press.
Levin, S. 1999. *Fragile Dominion: Complexity and the Commons*. Reading, MA: Perseus Books.
Lewis, H.T. and Ferguson, T.A. 1988. Yards, corridors and mosaics: how to burn a boreal forest. *Human Ecology* 16: 57–77.
Lundberg, J. and Moberg, F. in review. The role of mobile link species for ecosystem functions. Effects of global environmental change and the implications for management.
McIntosh, R.J. 2000. Social memory in Mande. In *The Way the Wind Blows: Climate, History, and Human Action*, pp. 141–80, ed. R.J. McIntosh, J.A. Tainter, and S.K. McIntosh. New York: Columbia University Press.
McIntosh, R.J., Tainter, J.A., and McIntosh, S.K., eds. 2000. *The Way the Wind Blows: Climate, History, and Human Action*. New York: Columbia University Press.

Muchagata, M. and Brown, K. 2000. Colonist farmers' perceptions on fertility and the frontier environment in eastern Amazonia. *Agriculture and Human Values* 17(4): 371–84.

Nabhan, G.P. 1997. *Cultures of Habitat: on Nature, Culture, and Story*. Washington DC: Counterpoint.

National Research Council 1998. *Sustaining Marine Fisheries*. Washington DC: National Academy Press.

Nelson, J.G. and Serafin, R. 1992. Assessing biodiversity: a human ecological approach. *Ambio* 21: 212–18.

Niamir-Fuller, M. 1998. The resilience of pastoral herding in Sahelian Africa. In *Linking Social and Ecological Systems. Management Practices and Social Mechanisms for Building Resilience*, pp. 250–84, ed. F. Berkes and C. Folke. Cambridge: Cambridge University Press.

Norgaard, R.B. 1994. *Development Betrayed: the End of Progress and a Coevolutionary Revisioning of the Future*. New York: Routledge.

Nyström, M. and Folke, C. 2001. Spatial resilience of coral reefs. *Ecosystems* 4: 406–17.

Nyström, M., Folke, C., and Moberg, F. 2000. Coral-reef disturbance and resilience in a human-dominated environment. *Trends in Ecology and Evolution* 15: 413–17.

Olsson, P. and Folke, C. 2001. Local ecological knowledge and institutional dynamics for ecosystem management: a study of Lake Racken Watershed, Sweden. *Ecosystems* 4: 85–104.

Ostrom, E. 1990. *Governing the Commons: the Evolution of Institutions for Collective Actions*. Cambridge: Cambridge University Press.

Ostrom, E. 1998. Scales, polycentricity, and incentives: designing complexity to govern complexity. In *Protection of Global Biodiversity: Converging Strategies*, pp. 149–67, ed. L.D. Guruswamy and J.A. McNeely. Durham, NC: Duke University Press.

Ostrom, E., Burger, J., Field, C.B., Norgaard, R.B., and Policansky, D. 1999. Sustainability – revisiting the commons: local lessons, global challenges. *Science* 284: 278–82.

Peterson, G., Allen, C.R., and Holling, C.S. 1998. Ecological resilience, biodiversity, and scale. *Ecosystems* 1: 6–18.

Pinkerton, E. 1998. Integrated management of a temperate montane forest ecosystem through wholistic forestry: a British Columbia example. In *Linking Social and Ecological Systems. Management Practices and Social Mechanisms for Building Resilience*, pp. 363–89, ed. F. Berkes and C. Folke. Cambridge: Cambridge University Press.

Polanyi, M. 1958. *Personal Knowledge: Towards a Post-Critical Philosophy*. Chicago: University of Chicago Press.

Posey, D.A. 1985. Indigenous management of tropical forest ecosystems: the case of the Kayapo Indians of the Brazilian Amazon. *Agroforestry Systems* 3: 139–58.

Ramakrishnan, P.S., Saxena, K.G., and Chandrashekara, U.M., eds, 1998. *Conserving the Sacred for Biodiversity Management*. New Delhi: Oxford and IBH Publishing Co. Pvt. Ltd.

Riedlinger, D. and Berkes, F. 2001. Contributions of traditional knowledge to understanding climate change in the Canadian arctic. *Polar Record* 37: 315–28.

Röling, N.G. and Wagemakers, M.A.E. 1998. *Facilitating Sustainable Agriculture: Participatory Learning and Adaptive Management in Times of Environmental Uncertainty*. Cambridge: Cambridge University Press.

Ruddle, K. 1988. Social principles underlying traditional inshore fishery management systems in the Pacific Basin. *Marine Resource Economics* 5: 351–63.
Scheffer, M., Carpenter, S., Foley, J., Folke, C., and Walker, B. 2001. Catastrophic shifts in ecosystems. *Nature* 413: 591–6.
Scoones, I. 1999. New ecology and the social sciences: what prospects for a fruitful engagement? *Annual Review of Anthropology* 28: 479–507.
Tainter, J. 1988. *The Collapse of Complex Societies*. Cambridge: Cambridge University Press.
Turner, B.L., Clark, W.C., and Kates, W.C., eds. 1990. *The Earth as Transformed by Human Action: Global and Regional Changes in the Biosphere over the Past 300 Years*. Cambridge: Cambridge University Press.
Turner, N.J., Boelscher Ignace, M., and Ignace, R. 2000. Traditional ecological knowledge and wisdom of aboriginal peoples in British Columbia. *Ecological Applications* 10: 1275–87.
van der Leeuw, S.E. 2000. Land degradation as a socio-natural process. In *The Way the Wind Blows: Climate, History, and Human Action*, pp. 357–83, ed. R.J. McIntosh, J.A. Tainter, and S.K. McIntosh. New York: Columbia University Press.
Walker, B.H. 1992. Biodiversity and ecological redundancy. *Conservation Biology* 6: 18–23.

Index

Page numbers in bold print indicate tables and figures.